T0073161

GAME THEORY

A Nontechnical Introduction to the Analysis of Strategy
(4th Edition)

GAME THEORY

A Nontechnical Introduction to the Analysis of Strategy
(4th Edition)

Roger A McCain

Drexel University, USA

 **World Scientific**

NEW JERSEY • LONDON • SINGAPORE • BEIJING • SHANGHAI • HONG KONG • TAIPEI • CHENNAI

Published by

World Scientific Publishing Co. Pte. Ltd.

5 Toh Tuck Link, Singapore 596224

USA office: 27 Warren Street, Suite 401-402, Hackensack, NJ 07601

UK office: 57 Shelton Street, Covent Garden, London WC2H 9HE

Library of Congress Cataloging-in-Publication Data

Names: McCain, Roger A., author.

Title: Game theory : a nontechnical introduction to the analysis of strategy /
    Roger A. McCain, Drexel University, USA.

Description: 4th edition. | New Jersey : World Scientific, [2023] | Includes index.

Identifiers: LCCN 2022032910 | ISBN 9789811262951 (hardcover) |
    ISBN 9789811262968 (ebook for institutions) | ISBN 9789811262975 (ebook for individuals)

Subjects: LCSH: Game theory.

Classification: LCC QA269 .M425 2023 | DDC 519.3--dc23/eng20221013

LC record available at https://lccn.loc.gov/2022032910

British Library Cataloguing-in-Publication Data

A catalogue record for this book is available from the British Library.

For any available supplementary material, please visit
https://www.worldscientific.com/worldscibooks/10.1142/13043#t=suppl

Desk Editors: Soundararajan Raghuraman/Pui Yee Lum

Typeset by Stallion Press
Email: enquiries@stallionpress.com

Printed in Singapore

Preface to the Fourth Edition

As with the earlier editions, this edition is intended to teach game theory principally by examples, relying generally on the Karplus learning cycle of examples followed by general observations, reinforced by further, and sometimes more advanced examples. New material includes more extensive and updated information on Bayesian Nash Equilibria, cooperative solutions in NTU games, and social mechanism design, in the last case reflecting some more recent Nobel Memorial Prizes. New examples arise from events associated with the COVID pandemic and global warming. Some new business examples have also been added, including beer branding and the make-or-buy decision. On the other side, this book has been reorganized into sections with more basic topics and examples in the earlier chapter and a clearer progression to the more advanced topics. Chapters on some topics that seem less important in recent work in game theory, such as auctions and bargaining, have been deleted so that this book is a little more compact.

About the Author

Roger A. McCain grew up in the State of Louisiana and obtained degrees in mathematics and economics at Louisiana State University. He joined the faculty of Drexel University in 1988. McCain is the author of over 100 scholarly articles and several books, mostly in economics and game theory.

Contents

PART I

Interactive Decisions

CHAPTER 1

Conflict, Strategy, and Games

What is game theory? And what does it have to do with strategy and conflict? Of course, strategy and conflict arise in many aspects of human life, including games. Conflicts may have winners and losers, and games often have winners or losers. This textbook is an introduction to a way of thinking about strategy, a way of thinking derived from the mathematical study of games. The first step, in this chapter, is to answer those questions — what is game theory and what does it have to do with strategy? But rather than answer the questions immediately, let's begin with some examples. The first one will be an example of the human activity we most often associate with strategy and conflict: War.

1. THE SPANISH REBELLION: PUTTIN' THE HURT ON HIRTULEIUS

Here is the story (as novelized by Colleen McCullough from the history of the Roman Republic):

In about 75 BCE, Spain (Hispania in Latin) was in rebellion against Rome, but the leaders of the Spanish rebellion were Roman soldiers and Spanish people who had adopted Roman culture. It was widely believed that the Spanish leader, Quintus Sertorius, meant to use Spain as a base to make himself master of Rome. Rome sent two armies to put down the rebellion: one commanded by the senior, aristocratic, and respected Metellus Pius, and the other commanded by Pompey, who was (as yet) young and untried but very rich and willing to pay for his own army. Pompey was in command over Metellus Pius. Pius resented his subordinate position since Pompey was not only younger but a social inferior. Pompey set out to relieve the siege of a small Roman garrison at New Carthage, but got no further west than Lauro, where Sertorius caught and besieged him. (See the map in Figure 1.1.) Thus, Pompey and Sertorius had stalemated one another in Eastern Spain. Metellus Pius and his army were in Western Spain, where Pius was governor. This suited Sertorius, who did not want the two Roman armies to unite, and Sertorius sent his second-in-command, Hirtuleius, to garrison Laminium, northeast of Pius's camp, and prevent Pius from coming east to make contact with Pompey.

Figure 1.1.   Spain, with Strategies for Hirtuleius and Pius.

Pius had two strategies to choose from. They are shown by the light gray arrows in the map. He could attack Hirtuleius and take Laminium, which, if successful, would open the way to eastern Spain and deprive the rebels of one of their armies. If successful, he could then march on to Lauro and unite with Pompey against Sertorius. But his chances of success were poor. Fighting a defensive battle in the rough terrain around Laminium, the Spanish legions would be very dangerous and would probably destroy Pius' legions. Alternatively, Pius could make his way to Gades and take ships to New Carthage, raise the siege of New Carthage that Pompey had been unable to raise, and march on to Lauro, raising the siege of Pompey's much larger forces. To Pius, this was the better outcome, since it would not only unite the Roman armies and set the stage for the defeat of the rebels, but would also show up the upstart Pompey, demonstrating that the young whippersnapper couldn't do the job without getting his army saved by a seasoned Roman aristocrat.

Hirtuleius, a fine soldier, faced a difficult problem of strategy choice to fulfill his mission to contain or destroy Pius. Hirtuleius could march directly to New Carthage, and fight Pius at new Carthage along with the small force already there. His chances of defeating Pius would be very good, but Pius would learn that Hirtuleius was marching for New Carthage, and then Pius could divert his own march to the north, take Laminium without a fight, and break out to the northeast. Thus, Hirtuleius would fail in his mission. Alternatively, Hirtuleius could remain at Laminium until Pius marched out of his camp, and then intercept Pius at the ford of the River Baetis. He would arrive with a tired army and would fight on terrain more favorable to the Romans, and so his chances were less favorable; but there would be no possibility of losing Laminium and the Romans would have to fight to break out of their isolation.

Thus, each of the two commanders has to make a decision. We can visualize the decisions as a tree diagram like the one in Figure 1.2. Hirtuleius must first decide whether to commit his troops to the march to new Carthage or remain at Laminium where he can intercept Pius at the Baetis. Begin at the left, with Hirtuleius'

HEADS UP!

Here are some concepts we will develop as this chapter goes along:

Game theory is the study of the choice of strategies by interacting rational agents, or in other words, *interactive decision theory.*

A key step in a game theoretic analysis is to discover which strategy is a person's **best response** to the strategies chosen by the others. Following the example of neoclassical economics, **we define the best response for a player as the strategy that gives that player the maximum payoff, given the strategy the other player(s) has chosen or can be expected to choose.**

Game theory is based on a scientific metaphor, the idea that many interactions we do not usually think of as games, such as economic competition, war and elections, can be treated and analyzed as we would analyze games.

decision, and then we see the decision Pius has to make depending on which decision Hirtuleius has made. What about the results? For Hirtuleius, the downside is the simple part. If he fails to stop Pius, he fails in his mission. If he intercepts Pius at new Carthage, he has a good chance of winning. If he intercepts Pius at the ford on the Baetis, he has at least a 50–50 chance of losing the battle. On the whole, Pius wins when Hirtuleius loses. If he breaks out by taking Laminium he is successful. However, if he raises the siege of New Carthage, he gets the pleasure of showing up his boss as well. But he cannot be sure of winning if he goes to New Carthage.

Figure 1.2 shows a tree diagram with the essence of Hirtuleius' problem.

If Hirtuleius goes to New Carthage, Pius will go to Laminium and win. If Hirtuleius stays at Laminium, Pius will strike for New Carthage. Thus, the best Hirtuleius can do is to stay at Laminium and try to intercept Pius at the river.

In fact, Pius moved more quickly than Hirtuleius expected, so that Hirtuleius' tired troops had to fight a rested Roman army. The rebels were badly beaten and ran, opening the way for Pius to

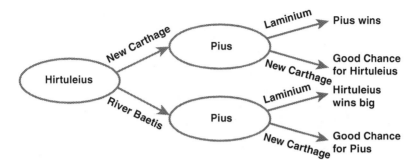

Figure 1.2. The Game Tree for The Spanish Rebellion.

continue to Gades and transport his legions by sea to New Carthage, where they raised the siege and moved on to raise the siege of Pompey in Lauro, and so Pius returned to Rome a hero. Pompey had plenty of years left to build his own reputation, and would eventually be First Man in Rome, only to find himself in Julius Caesar's headlights. But that's another story.[1]

In analyzing the strategies of Pius and Hirtuleius with the tree diagram, we are using concepts from **game theory**.

2. WHAT DOES THIS HAVE TO DO WITH GAMES?

The story about The Spanish Rebellion is a good example of the way we ordinarily think about strategy in conflict. Hirtuleius has to go first, and he has to try to guess how Metellus Pius will respond to his decision. Somehow, each one wants to try to outsmart the other one. According to common sense, that's what strategy is all about.

There are some games that work very much like the conflict between Metellus Pius and Hirtuleius. A very simple game of that kind is called Nim. Actually, Nim is a whole family of games, from smaller and simpler versions up to larger and more complex versions. For this example, though, we will only look at the very simplest version. Three coins are laid out in two rows, as shown in Figure 1.3. One coin is in the first row, and two are in the second. The two

---

[1]*Source*: Colleen McCullough, *Fortune's Favorites* (Avon PB, 1993), pp. 621–625.

Figure 1.3.   Nim.

players take turns, and on each turn a player must take at least one coin. At each turn, the player can take as many coins as they wish from a single row, but can never take coins from more than one row on any round of play. The winner is the player who picks up the last coin. Thus, the objective is to put the opponent in the position that they are required to leave just one coin behind.

There are some questions about this game that we would like to answer. What is the best sequence of plays for each of the two players? Is there such a best strategy at all? Can we be certain that the first player can win? Or the second? These are questions you might like to know the answer to, for example, if someone offered to make you a bet on a game of Nim.

Let's say that our two Nim players are Anna and Barbara. Anna will play first. Once again, we will visualize the strategies of our two players with a tree diagram. The diagram is shown in Figure 1.4. Anna will begin with the oval at the left, and each oval shows the coins that the player will see in case they arrive at that oval. Thus, Anna, playing first, will see all three coins. Anna can then choose among three plays at this first stage. The three plays are:

1.   Take one coin from the top row.
2.   Take one coin from the second row.
3.   Take both coins from the second row.

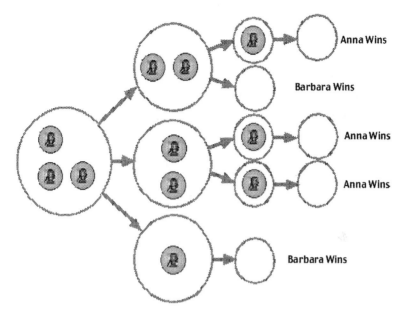

Figure 1.4. A Tree Diagram for Nim.

The arrows shown leading away from the first oval correspond from top to bottom to these three moves. Thus, if Anna chooses the first move, Barbara will see the two coins shown side by side in the top oval of the second column. In that case Barbara has the choice of taking either one or two coins from the second row, leaving either none or one for Anna to choose in the next round as shown in the top two ovals of the third column. Of course, by taking two coins, leaving none for Anna, Barbara will have won the game.

In a similar way, we can see in the diagram how Anna's other two choices leave Barbara with other alternative moves. Looking to strategy 3, we see that it leaves Barbara with only one possibility; but that one possibility means that Barbara wins. From Anna's point of view move 2, in the middle, is the most interesting. As we see in the middle oval, second column, this leaves Barbara with one coin in each row. Barbara has to take one or the other — those are her only choices. But each one leaves Anna with just one coin to take, leaving Barbara with nothing on her next turn, and thus winning the game for Anna. We can now see that Anna's best move is to take one coin from the second row, and once she has done that, there is nothing Barbara can do to keep Anna from winning.

Now we know the answers to the questions above. There is a best strategy for the game of Nim. For Anna, the best strategy is "Take one coin from the second row on the first turn, and then take whichever coin Barbara leaves." For Barbara, the best strategy is "If Anna leaves coins on only one row, take them all. Otherwise, take any coin." We can also be sure that Anna will win if she plays her best strategy.

3. GAME THEORY EMERGES

Early in the 20th century, mathematicians began to study gambling games. These studies were the beginning of game theory. The great mathematician John von Neumann extended the study to games like poker. Poker is different from Nim and Chess in a fundamental way. In Nim, each player always knows what moves the other player has made. That's also true in Chess, even though Chess is very much more complex than Nim. In poker, by contrast, you may not know whether or not your opponent is "bluffing." Games like Nim and Chess are called games of perfect information, since there is no bluffing, and every player always knows what moves the other player has made. Games like poker, in which bluffing can take place, are called games of imperfect information.

Von Neumann's analysis of games of imperfect information was a step forward in the mathematical study of games. But a more important

Definition: *Perfect Information* — A *game of perfect information* is a game in which every player always knows every move that other players have made that will influence the results of his or her own choice of strategies. A *game of imperfect information* is a game in which some players sometimes do not know the strategy choices other players have made, either because those choices are made simultaneously or because they are concealed.

departure came when von Neumann teamed up with the mathematical economist Oskar Morgenstern. In the 1940's, they collaborated on a book entitled *The Theory of Games and Economic Behavior*. The idea behind the book was that many aspects of life that we do not think of as games, such as economic competition and military conflict, can be analyzed *as if* they were games. Today, game theorists treat all kinds of human strategy choices as if they were strategies for games. Game theory is thought of as a *theory of interactive decisions*, according to the two game theorists who shared the 2005 Nobel Memorial Prize in Economics, Robert Aumann and Thomas Schelling.

As we have said, game theory studies the rational choice of strategies. This conception of rationality has a great deal in common with neoclassical economics. Thus, rationality is a key link between neoclassical economics and game theory. Of course, Morgenstern was an economist, but von Neumann was also well acquainted with neoclassical economics, so it was natural that they would draw from the neoclassical economic tradition.

Neoclassical economics is based on the assumption that human beings are absolutely rational in their economic choices. Specifically, the assumption is that each person maximizes their rewards — profits, incomes, or subjective benefits — in the circumstances that they face. This hypothesis serves a double purpose in the study of economics. First, it narrows the range of possibilities somewhat. Absolutely rational behavior is more predictable than irrational behavior. Second, it provides a criterion for evaluation of the efficiency of an economic system. If the system leads to a reduction in the rewards coming to some people,

A Closer Look: John von Neumann, 1903–1957

John von Neumann, born and educated in Hungary, was one of the leading mathematicians of the twentieth century, participated in the invention of computers, and among many other contributions, was the key person in the foundation of game theory, particularly in his collaboration with Oskar Morgenstern.

without producing more than compensating rewards to others (that is, if costs are greater than benefits, broadly speaking) then something is wrong. Pollution of air and water, the overexploitation of fisheries, and inadequate resources committed to research can all be examples of inefficiency in this sense.

A key step in a game theoretic analysis is to discover which strategy is a person's **best response** to the strategies chosen by the others. Following the example of neoclassical economics, *we define the best response for a player as the strategy that gives that player the maximum payoff, given the strategy the other player has chosen or can be expected to choose.* If there are more than two players, we say that the best response is the strategy that gives the maximum payoff, given the strategies all the other players have chosen. This is a very common concept of rationality in game theory, and we will use it in many of the chapters that follow in this book. However, it is not the only concept of rationality in game theory, and game theory does not always assume that people are rational. In some of the chapters to follow, we will explore some of these alternative views.

> **A Closer Look:** Oskar Morgenstern 1902–1977
>
> A noted mathematical economist, Morgenstern was born in Germany and worked in Vienna, Austria, before the NAZI takeover there. He then became a faculty member at Princeton and collaborated with von Neumann in writing the founding book of game theory, *The Theory of Games and Economic Behavior.* Morgenstern was also known for his work on the economics of national defense and space travel and on economic forecasting.

4. GAME THEORY, NEOCLASSICAL ECONOMICS AND MATHEMATICS

In neoclassical economics, the rational individual faces a specific system of institutions, including property rights, money, and highly competitive markets. These are among the "circumstances" that the person takes into account in maximizing rewards. The implication of property rights, a money economy and ideally competitive markets is that the individual needs not consider her or his interactions with other individuals. She or he needs to consider only his or her own situation and the "conditions of the market." But this leads to two problems. First, it limits the range of the theory. Whenever competition is restricted (but there is no monopoly), or property rights are not fully defined, consensus neoclassical economic theory is inapplicable, and neoclassical economics has never produced a generally accepted extension of the theory to cover these cases. Decisions taken outside the money economy were also problematic for neoclassical economics. Game theory was intended to confront just this problem: to provide a theory of economic and strategic behavior when people interact directly, rather than "through the market."

In neoclassical economic theory, to choose rationally is to maximize one's rewards. From one point of view, this is a problem in mathematics: choose the activity that maximizes rewards in given circumstances. Thus, we may think of a rational economic choice as the "solution" to a problem of mathematics. In game theory, the case is more complex, since the outcome depends not only on my own strategies and the "market conditions," but also directly on the strategies chosen by others. We may still think of the rational choice of strategies as a mathematical problem — maximize the rewards of a group of interacting decision makers — and so we again speak of the rational outcome as the "solution" to the game.

5. THE PRISONER'S DILEMMA

John von Neumann was at the Institute for Advanced Study in Princeton. Oskar Morgenstern was at Princeton University. As a result of their collaboration, Princeton was soon buzzing with game theory. Alfred Tucker, a faculty member in the mathematics department at Princeton, was visiting at Stanford University, and wanted to give a group of psychologists some idea of what all the buzz was about, without using much mathematics. The example that he gave them is called the "Prisoner's Dilemma."[2] It is the most studied example in game theory and possibly the most influential half a page written in the 20th century. You may very well have seen it in some other class. The Prisoner's Dilemma is presented a little differently than the two previous examples, however.

Tucker began with a little story, like this: two burglars, Bob and Al, are captured near the scene of a burglary and are given the "third degree" separately by the police. Each has to choose whether or not to confess and implicate the other. If neither man confesses, then both will serve 1 year on a charge of carrying a concealed weapon. If each confesses and implicates the other, both will go to prison for 10 years. However, if one burglar confesses and implicates the other, and the other burglar does not confess, the one who has collaborated with the police will go free, while the other burglar will go to prison for 20 years on the maximum charge.

The strategies in this case are: confess or don't confess. The payoffs (penalties, actually) are the sentences served. We can express all this compactly in a "payoff table" of a kind that has become pretty standard in game theory. Table 1.1 gives the payoff table for the Prisoners' Dilemma game.

The table is read like this: Each prisoner chooses one of the two strategies. In effect, Al chooses a column and Bob chooses a row. The two numbers in each cell tell the outcomes for the two prisoners when the corresponding pair of strategies is chosen. The number to the left of the comma tells the payoff to the person who chooses the rows (Bob) while the number to the right of the cell tells the payoff to the person who chooses the columns (Al). Thus (reading down the first column) if they both confess, each gets 10 years, but if Al confesses and Bob does not, Bob gets 20 and Al goes free.

So: how to solve this game? What strategies are "rational" if both men want to minimize the time they spend in jail? Al might reason as follows: "Two things can happen: Bob can confess or Bob can

---

[2] *Source:* S. J. Hagenmayer, Albert W. Tucker, 89, Famed Mathematician, *The Philadelphia Inquirer* (Thursday, February 2, 1995), p. B7.

Table 1.1. The Prisoner's Dilemma.

		Al	
		Confess	Don't
Bob	Confess	10 years, 10 years	0, 20 years
	Don't	20 years, 0	1 year, 1 year

keep quiet. Suppose Bob confesses. Then I get 20 years if I don't confess, 10 years if I do, so in that case it's best to confess. On the other hand, if Bob doesn't confess, and I don't either, I get a year; but in that case, if I confess I can go free. Either way, it's best if I confess. Therefore, I'll confess."

But Bob can and presumably will reason in the same way — so that they both confess and go to prison for 10 years each. Yet, if they had acted "irrationally," and kept quiet, they each could have gotten off with 1 year each.

6. ISSUES WITH RESPECT TO THE PRISONERS' DILEMMA

This remarkable result — that self-interested and seemingly "rational" action results in both persons being made worse off in terms of their own self-interested purposes — is what has made the wide impact in modern social science. For there are many interactions in the modern world that seem very much like that, from arms races through road congestion and pollution to the depletion of fisheries and the overexploitation of some subsurface water resources. These are all quite different interactions in detail, but are interactions in which (we suppose) individually rational action leads to inferior results for each person, and the Prisoners' Dilemma suggests something of what is going on in each of them. That is the source of its power.

Having said that, we must also admit candidly that the Prisoners' Dilemma is a very simplified and abstract — if you will, "unrealistic" — conception of many of these interactions. A number of critical issues can be raised with the Prisoners' Dilemma, and each of these issues has been the basis of a large scholarly literature:

- The Prisoners' Dilemma is a two-person game, but many of the applications of the idea are really many-person interactions.
- We have assumed that there is no communication between the two prisoners. If they could communicate and commit themselves to coordinated strategies, we would expect a quite different outcome.
- In the Prisoners' Dilemma, the two prisoners interact only once. Repetition of the interactions might lead to quite different results.
- Compelling as the reasoning that leads to this conclusion may be, it is not the only way the problem might be reasoned out. Perhaps it is not really the most rational answer after all.

7. GAMES IN NORMAL AND EXTENSIVE FORM

Definition: *Extensive and Normal Form* — A game is *represented in extensive form* when it is shown as a tree diagram in which each strategic decision is shown as a branch point. A game is *represented in normal form* when it is shown as a table of numbers with the strategies listed along the margins of the table and the payoffs for the participants in the cells of the table.

There are both important similarities and contrasts between this example and the previous two. A contrast can be seen in the way the examples have been presented. The Prisoner's Dilemma has been presented as a table of numbers, not as a tree diagram. These two different ways of presenting a game will play important and different roles in this book, as they have in the history of game theory.

When a game is represented as a tree diagram, we say that the game is represented in "extensive form." The extensive form, in other words, represents each decision as a branch point in a tree diagram. One alternative to the extensive form is the representation we see in the Prisoner's Dilemma. This is called the "normal form." In a normal form representation, the game is shown as a table of numbers with the different strategies available to the players enumerated at the margins of the table.

The normal form representation is probably less intuitive than the extensive form. Nevertheless, it has been very influential and we will rely mostly on the normal form in the next few chapters of this book.

8. A BUSINESS CASE

So far, we have seen three examples — cases from war, concealment of a crime, and a recreational game. There are many applications to business, so let us consider a business case before concluding this chapter.[3] We will apply the game metaphor and the representation of the game in normal form. The business example will be very much like the Prisoner's Dilemma.

Before 1964, television advertising of cigarettes was common. Following the Surgeon General's Report in 1964, the four large tobacco companies, American Brands, Reynolds, Philip Morris, and Ligget and Myers, negotiated an agreement with the federal government. The agreement came into effect as of 1971 and included a pledge not to advertise on television. Can this be explained by means of game theory?

Here is a two-person advertising game much like the situation faced by the tobacco companies. Let's call the companies Fumco and Tabacs. The strategies for each firm are don't advertise or advertise. We assume that if neither of them advertises, they will divide the

---

[3] Game theory is important for business and economics, and is valuable also as a link across the disciplines to the other social sciences and philosophy. Thus, part of the plan for this book is that every chapter will include at least one major business case, but also at least one major case from another discipline. The exceptions will be the chapter on industry strategy and prices, which will be pretty nearly all business and economics, and the chapter on games and politics, which will not include a business application.

Table 1.2.   The Advertising Game.

		Fumco	
		Don't advertise	Advertise
Tabacs	Don't advertise	8,8	2,10
	Advertise	10,2	4,4

market and their low costs (no advertising costs) will lead to high profits in which they share equally. If both advertise, they will again divide the market equally, but with higher costs and lower profits. Finally, if one firm advertises and the other does not, the company that advertises gets the largest market share and substantially higher profits. Table 1.2 shows payoffs rating profits on an arbitrary scale from 1 to 10 — with 10 best. The table is read as the Prisoner's Dilemma table is: Fumco chooses the column, Tabacs chooses the row, and the first payoff is to Tabacs, the second to Fumco.

We will find that this game is very much like the Prisoner's Dilemma. Each firm can reason as follows: "If my rival does not advertise, then I am better off to advertise, since I will get profits of 10 rather than 8. On the other hand, if my rival does advertise, I am better off to advertise, since I will get profits of 4 rather than 2. Either way, I had better advertise." Thus, both advertise and get profits of 4 rather than 8.

This is like the Prisoner's Dilemma in that rational, self-seeking behavior leads the two companies to a result that both dislike. But it may be difficult for competitive companies, as it is for prisoners in different interrogation rooms, to trust one another and choose the strategy that is better for both. However, when a third party steps in — as the Federal Government did in the tobacco case — they are happy to agree to restrain their advertising expenditure.

9. A SCIENTIFIC METAPHOR

Now, let's return to the question, "what is game theory?" Since the work of John von Neumann, "games" have been a scientific metaphor for a much wider range of human interactions in which the outcomes depend on the interactive strategies of two or more persons, who have opposed or at best mixed motives. **Game theory is a distinct and interdisciplinary approach to the study of human behavior, an approach that studies rational choices of strategies and treats the interactions among people as if it were a game, with known rules and payoffs and in which everyone is trying to "win."** The disciplines most involved in game theory are mathematics, economics and the other social and behavioral sciences. Increasingly, engineers and biologists also make use of game theory. Among the issues discussed in game theory are:

(1) What does it mean to choose strategies "rationally" when outcomes depend on the strategies chosen by others and when information is imperfect?
(2) In "games" that allow mutual gain (or mutual loss) is it "rational" to cooperate to realize the mutual gain (or avoid the mutual loss) or is it "rational" to act aggressively in seeking individual gain regardless of mutual gain or loss?
(3) If the answers to (2) are "sometimes," in what circumstances is aggression rational and in what circumstances is cooperation rational?
(4) In particular, do ongoing relationships differ from one-off encounters in this connection?
(5) Can moral rules of cooperation emerge spontaneously from the interactions of rational egoists?
(6) How does real human behavior correspond to "rational" behavior in these cases?
(7) If it differs, in what direction? Are people more cooperative than would be "rational?" More aggressive? Both?

10. SUMMARY

In this chapter, we have addressed the questions "What is game theory? And what does it have to do with strategy and conflict?" We have seen from some examples that game theory is a distinct and interdisciplinary approach to the study of human behavior, based on a scientific metaphor. The metaphor is that conflicts and choices of strategy, as in war, deception, and economic competition, can be treated "as if" they were games. We have seen two major ways that these "games" can be represented:

In normal form
- As a table of numbers with the different strategies available to the players enumerated at the margins of the table.

In extensive form
- As a "tree" diagram with each strategic decision as a branch point.

We have seen that game theory often assumes that people act rationally in the sense that they adopt a best response strategy. Like the neoclassical conception of rational behavior in economics, the assumption is that people are acting rationally when they act as though they are maximizing something: profits, winnings in the game, or subjective benefits of some kind — or, perhaps, minimizing a penalty, such as the number of years in jail. The "best response" is the strategy that gives a player the maximum payoff, given the strategies the other player has chosen or can be expected to choose. These concepts are the beginning point for a study of game theory. In the next chapter, we will explore the relationships among some of them, especially between games in normal and extensive form.

Q1. PROBLEMS AND DISCUSSION QUESTIONS

Q1.1. The Spanish Rebellion

In her story about the Spanish Rebellion, McCullough writes "There was only one thing Hirtuleius could do: march down onto the easy terrain ... and stop Metellus Pius before he crossed the Baetis." Is McCullouch right? Discuss.

Q1.2. Nim

Consider a game of Nim with three rows of coins, with one coin in the top row, two in the second row, and either one, two or three in the third row. (A) Does it make any difference how many coins are in the last row? (B) In each case, who wins?

Q1.3. Matching Pennies

Matching pennies is a school-yard game. One player is identified as "even" and the other as "odd." The two players each show a penny, with either the head or the tail showing upward. If both show the same side of the coin, then "even" keeps both pennies. If the two show different sides of the coin, then "odd" keeps both pennies. Draw a payoff table to represent the game of matching pennies in normal form.

Q1.4. Happy Hour

Jim's Gin Mill and Tom's Turkey Tavern compete for pretty much the same crowd. Each can offer free snacks during happy hour, or not. The profits are 30 to each tavern if neither offers snacks, but 20 to each if they both offer snacks, since the taverns have to pay for the snacks they offer. However, if one offers snacks and the other does not, the one who offers snacks gets most of the business and a profit of 50, while the other loses 20. Discuss this example using concepts from this chapter. How is the competition between the two tavern owners like a game? What are the strategies? Represent this game in normal form.

CHAPTER 2

Some Foundations

In the previous chapter, we saw two rather different kinds of examples. The example of Hirtuleius and Pius and the game of Nim were represented in extensive form, that is, as a tree diagram. The Prisoner's Dilemma and the Advertising Game were represented in normal form, that is, in tabular form. There are some other differences between those games, and the representation as a tree diagram or as a table is partly a matter of convenience.

In the early development of game theory, the representation of games in normal form was more common and was very influential. In some more recent work, the representation of games in extensive form has played a key role. Following this history, the next few chapters of this book will focus mainly on games in normal form, and later chapters will return and analyze games in extensive form.

1. REPRESENTATION IN NORMAL FORM: A BUSINESS CASE

We have seen that games can be represented in two different ways: extensive and normal form. Although it is sometimes more convenient to represent a particular game in one way or another, there is nothing absolute about this. Any game can be represented in either form. This is not obvious, of course. It was one of von Neumann's key discoveries. And there is a trick to it. Here is an example, which is also a business case.

According to a study[1] by the McKinsey business consulting organization, deregulation can create a difficult transition for formerly regulated companies. Often these companies have been "public utility" monopolies. During most of the twentieth century, "public utilities" in the United States were allowed to operate as monopolies, with their prices regulated and profits limited to a "fair rate of return;" but with new competition prohibited by law. However, under deregulation they face the entry of new competition. The monopoly will be tempted to respond to the new entry with a price war, although, according to the McKinsey organization, this is usually an unprofitable strategy.

Definitions: *Contingency* — A contingency is an event that may or may not occur, such as the event that another player adopts a particular strategy.

Contingent Strategy — A contingent strategy is a strategy to be adopted only when it is known that the contingent event has occurred.

Contingency Plan — A contingency plan is a plan to be put into operation only when it is known that the contingent event has occurred.

Let's illustrate that with an example. Goldfinch Corp. provides telecommunications services in Gentilia City, but the telecommunications market has been deregulated. Bluebird Communications is considering entering the Gentilia market. If Bluebird does enter the market, Goldfinch has two choices: Goldfinch can cut prices, entering into a price war, to retain their market share so far as possible and perhaps to punish Bluebird for entering the market and try to drive them out. (We ignore the possibility that Goldfinch might run into legal problems if they do this.) Alternatively, Goldfinch can cut back on their own output, "accommodating" the new entering firm and keeping the

---

[1]Florissen, Andreas, Boris Maurer, Bernhard Schmidt, and Thomas Vahlenkamp (2001), "The Race to the Bottom," (McKinsey Quarterly).

price up. Either way, Goldfinch will expect decreased profits. Rating profits on a scale of 10 for best, payoffs for Goldfinch are 10 if Bluebird does not enter, 5 if Bluebird enters and Goldfinch shares the market, and 2 if Bluebird enters and Goldfinch starts a price war. The payoff of 2 includes the present value of any monopoly profits gained if Bluebird is driven out of the market. Bluebird's payoffs are 0 if they do not enter, 3 if they enter and Goldfinch shares the market, and –5 if there is a price war. As the weaker firm, in a financial sense, Bluebird does not profit as much as Goldfinch if the market is shared and will lose out in a price war. (In case the market is shared, total profits in the industry are reduced from 10 to 8, while consumers, who are not part of the game, benefit from lower prices.)

This "game" is different from the Prisoner's Dilemma and the Advertising Game (but similar to the Spanish Rebellion and Nim) in that one participant, Bluebird, Ltd, has to go first in choosing their strategy and the other participant, Goldfinch, can wait and see what Bluebird will do before choosing their strategy. It seems natural to

---

HEADS UP!

Here are some concepts we will develop as this chapter goes along:

Games in Normal and Extensive Form: A game is *represented in extensive form* when it is shown as a tree diagram in which each strategic decision is shown as a branch point. A game is *represented in normal form* when it is shown as a table of numbers with the strategies listed along the margins of the table and the payoffs for the participants in the cells of the table.

Contingency: A contingency is an event that may or may not occur, such as the event that another player adopts a particular strategy.

Contingent Strategy: A contingent strategy is a strategy to be adopted only when it is known that the contingent event has occurred.

Information set: In a game in extensive form (tree diagram) a decision node with more than one branch included in it is called an "information set."

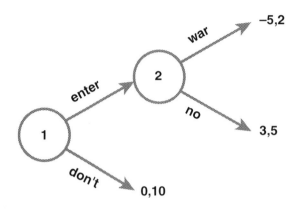

Figure 2.1. The Game of Market Entry in Extensive Form.

represent this game in extensive form, as a tree diagram, as shown in Figure 2.1.

In the figure, Bluebird's choice is shown at node 1 and Goldfinch's choice is shown at node 2. The numbers at the right side show the payoffs, with Bluebird's payoff first. In this case, Bluebird will want to "think strategically," that is, to anticipate how Goldfinch will respond if Bluebird does decide to enter; and Goldfinch will want to plan for the "scenario" or "contingency" that Bluebird will enter. Looking just at decision node 2, we see that Goldfinch gets a payoff of 5 (hundred million) from the strategy of accommodation, and a payoff of only 2 from the strategy of a price war. Thus, if Goldfinch maximizes its payoffs, Goldfinch will accommodate the new entry, and Bluebird can anticipate that, and will choose to enter for a payoff of 3 rather than staying out for zero. (We will see in a later chapter that this sort of commonsense reasoning is very central to game theory, but does not always give quite such commonsense results in more complex examples.)

The tree diagram is called the "extensive form" and here we are examining the game of market entry in extensive form. It seems natural to express a game like this in extensive form, and that probably agrees with our commonsense way of thinking of a "game" and of strategy. But, as von Neumann and Morgenstern pointed out in their founding book, this game can be represented in "normal form," that is as a table of numbers like the Prisoner's Dilemma. There is something to be gained by looking at all games in the same way, and von Neumann and Morgenstern selected the normal form as their common framework for looking at all games.

But there is a trick to it. In the game of market entry, Goldfinch's strategies are *contingent* strategies:

(1) *If Bluebird enters then* accommodate.
(2) *If Bluebird enters then* initiate price war.

A "contingent" strategy is a strategy that is only adopted if a particular "contingency" arises. The phrases in italics in (1) and (2) indicate the contingencies in which Bluebird's strategies are relevant. Like the Market Entry Game, any game can be represented in normal form, but we may have to treat some of its strategies as contingent strategies in order to do it.

In the game of market entry, as in chess, each participant knows all decisions by the opponent that are relevant to his own decision. For example, Bluebird must choose first whether to enter or not, and Goldfinch knows what that decision is when they make their own decision to retaliate or not.

Bluebird's strategies are:

(1) Enter.
(2) Don't enter.

Thus, the game in normal form is as shown in Table 2.1. However, game theorists often use a "shorthand," leaving out the contingent phrase such as "if Bluebird enters then," and showing Goldfinch's strategies simply as "initiate price war" and "accommodate." This does not cause confusion in a simple case like this, but it can cause confusion in more complex examples, so it is best to be more careful and to keep in mind that the strategies used when we represent games in normal form are contingent strategies. "Accommodate," "price war," without the "If" statements, are often

Table 2.1. The Market Entry Game in Normal Form.

		Goldfinch	
		If *Bluebird enters* then accommodate; *if Bluebird does not enter,* then do business as usual.	If *Bluebird enters* then initiate price war; *if Bluebird does not enter,* then do business as usual.
Bluebird	Enter	3,5	−5,2
	Don't	0,10	0,10

called *behavior strategies*; and "enter" and "don't" would also be behavior strategies.

How will the game come out? First, what is the "best response" for each company? If Bluebird does enter, "accommodate" is the better behavior strategy — so there will be no price war, and Bluebird, anticipating this, will enter. But the McKinsey study warns that deregulated companies, inexperienced in competitive markets, may make the wrong decision in these circumstances, which will harm the profits of both companies.

Contingent strategies play an important role in many business situations. For example, consider this quotation[2]: "All airlines have the grim task of preparing contingency plans in case one of their jets should fall out of the sky." A contingency plan is a plan to be put into effect if a particular contingency occurs. Contingency planning is not only important in business, but in many other fields as well, such as military affairs (from which the idea originally came) and government. Whenever two or more decision makers are making contingency plans to deal with one another, we have contingent strategies.

2. THE NORMAL FORM IN GENERAL

The "Prisoner's Dilemma" and "Advertising Game" examples from Chapter 1 illustrate why game theory has been influential, and some

---

[2] From "Airline Management Style Honed By Catastrophe," by Laurence Zuckerman, *The New York Times* (Thursday, November 15, 2001), p. C1.

of the issues that arise in more complicated cases. But the way those examples were presented — as tables of numbers — probably is not the way that most of us are accustomed to thinking of games. The table-of-numbers presentation is what the founders of game theory

> **Definition:** *Normal Form* — The game in normal form is a tabular list of strategies available to each participant, with the payoffs that result if the participants choose each pair (or triple, etc.) of strategies.

called "the game in normal form." The game in normal form puts stress on one important aspect of games and other game-like interactions: each participant's payoffs depend not only on his own strategy but also on the strategy chosen by the other. However, many games seem, on first impression, to be too complex for this treatment — although they may not be very complex games!

A key discovery in the book by von Neumann and Morgenstern is that all games can be represented in normal form. To do this, it will be necessary to treat the strategies as contingent strategies. In complex games such as chess and war, with many steps of play, almost all strategies will naturally be thought of as contingent strategies, relevant only if the opponent has already made certain strategic commitments. The number of contingencies that can arise in chess, and thus the number of distinct contingent strategies, are literally inconceivably large, and beyond the computational capacity of any existing computer or any computer in the foreseeable future. Nevertheless, *in principle,* even chess could be represented in normal form.

In principle, all strategies are contingent strategies. In the market entry game, for example, even Bluebird's strategies are contingent: "Whether Goldfinch retaliates or not, enter," and "whether Goldfinch retaliates or not, don't enter." Since Bluebird goes first, and isn't informed whether Goldfinch will retaliate or not, Bluebird has only one contingency for each strategy — so we usually ignore the contingencies for the person who goes first. But, to be quite thorough, we should remember that the contingency is there.

This sort of thoroughness is a characteristic that game theory inherits from its mathematical origins — and shares with computer programming. We could put it this way: a strategy is an if-then rule, and there must always be something in the "if" part, even if it doesn't make any difference in a particular example. You probably know from experience what a computer does to you if you do not follow the rules *thoroughly*.

For some games that are conveniently shown in extensive form, we will need to know how to convert them to normal form. For example, let's think again about the Spanish Rebellion. If Hirtuleius were to march toward New Carthage, Pius would know, and could march on Laminium unopposed. But Pius would only march toward Laminium if he knew that Hirtuleius had already committed himself to go to New Carthage. Thus, Pius' contingent strategy is "if Hirtuleius marches to New Carthage then attack Laminium, but otherwise go to New Carthage." In this case "attack Laminium" and "go to New Carthage" are behavior strategies.

Let's convert the Spanish Rebellion into normal form. As a first step, how many strategies do the two generals have? Clearly, Hirtuleius has only two. Looking at the tree diagram, it is not quite so clear how many strategies Pius has. In fact, Pius has four contingent strategies. There are two strategies Pius may choose if Hirtuleius goes to New Carthage, and two others that he may choose if Hirtuleius marches towards the River Baetis.

Using this information, the Spanish Rebellion game is shown in normal form in Table 2.2.

In this case, if we were to leave out the contingent phrases *If Hirtuleius goes to New Carthage,* and *If Hirtuleius marches for the River Baetis,* we might suppose that Pius has only two strategies, and not take into account his superior information. That could lead to confusion in this slightly more complex game.

While it is not essential, it is often useful in game theory to indicate the results of the game in terms of numbers, as we did with the Prisoner's Dilemma and the Advertising Game. We will often choose numbers more or less arbitrarily to express the relative desirability

Table 2.2. Spanish Rebellion in Normal Form.

		Hirtuleius	
		(Regardless of Pius' strategy) go to R. Baetis	*(Regardless of Pius' strategy)* go to New Carthage
Pius	*If Hirtuleius goes to New Carthage, then go to Laminium; if Hirtuleius marches for the River Baetis, then go to Laminium*	Hirtuleius wins big	Pius wins
	If Hirtuleius goes to New Carthage, then go to New Carthage; if Hirtuleius marches for the River Baetis, then go to New Carthage	Good chance for Pius	Good chance for Hirtuleius
	If Hirtuleius goes to New Carthage, then go to Laminium; if Hirtuleius marches for the River Baetis, then go to New Carthage	Good chance for Pius	Pius wins
	If Hirtuleius goes to New Carthage, then go to New Carthage; if Hirtuleius marches for the River Baetis, then go to Laminium	Hirtuleius wins big	Good chance for Hirtuleius

or undesirability of the outcomes. Let's see how we might do that for the example of Hirtuleius and Pius.

In the upper right cell of Table 2.2, the outcome is "Pius wins." This is the worst outcome for Hirtuleius and the best for Pius. Let's assign that a payoff of 5 for Pius and minus 5 for Hirtuleius. This is shown in Table 2.3, upper right cell, with the payoff for Pius to the left, and the payoff for Hirtuleius to the right. As usual, the payoff to the player who chooses the row is to the left, since that player's strategies are to the left. In the next cell, in Table 2.2, the outcome is "good chance for Hirtuleius." We translate that (on a relative scale) as 3 for Hirtuleius and –3 for Pius. In the third and fourth rows, these estimates are more or less reversed, giving us the table of numerical payoffs shown in Table 2.3.

Table 2.3.   Spanish Rebellion in Normal Form with Numeric Outcomes.

		Hirtuleius	
		(Regardless of Pius' strategy) go to R. Baetis	*(Regardless of Pius' strategy)* go to New Carthage
Pius	*If Hirtuleius goes to New Carthage, then go to Laminium; if Hirtuleius marches for the River Baetis, then go to Laminium*	−5, 5	5, −5
	If Hirtuleius goes to New Carthage, then go to New Carthage; if Hirtuleius marches for the River Baetis, then go to New Carthage	3, −3	−3, 3
	If Hirtuleius goes to New Carthage, then go to Laminium; if Hirtuleius marches for the River Baetis, then go to New Carthage	3, −3	5, −5
	If Hirtuleius goes to New Carthage, then go to New Carthage; if Hirtuleius marches for the River Baetis, then go to Laminium	−5, 5	−3, 3

Let's try one more example, a new kind of game. The example will be a simple version of the Dictator Game. The Dictator Game has been studied in experimental research in game theory, but we will leave the discussion of that research for a later chapter. Amanda is the dictator. She has a candy bar that she has to share with her little sister Barbara. (Mother will be angry if she doesn't share at all.) Amanda can share 50–50, or she can keep 90% for herself. Barbara's only choice is to accept or reject what she is offered. The tree diagram for this game is shown in Figure 2.2.

In the Dictator game, Amanda's strategies are 50–50 and 90–10. Amanda's decision is shown by the circle marked A. Barbara's strategies in each case are accept or reject. Her choices are shown by the circles marked B. The first payoff is to Amanda, and the second is to Barbara. The game in normal form is shown in Table 2.4. This doesn't look very good for Barbara — perhaps you can see why it is

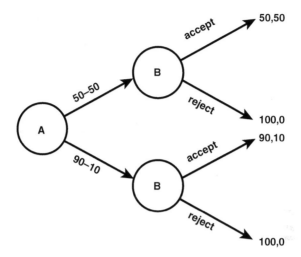

Figure 2.2. The Dictator Game.

Table 2.4. The Dictator Game in Normal Form.

		Amanda	
		(Whether Barbara accepts or rejects) offer 50–50	(Whether Barbara accepts or rejects) offer 90–10
Barbara	If *Amanda offers 50–50,* then accept; if *Amanda offers 90–10,* then accept	50, 50	10, 90
	If *Amanda offers 50–50,* then accept; if *Amanda offers 90–10,* then reject	50, 50	0, 100
	If *Amanda offers 50–50,* then reject; if *Amanda offers 90–10,* then accept	0, 100	10, 90
	If *Amanda offers 50–50,* then reject; if *Amanda offers 90–10,* then reject	0, 100	0, 100

called the Dictator Game. But the main point here is to see how we can translate the game from extensive form to normal form by using the contingent expression of the strategies.

3. THE PRISONER'S DILEMMA IN EXTENSIVE FORM

For an example of converting a game from normal to extensive form, let's take a look at our old friend, the Prisoner's Dilemma, in extensive form. In Prisoner's Dilemma, both prisoners make their decisions simultaneously, and we have to allow for that. The Prisoner's Dilemma is shown in normal form in Table 2.5. It is shown in extensive form in Figure 2.3, and we may suppose that Al makes his decision to confess or not confess at 1, and Bob makes his decision at 2. Notice one difference from the examples so far. Bob's decision, in the two different cases, is enclosed in a single oval. That is game-theory code, and what it tells us is that when Bob makes his decision, he doesn't know what decision Al has made. Bob has to make his decision without knowing Al's decision. Graphically, Bob doesn't know whether he is at the top of the oval or the bottom.

Contrast this example with the Market Entry Game, Figure 2.1. In Figure 2.1, the fact that node 2 gives only one set of arrows means, in game theory code, that Bluebird knows Goldfinch's decision before Bluebird makes its decision. Recall, Bluebird knows what Goldfinch has decided, but, but Bob does not know what Al has decided (or will decide).

Because it tells us something about the (limited) information Bob has, node 2 is called an "information set." Conversely, the extensive form of the game can be useful as a way of visualizing the information available to a participant at each stage of the game.

Table 2.5. The Prisoner's Dilemma (Repeats Table 1.1, Chapter 1).

		Al	
		Confess	Don't
Bob	Confess	10, 10	0, 20
	Don't	20, 0	1, 1

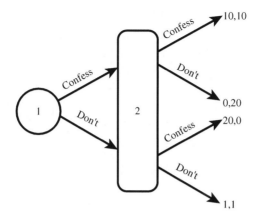

Figure 2.3. The Prisoner's Dilemma in Extensive Form.

Where information availability is important, we will often want to use the game in extensive form. But for games like the Prisoner's Dilemma, there is no available information, and the game in normal form contains all we need to know about the game.

> **Definition:** *Information set* — A decision node with more than one branch included in it is called an "information set."

The Prisoner's Dilemma can be represented in extensive form in two different ways. We have shown it here with Al going first, and Bob following, although not knowing what decision Al made. Since the decisions are made simultaneously, it is equally correct to represent it with Bob going first and Al following. The important thing is the information that Al and Bob have, or rather, the information they lack. Since each representation shows both men making their decision without knowledge of the decision the other makes, both are equally correct.

The idea here is nothing more than a visual code. If for any reason the player doesn't know which decision the other player has made, then he does not really know which branch in the tree he is taking. We express that by putting both branches, or all of the branches that he might be making, within a single node in the decision tree. When a game theorist sees two or more branches grouped within a single node, the game theorist knows that the player lacks information — the player doesn't know which branch he is really at.

4. AN EXAMPLE FROM MILITARY HISTORY

Let's look at another example, this time an example drawn from military history. By the early 20th century heavy, long-range artillery had become a decisive influence in European war. The big cannons were transported by rail, sometimes using special railroads built for the purpose, but also relying on ordinary freight railroads. As a result, in a crisis it was important, if possible, to be the first to mobilize. For suppose one country moved its cannon into position first. That country would be able to destroy the other country's rail approaches and thus prevent the enemy from moving its cannon into place. The result would be an immediate advantage in the war to follow.

This combination of artillery dominance and limited mobility contributed to the sudden outburst of full-scale war at the beginning of World War I. The crisis came when Gavrilo Princip, a Yugoslav nationalist, assassinated Archduke Franz Ferdinand of Austria. This led, over a few months, to a general crisis. Knowing that a war might follow, Austria, Germany, France and their allies all rushed to mobilize their forces, rather than taking a chance that their enemies might mobilize first and leave them at the disadvantage.

The game of Mobilization is shown in normal form in Table 2.6, where the two countries are France and Germany. Rating disasters on a scale of 1 to 10, the bloody European war from 1914 to 1918 certainly was a 10. Accordingly, we show a payoff of negative 10 to each country if both countries mobilize. If neither country mobilizes, peace continues, and the payoff is zero to each country. If one

Table 2.6. Mobilization.

		Germany	
		Mobilize	Don't
France	Mobilize	−10, −10	−9, −11
	Don't	−11, −9	0, 0

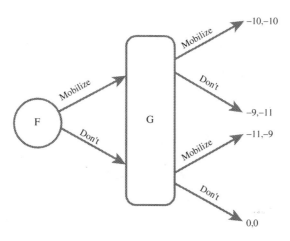

Figure 2.4. Mobilization.

country mobilizes and the other does not, the country that mobilizes does slightly less badly, with a payoff of negative 9; while a country that does not mobilize does slightly worse, with the payoff of negative 11.

As in the Prisoner's Dilemma, the two countries have to make their decisions simultaneously. Each one has to decide whether to mobilize without knowing whether the other country will mobilize or not. This lack of information is the crucial point in constructing the diagram of the game in extensive form.

At the same time, and again like the Prisoner's Dilemma, it doesn't matter whether we show France or Germany going first, since they are choosing simultaneously. Let's show the game with France going first. Then we show Germany's choices in a single oval, to express the fact that Germany doesn't know whether France is mobilizing or not. The diagram is shown in Figure 2.4.

In general, whenever one of the players has to choose without knowledge of choices the other player has made, or is making at the same time, we will represent that lack of information by putting all of the branches that the player may be taking in the same node in the tree diagram. This is how the game in extensive form shows information and the lack of information in games.

5. ZERO AND NON-CONSTANT SUM GAMES

Looking back at Table 2.3, we notice that the payoffs in each cell of the Spanish Rebellion Game add up to zero. It makes sense to treat the Spanish Rebellion in this way, since whatever Hirtuleius wins, Pius loses, and vice versa. The interests of the two generals are absolutely opposed, and we express that by showing the payoffs to one of them as just the negative of the payoffs of the other: thus, the payoffs always add up to zero. Accordingly, we may think of the Spanish Rebellion as a *zero-sum game*.

> **Definition:** *Zero-Sum Game* — A game in which the payoffs for the players always add up to zero is called a zero-sum game.

Before he teamed up with Morgenstern, John von Neumann had done some work on the mathematical analysis of gambling games, such as poker. This was game theory in a non-metaphorical sense. For these games, it was natural to assume that the payoffs would add up to zero: whatever one person wins, another loses. Thus, von Neumann made the assumption of a zero sum one of his basic assumptions for the theory of gambling games.

However, when von Neumann and Morgenstern began to apply the same sort of reasoning to "economic behavior" and other human decisions outside of gambling, recreational games and war, it seemed that there were many important examples in which the payoffs would not add up to zero, and in which the total payoffs to the players could depend on just which strategies they choose. We have seen several examples already: the Prisoner's Dilemma, the Advertising Game, and the Market Entry Game are all games of this kind. They are called *non-zero-sum games* or *non-constant-sum games*. They are games with win-win, or lose-lose, or other, still more complicated outcomes.

Zero-sum games are simpler in some ways than non-constant sum games. Since they represent interactions in which the players' interests are absolutely opposed, we need not consider how win-win outcomes, or similar issues, might affect the rational play of the game. It is not only the zero-sum games that are simple in this way. Refer to Table 2.4, the Dictator Game in normal form. We see that the payoffs always add up to 100%. Thus, as in a zero-sum game, the interests of the players are absolutely opposed. There are no win-win or lose-lose possibilities. Thus, the Dictator Game is an example of a *constant-sum game*.

Definition: *Constant-sum and non-constant-sum game* — If the payoffs to all players add up to the same constant, regardless which strategies they choose, then we have a constant-sum game. The constant may be zero or any other number, so zero-sum games are a class of constant-sum games. If the payoffs do not add up to a constant, but vary depending on which strategies are chosen, then we have a non-constant sum game.

Although zero-sum and constant-sum games are simpler than non-constant sum games, the non-constant-sum games are the general case. Anything we can apply to non-constant sum games will also apply to constant-sum games, including zero-sum games. Thus, in the next few chapters we will focus on non-constant-sum games, although we will return to zero-sum games, and consider some special methods for those games, in the next section.

6. THE MAXIMIN APPROACH

Here is another example: Olga and Pamela are contestants in a quiz show, but the question for this round of the game is so easy that both women are confident that they can answer it. Of course, that isn't the whole story. The rules of the game are: to answer a question a contestant has to sound a buzzer. If one contestant buzzes, she gets to answer the question. A correct answer scores one point. The contestant who does not buzz is penalized one point. If neither contestant buzzes, the question is passed, with no score for anyone. Also (this is the twist) if both contestants buzz, the question is passed with no score for anyone. The contestants have two strategies: Buzz or don't buzz. The payoff table is shown in Table 2.7.

Of course, this is a zero-sum game. How will rational, self-interested individuals decide their strategies for such a game? We have the answer, thanks to von Neumann. To illustrate the point, let's look back at this game strictly from Pamela's point of view. Table 2.8 shows the payoffs just for Pamela. But all of the information in

Table 2.7. The Buzzer Game.

		Pamela	
		Buzz	Don't
Olga	Buzz	0, 0	1, −1
	Don't	−1, 1	0, 0

Table 2.8. Pamela's Payoffs in The Buzzer Game.

		Pamela	
		Buzz	Don't
Olga	Buzz	0	−1
	Don't	1	0
Minimum		0	−1

Table 2.7 is really there, because Pamela knows that whatever makes Olga better off will make her, Pamela, worse off. Thus, Pamela has enough information in this table to figure out what Olga's payoffs and best responses will be. Whatever strategy Pamela chooses, Olga's best response is the strategy that gives Pamela the worst, the *minimum* payoff. This is shown in the last row in Table 2.8. Thus, to choose her best strategy, all Pamela needs to do is choose the strategy that gives the biggest minimum payoff, namely zero. That is, Pamela chooses the strategy that *maximizes the minimum payoff*, the *maximin* strategy for short. And since Olga can reason in just the same way, she too chooses the strategy that maximizes her minimum payoff, but it is equally the strategy that minimizes Pamela's maximum payoff (the *minimax* payoff).

For another example, let us return to the Spanish Rebellion in normal form with numerical payoffs as shown in Table 2.3. Hirtuleius, we recall, has two strategies. The minimum payoff for "go to R. Baetis" is −3 whereas it is −5 for "go to New Carthage," so "go to R. Baetis" is his maximin strategy. Pius has four contingent strategies, but of the four, the third, "*If*

Definition: *Maximin strategy* — If we determine the least possible payoff for each strategy, and choose the strategy for which this minimum payoff is largest, we have the maximin strategy. Since this also minimizes the opponent's maximum payoff, it can also be called the minimax strategy.

Hirtuleius goes to New Carthage, then go to Laminium; *if Hirtuleius marches for the River Baetis,* then go to New Carthage," gives a minimum payoff of 3, while all others have negative minimum payoffs, so that is his maximin strategy. In real history, both Hirtuleius and Pius chose their maximin strategies.

The maximin approach can be applied whenever the payoffs for the two contestants add up to the same *constant* number regardless of the strategies chosen. In this case the constant number happens to be zero. If it were a constant other than zero, we would say that the game is a *constant-sum game*. Every zero-sum game is a constant-sum game but not every constant-sum game is a zero-sum game. As we saw, the dictator game is a constant-sum game where the total payoff is 100%. In all two person constant-sum games, regardless whether they are zero-sum games or not, the rational solution is the maximin strategies.

For an example of a game that is not a constant-sum game, we do not need to go any further than the Prisoner's Dilemma. In the Prisoner's Dilemma as shown in Chapter 1, the total payoffs to the two prisoners can range from 2 to 20 (years in prison). Thus, we would say that the Prisoner's Dilemma is a *non-constant sum game.*

Thus, we will contrast "zero-sum games" with "non-constant sum games" like the Prisoner's Dilemma. This does leave something out — the non-zero constant sum games — and so far as logic is concerned, that group is just as important as any other. We have seen a few applications of non-zero constant sum games, such as the election game at Chapter 3, Section 8.

7. THE SIGNIFICANCE OF ZERO-SUM GAMES

Zero-sum games are simple, but, as we have seen, may not be applicable to many real-world interactions. Most of the examples we have seen are in recreational games and war — and even in war, we have the Mobilization Game as an example of a non-constant sum war game.[3] Apart from their simplicity, what is their significance for the application of game theory to serious human interactions?

---

[3] And even in recreational games, the zero-sum scoring may not be the whole story. A personal reminiscence can illustrate the point. I played some playground basketball when I was in grade school. I was tall and lanky for my age, but I was also a terrible shooter and medium-slow. As a result, I was a lot better on defense than on offense, so my idea was to do what I could do best and play a really sticky defense. But even my teammates didn't want me to do that. They wanted to have fun running and shooting as much as they wanted to win, if not more. So, when I played defense, it actually cut down on the subjective payoffs for both sides. And subjective payoffs are the ones that really count. I have to conclude that my schoolyard basketball was not a zero-sum game, even though there could only be one winner and one loser.

By coincidence there was a good illustration in the morning newspaper when the first draft for this chapter was being written, the Philadelphia Inquirer for Presidents' Day, February 18, 2002. The lead opinion column in the business section dealt with proposals for tax reform in Philadelphia and said that the proposed reforms would be a win-win game for Philadelphia government and business.[4] The headline is "Cutting wage tax wouldn't be a zero-sum game for city," and the columnist is warning us not to oversimplify. Some people may assume that, in order for tax reform to benefit some people, it must harm others. That would be true in a zero-sum game, but not in a win-win game, since any win-win game is non-constant sum. Similarly, economist Lester Thurow wrote a book,[5] *The Zero-Sum Society*, focused on the negative results of this sort of oversimplification.

What these examples suggest is that the most important significance of zero-sum games is negative. It is the contrast with the much richer potential we can find in much of our day-to-day business. It is the warning not to oversimplify. On the other hand, this solution for constant-sum games provides us with a model for the solution of more complex games. This generalization of the maximin solution was a major objective for von Neumann and Morgenstern and, a few years later, of John Nash, although their approaches went in different directions (Chapters 3–6 and 12–13 respectively.)

8. SUMMARY

One of the early discoveries in game theory was that all kinds of games can be represented in normal form; that is, as a table of numbers with the strategies listed at the margin and the payoffs to the participants in the cells. For some games, we have to understand

---

This may tell us why so many coaches stress the importance of "D" — of defense. The coach doesn't get to have fun running and shooting anyway, so he just wants to win.

[4]Andrew Cassel, "Commentary: Cutting wage tax wouldn't be a zero-sum game for city," *The Philadelphia Inquirer* (Monday, February 18, 2002), Section C, pp. C1, C9.

[5]Lester Thurow, *The Zero-Sum Society* (Basic Books, 1980).

that some of the strategies are contingent strategies, strategies that are considered only if the other player has already taken some specific action. Contingent strategies and contingency planning are important in themselves, but doubly important in game situations where players may get an advantage from the information they have. On the other hand, when we represent games like the Prisoner's Dilemma (games in which information on the others' moves may be missing) in the extensive form, we have to show what information each decision maker has at each step. When one decision maker does not know what decision, the other has made or is making, we put the results of both decisions within a single decision node. In a tree diagram, this will be shown by including two or more branches and a single oval. Such an oval is called an "information set," since it illustrates what information the decision maker has. To be more exact, it illustrates the information the decision maker does not have — they cannot tell which of the decisions within the oval they are making.

Zero-sum or constant-sum games are a category that is usually represented in normal form. These are games in which the total payoffs add up to a constant, so that whatever one wins the other loses. For these games we have a "solution" for rational self-interested play: maximize the minimum payoff.

Our conclusion for this chapter is that the extensive form of a game and the normal form are just alternative ways of looking at games, and each can be applied to any game, so we can use the representation that works best in a particular case. For the next few chapters, we will focus on non-constant sum games in the normal form, with special attention to non-constant sum games.

Q2. PROBLEMS AND DISCUSSION QUESTIONS

Q2.1. Sibling Rivalry

Two sisters, Iris and Julia, are students at Nearby College, where all the classes are graded on the curve. Since they are the two best students in their class, each of them will top the curve unless they

Table 2.9. Grade Point Averages
for Iris and Julia.

		Iris	
		Math	Lit
Julia	Math	3.8, 3.8	4.0, 4.0
	Lit	3.8, 4.0	3.7, 4.0

enroll in the same class. Iris and Julia each have to choose one more class this term, and each of them can choose between math and literature. They're both very good at math, but Iris is better at literature. Each wants to maximize her grade point average. The grade point averages are shown in Table 2.9, which treats their friendly rivalry as a game in normal form.

a. What are the strategies for this game?
b. Express this game in extensive form, assuming that the sisters make their decisions at the same time.
c. Express this game in extensive form, assuming that Iris makes her decision first, and Julia knows Iris's decision when she chooses her strategy.

Q2.2. The Great Escape

A prisoner is trying to escape from jail. He can attempt to climb over the walls or dig a tunnel from the floor of his cell. The warden can prevent him from climbing by posting guards on the wall, and he can prevent the con from tunneling by staging regular inspections of the cells, but he has only enough guards to do one or the other, not both.

a. What are the strategies and payoffs for this game?
b. Express the payoffs in both non-numerical and numerical terms.
c. Express this game in normal form.

d. Express this game in extensive form, assuming that the prisoner and the warden make their decisions at the same time.
e. Express this game in extensive form, assuming that the warden makes his decision first, and the prisoner knows the warden's decision when he chooses his strategy.

Q2.3. Checkers

Here is a very simplified version of the familiar game of checkers:

Mini-checkers is a small scale version of the familiar game of checkers. It is played on a checkerboard just for squares wide and three deep. Each player has just two pieces. The checkerboard and the initial placement of the pieces in shown in Figure 2.5. In color, the squares would be a red and black, and so would the pieces; but here, red is shown by the lighter gray, and black by the darker gray. As in ordinary checkers, the pieces move diagonally, only on the red squares. Black goes first. As in ordinary checkers again, a piece can "jump" an enemy piece, provided there is an open red square beyond the enemy piece. This is shown in Figure 2.6.

The game ends when either (1) the player whose turn it is cannot move; in which case the player who cannot move loses, or (2) a player advances one of his pieces to the opponent's E end of the

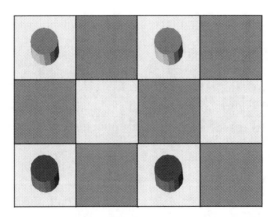

Figure 2.5. Mini-Checkers.

Figure 2.6. Black Jumps and Wins.

checkerboard, so that the piece is "kinged"; in which case the player whose piece is kinged is the winner. In Figure 2.6, Black has won by having his piece kinged.

a. Draw a diagram for the game of mini-checkers in extensive form.
b. Draw the tabular representation of the game of mini-checkers in normal form.
c. Can you determine whether black or red will win at mini-checkers if both sides play their best strategies? Which?

Q2.4. Water Game

Eastland and Westria share the valley of Southflowing River, which forms the boundary between them. Each country controls some of the northern tributaries of the river, and could divert water from the tributary streams for their own use. However, any diversion from the tributaries of the river will divert water that the citizens of the southern regions of both countries use for irrigation and other purposes, and if both countries divert the water of the tributaries, the flow in the south will be so reduced that silting and problems of navigation also will occur. Reliable cost-benefit studies have provided the following figures: if just one country diverts water from the tributaries, the net benefit to that country will be 3 billion Euros, but the other country will lose 4 billion. However, if both countries divert water from the tributaries, each country will suffer a net loss of 2 billion. The two countries do not trust one another and keep their decisions strictly secret from one another as long as possible, so each country can only conjecture as to what the other country will decide and feels no possibility of influencing the decision of the other. Explain how this interaction can be interpreted as a game. Will the normal form or the extensive form be more appropriate? Show both the normal and extensive forms of the Water Game.

Q2.5. Nim strikes again

Translate the game of Nim (Chapter 1) from extensive to normal form.

Q2.6. Beggars are Choosers

Vick and Will beg for donations on a street that leads to the superhighway onramp, where cars back up to wait and some will give some money. The total contributions per day are 100. The two beggars can each choose between the northern end of the street, where cars enter, or the south end, where they wait for a stoplight to let them on the highway and so the queues are more frequent. If both take the same end, they divide the contributions equally, but if one takes the south end, he gets it all. The payoffs are shown in Table 2.10.

Table 2.10. Contributions.

		Will	
		North	South
Vick	North	50, 50	0, 100
	South	100, 0	50, 50

a. What kind of game is this?
b. Can the maximin approach give a solution for this game? What is it? Why?

Q2.7. Beggars Pay Admission

A local gangster has taken control of the begging at the approach street to the superhighway (from the previous problem), and he requires a beggar to pay him 50 to beg at either end. Thus, every payoff is reduced by 50.

a. Construct the payoff table for this modified game.
b. What kind of game is this?
c. Can the maximin approach give a solution for this game? What is it? Why?

PART II

Equilibrium in Normal Form Games

CHAPTER 3

Dominant Strategies and Social Dilemmas

The first two chapters have focused mainly on concepts, pointing out that many problems of strategy can be studied as if they were games, and exploring the way the strategies and games are represented. But game theory is particularly interested in the interactions among the people involved in the game. From this point on, we will focus very strongly on the interactions.

We want to discover patterns of interaction that are stable and predictable. In economics, a stable and predictable pattern of interaction is usually called "an equilibrium." Game theorists follow this example so we can say that we are investigating "equilibrium" patterns of play in games.

As a first example, we will use a case from environmental policy and environmental economics. This is especially appropriate. Both of the words "economics" and "ecology" come from the Greek root word "oikos," which means a household. Here is the idea: Within a household, the members of the household are in constant interaction, and their interaction is what makes the household work. When we use the words "economic" and "ecological" we are implying that in economic and ecological matters, the interaction is just as constant and crucially important as it is within a household.

1. THE DUMPING GAME

For this game, we will begin with a story, as we so often do. In this story the characters are two property owners. We will call them Mr. Jones and Mr. Smith. Mr. Jones and Mr. Smith own weekend homes on side by side plots of land in a remote area that has no routine garbage collection. They can contract to have their garbage picked up by a trucking firm, but that is rather costly. It would cost either property owner $500 per year to arrange for the garbage pickup. Each one has another possible strategy to get rid of his garbage. Mr. Jones can dump his garbage near Mr. Smith's house, but far from his own; and similarly, Mr. Smith can dump his garbage near Mr. Jones's house. (Figure 3.1 shows a rough plot of their properties and these strategies.)

The two landowners will make their decisions simultaneously, so neither will know what strategy the other person is choosing. Therefore, we will want to represent this game in normal form. Each

HEADS UP!

Here are some concepts we will develop as this chapter goes along:

Dominant Strategy: Whenever one strategy yields a higher payoff than a second strategy, regardless which strategies the other players choose, the first strategy dominates the second. If one strategy dominates all other strategies (for a particular player in the game) it is said to be a dominant strategy (for that player).

Dominant Strategy Equilibrium: If, in a game, each player has a dominant strategy, then that combination of (dominant) strategies and the corresponding payoffs are said to constitute the dominant strategy equilibrium for that game.

Cooperative and non-cooperative solutions: the cooperative solution of a game is the list of strategies and payoffs that the participants would choose if they could commit themselves to a coordinated choice of strategies: For example, by signing an enforceable contract. The strategies and payoffs they would choose if there are no enforceable agreements is the non-cooperative solution.

Social dilemma: A social dilemma is a game that has a dominant strategy equilibrium and the dominant strategy solution is different than the cooperative solution to the game.

Figure 3.1. Mr. Smith and Mr. Jones' Dumping Strategies.

landowner chooses between two strategies: Pay for a garbage pickup or dump the garbage. But what are the payoffs in this game?

The benefits the two landowners get from their weekend property are subjective benefits. They consist of the enjoyment the two get from spending their time in this remote and scenic location. But we need to express the benefits in money terms, if only in order to compare them with the dollar cost of the garbage pickup. We can use an idea from economics to do this. We know that the property owners could benefit in other, monetary ways from their property. For example, each one could rent his property out for the season, rather than occupying it himself. We can estimate the money value of a person's subjective enjoyment of his property — it is the smallest amount of rent that he would accept in return for giving up his own occupation of the property.

Of course, the subjective benefits depend on whether someone is dumping his garbage near your property or not. Taking that into account, we will say that each person values his experience at his weekend home at $5,000 per year if there is no dumping, but at $4,000 per year if there is. In other words, if there were no dumping, neither Mr. Jones nor Mr. Smith would give up a year of occupancy for less rent than $5,000; but if there is dumping on the property each of them would give up the year of occupancy for $4,000.

Using this information, we can represent the Dumping Game in normal form as shown in Table 3.1.

Now remember, in game theory we focus on "rational" behavior, which means that each player chooses his "best response" to the strategy the other player has chosen or can be expected to choose. First, let's think of Mr. Smith's best response to the strategies that Mr. Jones might choose. These are shown in Table 3.2.

What we see in Table 3.2 is that "dump" is always the best response, regardless of which strategy Mr. Jones chooses. Now let's think through Mr. Jones's best responses to the strategies that Mr. Smith might choose. Since the game is symmetrical, it should come as no surprise that the best response is the same for Mr. Jones as it is for Mr. Smith. That is what we see in Table 3.3.

It seems that both players in the Dumping Game have rather easy decisions. That is because the strategy "dump" is an example of a *dominant strategy*. A dominant strategy is a strategy that is the best response to any strategy that the other player or players might choose. In the Dumping Game, "dump" is a dominant strategy for both players. Since both players in the Dumping Game have dominant strategies, the Dumping Game also gives us a good example of a dominant strategy equilibrium. When each of the players in a game chooses his dominant strategy, the result is a *dominant strategy equilibrium*. We can also say that the strategy "dump" dominates the strategy "hire a truck." Whenever one strategy yields a higher payoff than a second strategy, regardless which strategies the other players choose, the

Definition: *Dominated Strategy* — Whenever one strategy yields a higher payoff than a second strategy, regardless which strategies the other players choose, the first strategy dominates the second. In this case the second strategy is said to be a dominated strategy.

Definition: *Dominant Strategy* — If one strategy dominates all other strategies (for a particular player in the game) it is said to be a dominant strategy (for that player).

Table 3.1. The Dumping Game in Normal Form.

		Mr. Smith	
		Dump	Hire truck
Mr. Jones	Dump	4,000, 4,000	5,000, 3,500
	Hire truck	3,500, 5,000	4,500, 4,500

Table 3.2. Best Responses for Mr. Smith.

If Mr. Jones's strategy is	The best response for Mr. Smith is
Dump	Dump
Hire truck	Dump

Table 3.3. Best Responses for Mr. Jones.

If Mr. Smith's strategy is	The best response for Mr. Jones is
Dump	Dump
Hire truck	Dump

second strategy is dominated by the first, and is said to be a dominated strategy. In the Dumping Game, the strategy "hire a truck" is a dominated strategy.

2. DOMINANT STRATEGIES

When dominant strategies exist in a game, they provide a very powerful reason for choosing one strategy rather than another. You may have noticed that the Dumping Game is very much like the Prisoner's Dilemma. In the Prisoner's Dilemma, "confess" is a dominant strategy both for Al and for Bob; and (confess,confess) is a dominant strategy equilibrium. Similarly, the Advertising Game, in Chapter 1, is an example of a dominant strategy equilibrium. All of these games — the Dumping Game and the other two — are examples of social dilemmas. A *social dilemma* can be defined as a game with a dominant strategy equilibrium in which all players do worse than they would if they had all adopted non-equilibrium strategies.

3. SOCIAL DILEMMAS AND COOPERATIVE SOLUTIONS

From a mathematical point of view, the dominant strategy equilibrium is a "solution" to the game. That is, it tells us what strategies will be chosen and what the results will be if participants in the game make choices that are "rational" in a certain sense. But from the point of view of the participants in a social dilemma, the dominant strategy equilibrium is more the problem in itself. Let's return to the example of the dumping game. We can be pretty certain that both Mr. Smith and Mr. Jones would prefer the situation where each person chooses to hire a truck rather than the dominant strategy equilibrium. Since both are better off when both hire a truck, we can describe the (hire a truck, hire a truck) outcome as the "cooperative solution" to the Dumping Game.

Definition: *Dominant Strategy Equilibrium* — If, in a game, each player has a dominant strategy, then that combination of (dominant) strategies and the corresponding payoffs are said to constitute the dominant strategy equilibrium for that game.

Definition: *Cooperative Solution* — The *cooperative solution* of a game is the list of strategies and payoffs that the participants would choose if they could commit themselves to a coordinated choice of strategies: for example, by signing an enforceable contract.

Suppose, to continue the example, that Mr. Smith and Mr. Jones come together and negotiate a contract. The contract states that each of them will hire a truck, and that there will be no dumping. After they have signed the contract, they are committed to the strategy of hiring a truck. If either of them should "cheat," the other player can file a lawsuit and force the cheater to comply with the contract. Thus, the institution of contracts provides a solution for social dilemmas in some cases. In this instance it enables the two homeowners to arrive at the cooperative solution rather than the dominant strategy equilibrium.

> **Definition:** *Non-cooperative Solution* — The *non-cooperative solution* of a game is the list of strategies and payoffs that the participants would choose if there is no possibility to commit themselves to a coordinated joint strategy, so that each assumes the other will choose a best response strategy.

In fact, contracts of this kind among homeowners are quite common. They are called "covenants." Many suburban settlements have covenants against dumping and similar nuisances. Of course, legislation serves the same purpose in many incorporated settlements.

In general, we will define the cooperative solution of a game as the list of strategies and payoffs that the participants would choose if they could commit themselves to a coordinated choice of strategies, whether by means of a contract or by any other form of commitment.

By contrast, the dominant strategy equilibrium we have been considering is a non-cooperative solution. A non-cooperative solution of a game is the list of strategies and payoffs that the participants would choose if there is no possibility of a binding

> **Definition:** *Social dilemma* — A social dilemma is a game that has a dominant strategy equilibrium and the dominant strategy solution is different than the cooperative solution to the game.

commitment to coordinate strategies. In a non-cooperative solution, each player chooses his best response to the strategies chosen by the others, and assumes that they do the same, so each player chooses a best response to the best response strategies of the others. This is true in the Dumping Game, for example — each player assumes that the other player will choose "dump" and each player chooses the best response to "dump."

What defines a social dilemma is the fact that a dominant strategy equilibrium exists and is contrary to the cooperative solution. Thus, for example, in the Prisoner's Dilemma, the two prisoners would certainly prefer to coordinate their strategies, refuse to confess, then serve only 1 year in prison. The "third degree" treatment they receive is specifically designed to prevent them from coordinating their strategies. In the advertising game, again, not to advertise is the cooperative solution. When the government threatened to enforce a no advertising policy, the tobacco companies were in a position to profit by complying. They could profit because, with government enforcement, the choice of a no advertising strategy would be coordinated — both would adopt it simultaneously.

4. A PRICING DILEMMA

One of the most important applications of game theory in economics is to the study of oligopoly. In an oligopoly, that is, an industry with only a few sellers, the sellers may be able to maintain the monopoly price, and so maximize the profits for the group as a whole. This creates an incentive to "cheat" by offering a cheaper price and thus taking business away from the rival firms that continue to charge the monopoly price. For an example, consider the pricing dilemma shown in Table 3.4. The payoffs are profits in millions. The duopolists, Magnacorp and Grossco, can each benefit by maintaining prices, but each can gain an even bigger advantage by cutting price when the other does not. Since they produce a homogenous product, the game is symmetrical. We see that, in this oligopoly game, price cutting is a dominant strategy. Thus, the game among the sellers is a social dilemma. In this case there are others

Table 3.4. Payoffs in the Pricing Dilemma.

		Grossco	
		Maintain price	Cut
Magnacorp	Maintain price	5, 5	0, 8
	Cut	8, 0	1, 1

affected by the decisions of the sellers: Customers. The customers are not treated as players in the game. Thus, although the monopoly price is the cooperative solution for the two sellers, it can be inefficient, when the benefits to customers are also taken into account, and economists have considered the lower price to the more efficient. "Cooperative" is not always necessarily "good," and antitrust policies have aimed at encouraging the efficient low price. But the example suggests that antitrust policies may not be needed. (We will reconsider this in Chapter 11.) This approach to oligopoly pricing is due to Prof. Warren Nutter (1923–1979) of the University of Virginia.

Social dilemmas are a very important category of games, but dominant strategies and dominated strategies will be important in other kinds of games as well. Some games have dominant strategy equilibria and others do not. Also, some games that do have dominant strategy equilibria are not social dilemmas. A dominant strategy equilibrium does not have to be inferior. Here is a business example that illustrates the possibility of a dominant strategy equilibrium that is not inferior to other strategy combinations.

5. COLLABORATIVE PRODUCT DEVELOPMENT

Omnisoft Corp. and Microquip, Ltd. are considering a collaborative project of research and product development. Each company has two strategies to choose between: To commit plenty of resources to the project, or to hold back, and only commit minimal resources to the project. A difficulty that can arise in this sort of game is that

Table 3.5. Collaborative Product Development.

		Omnisoft	
		Commit	Hold back
Microquip	Commit	5, 5	2, 3
	Hold back	3, 2	1, 1

neither partner can monitor or enforce the commitment of effort and resources by the other. In this case, however, we are assuming that the project has "spinoff" technologies, that is, technologies that the two companies can put to use profitably even if the collaborative project does not work out, and that will make a difference. The payoffs (in billions) are shown in Table 3.5.

When we examine the game to determine whether or not there are dominant strategies, we find that there are. Suppose Omnisoft's strategy is to commit. Microquip can then earn 5 billion by choosing "commit" as its own strategy, but only three by choosing "hold back." If Omnisoft's strategy were to hold back, then Microquip can earn 2 billion by choosing "commit" and only 1 billion by choosing "hold back." Thus, "commit" is the dominant strategy for Microquip. By symmetrical reasoning, "commit" is the dominant strategy for Omnisoft as well. Since both players have dominant strategies, this game has a dominant strategy equilibrium. The dominant strategy equilibrium is for each of the two firms to choose commit. This leads to the best possible outcome, with each firm earning 5 billion of profits.

This example contrasts with some of the other ones in that the dominant strategy equilibrium is just the outcome that the two companies want. They could not improve on this outcome even if they merged. The equilibrium at (commit,commit) is not only the dominant strategy equilibrium in this game. It is also the "cooperative solution," that is, the outcome the players would choose if they could choose any pair of strategies at all. Games like this — in which the dominant strategy equilibrium is also the cooperative solution — do not play a large part in the literature of game theory. Most likely that is because they do not present any problems for people and for society. Game theory is a pragmatic study, oriented toward finding and solving problems. But games in which the cooperative solution is a dominant strategy equilibrium are logically possible, and may even be fairly common in business, since, after all, non-cooperative equilibria are barriers to increasing the profits, and (as economist George Stigler observed) "business" consists of all the methods we know for eliminating barriers to increased profits.

6. GAMES WITH MORE THAN TWO STRATEGIES

Most of our examples so far in this textbook have been "two-by-two games," that is, games with just two participants each of whom must choose between just two strategies. However, most real world interactions involve more than two participants or more than two strategies or both, and sometimes the number of participants or strategies or both is very large indeed. In future chapters, we will consider games with more than two participants. Games with more than two strategies are only a little more complicated than two-by-two games, when they are represented in normal form. Only the table has to have more than two rows and columns.

For another example, let's go back to the Dumping Game. We will change it by adding a third strategy: Burning the garbage. So now each of the two homeowners will have to choose among three strategies: Hire a truck, dump, or burn. Burning will have the same effect on the value of both tracts of land. If one landowner burns, it will reduce the value of both pieces of land by $250, and if both burn, by $350, from what it would be otherwise. The payoff table is shown in Table 3.6. It differs from our previous examples only in that it has one more row and one more column.

As usual, we want to investigate the best responses of each player to the strategies that might be chosen by the other player. The best responses for each are shown in Table 3.7. Once again, we see that "dump" is a dominant strategy for Mr. Smith. By symmetrical reasoning, it is also a dominant strategy for Mr. Jones. So, we will again have a dominant strategy equilibrium and a social dilemma.

Table 3.6.   The Dumping Game with Three Strategies.

		Mr. Smith		
		Dump	Hire truck	Burn
Mr. Jones	Dump	4,000, 4,000	5,000, 3,500	4,750, 3,750
	Hire truck	3,500, 5,000	4,500, 4,500	4,250, 4,750
	Burn	3,750, 4,750	4,750, 4,250	4,650, 4,650

Table 3.7.   Best Responses for Mr. Smith.

If Mr. Jones's strategy is	The best response for Mr. Smith is
Dump	Dump
Hire truck	Dump
Burn	Dump

7. A POLITICAL GAME

Strategic choices have to be made not only in business, waste disposal, recreational games, and war — but also in routine, peace-time politics. Let's consider a political example. Richard Nixon, a highly successful Republican politician, said that the way for a Republican to be elected to public office was to "run to the right in the primary election, but run to the center in the general election." The next example gives an explanation for the advice to "run to the center in the general election."

For this example, we have two candidates: Senator Blank and Governor Gray. Although the candidates have no personal preferences for one ideology over another, Senator Blank, as a Democrat, can take a position on the political left more credibly than Governor Gray, who is a Republican. Conversely, Governor Gray can take a

Table 3.8. Vote Payoffs for Two Candidates.

		Governor Gray		
		Left	Middle	Right
Senator Blank	Left	55, 45	30, 70	50, 50
	Middle	75, 25	50, 50	70, 30
	Right	50, 50	25, 75	45, 55

position on the political right more credibly than Senator Blank. Adopting the left and right political positions are two of the strategies that they can choose. But they both have a third strategy available. Either or both of them can adopt a "middle of the road" political position.

What are the payoffs in this game? We are assuming that the two candidates don't particularly care what positions they take. Their objective is not to advance a particular ideology, but just to get elected. Some German speaking political economists express this motivation as "Stimmungsmaximieren" — maximizing the vote. That's probably a little bit of an exaggeration. What the candidate needs is not the largest possible vote but one vote over 50%. On the other hand, a "landslide victory" by a large margin can be a big advantage to the winning candidate. In any case, we can express the payoffs as the percentage of the vote the candidate can expect to receive.

Of course, that will depend where the voters are. We shall assume that the voters are distributed symmetrically, with 30% favoring the political right, 30% favoring the political left, and 40% preferring to vote for a candidate with a middle-of-the-road position. The payoffs are shown in Table 3.8. We are assuming that if both candidates adopt the same "middle of the road" position, they split the vote 50–50, and the winner is determined by random errors in counting the vote. In that case, each has an equal chance of victory. If the two candidates take different positions, the voters will vote for

Table 3.9. Best Responses for Senator Blank.

If Gray's strategy is	The best response for Blank is
Left	Middle
Middle	Middle
Right	Middle

the one nearest their position, with two exceptions. First, not all of the voters on the left will choose the Republican candidate even if he takes the left position; and similarly, not all of the voters on the right will choose the Democrat if he takes the right position. Second, if nobody takes the middle position, the middle vote divides equally.

The best responses for Senator Blank are shown in Table 3.9. As we see, the "middle-of-the-road" strategy is a dominant strategy for Senator Blank. And even though the game is not perfectly symmetrical, we can reason symmetrically, and see that the "middle-of-the-road" strategy is also a dominant strategy for Gov. Gray. Since both candidates have dominant strategies, we have a dominant strategy equilibrium.

This is not a social dilemma. This is a constant-sum game, and a constant-sum game can have no cooperative solution. Since the total vote is the same, 100%, regardless of which strategies the two candidates choose, there is no conflict between the dominant strategy equilibrium and the cooperative solution in this game. But that is looking at it from the point of view of the candidates, and not from the point of view of the voters. This dominant strategy solution might not be a very good one from the point of view of the voters. Only one political point of view, the "middle-of-the-road" viewpoint, is ever expressed. Nevertheless, in this example, the middle group are the largest group. That probably was the case in the mid-Twentieth century when Nixon was successful, but American politics seems to have changed in ways that make the middle ground less predominant.

8. A TEXTBOOK-WRITING GAME

The concept of a dominant strategy equilibrium is a powerful one, but we will find that not all games have dominant strategy equilibria. Here is an example of a game of three strategies in which there is no dominant strategy equilibrium.

Professor Heffalump and Dr. Boingboing are the authors of rival textbooks of game theory. Their books are of equal quality in every way except length.[1] Both authors know that professors will usually choose the longer book if one is longer than the other. Each would like to get the larger audience, but writing a longer book is a bigger effort, so neither author wants to write a book longer than it needs to be to capture the bigger audience. Each of the two authors can choose among the following three strategies: Write a book of 400 pages, 600 pages, or 800 pages. The payoffs are shown in Table 3.10. The payoffs might be royalties, in thousands of dollars per year, or might also reflect some subjective benefits from being in first place — whatever it is that the authors want to get out of their writing.

Table 3.11 shows the best responses for Dr. Boingboing, depending on Prof. Heffalump's strategy choice. What we see is that Dr. Boingboing will want to choose a different strategy when Dr. Heffalump chooses a 400 page text than he will want to choose if Professor Heffalump chooses a 600 or 800 page text. Dr. Boingboing's idea is to write a text just one step longer than Dr. Heffalump's text, if he can. It follows that there is no one strategy that is Dr. Boingboing's best response to each of the different strategies Professor Heffalump might choose. In other words, there is no dominant strategy for Dr. Boingboing. And since the game is symmetrical, we can reason in

---

[1]This is an unrealistic simplifying assumption, since, of course, this book is far better than any other game theory textbook, page for page, in every way. By now you have probably noticed that we do not hesitate to make unrealistic simplifying assumptions in order to get the point across. Of course, textbooks vary in many ways, including the quality of the writing, the production, and whether or not the author has a sense of humor.

Table 3.10.   Writing a Game Theory Textbook.

		Prof. Heffalump		
		400 p.	600 p.	800 p.
Dr. Boingboing	400 p.	45,45	15,50	10,40
	600 p.	50,15	40,40	15,45
	800 p.	40,10	45,15	35,35

Table 3.11.   Best Responses for Dr. Boingboing.

If Prof. Heffalump's strategy is	The best response for Dr. Boingboing is
400	600
600	800
800	800

just the same way and find that there is no dominant strategy for Professor Heffalump either.

So, the textbook writing game gives us an example of a game in which there is no dominant strategy. If we are to find a "solution" for a game like this one, it will have to be a different kind of solution. We will go on to investigate that in the next chapter. So, we set this example aside for now, and will return to it in Chapter 4.

9. SUMMARY

One objective of game theoretic analysis is to discover stable and predictable patterns of interactions among the players. Following the example of economics, we call these patterns "equilibria."

Since we assume that players are rational, their choices of strategies will only be stable if they are best response strategies — the player's best response to the other players' strategies. If there is one strategy that is the best response to every strategy the other player or players might choose, we call that a dominant strategy. If every player in the game has a dominant strategy, then we have a dominant strategy equilibrium.

A dominant strategy equilibrium is a non-cooperative equilibrium, which means that each player acts independently, not coordinating the choice of strategies. If the players in the game are able to commit themselves to a coordinated choice of strategy, the strategies they choose are called a cooperative solution. It is possible that the cooperative solution may be the same as a dominant strategy equilibrium, but then again it may not be.

One important class of games with dominant strategy equilibria are the social dilemmas. The familiar Prisoner's Dilemma is a typical example of this class. What the Prisoner's Dilemma has in common with every other social dilemma is that it has a dominant strategy equilibrium that conflicts with its cooperative solution.

We have seen applications to environmental management, advertising, business, partnership and politics. It seems clear that dominant strategy equilibrium has a wide range of application. Nevertheless, we have also seen that not all games have dominant strategies or dominant strategy equilibria.

Q3. EXERCISES AND DISCUSSION QUESTIONS

Q3.1. Solving the Game

Explain the advantages and disadvantages of the dominant strategy equilibrium as a solution concept for non-cooperative games.

Q3.2. Effort Dilemmas

One family of "social dilemmas" arises where a group of people are involved in some task that depends on the efforts of each of them. The strategy choices are "work" and "shirk." In an effort dilemma, one person's shirking places the burden of increased effort on the other(s). Then the payoff table could be something like Table 3.12.

Table 3.12. An Effort Dilemma.

		Mr. Jones	
		Work	Shirk
Mr. Smith	Work	10, 10	2, 14
	Shirk	14, 2	5, 5

a. Are there any dominant strategies in this game? What?
b. Has this game a dominant strategy equilibrium?
c. What would be the cooperative solution in this game?
d. In the Pacific Northwest of the United States, in the middle of the twentieth century, between two and three dozen plywood companies were worker owned. The worker-owners controlled the companies by majority rule and hired and fired the managers. However, the managers typically had discretion to assign and discipline work, and were, if anything, more powerful than managers in similar profit-seeking companies. Does the example suggest an explanation of this?

Q3.3. Public Goods

In economics, a "public good" is defined as a good or service with two characteristics: (1) It is not practically possible to charge the agents who benefit from the good and make their payment a condition for them to get the benefit of the good (the good is "non-exclusive"). (2) The cost of providing the good is the same regardless of the number of beneficiaries there are (the good is "non-rivalrous"). Here is an example in the form of a two-person game. Joe and Irving each can choose between the strategy of producing one unit of a public good or no units of a public good. If the public good is not produced, each gets a payoff of 5. If either Joe or Irving produces a unit of the public good, both players' payoffs are

Table 3.13. Gas.

		Island	
		Gas	No
Mainland	Gas	−8, −8	3, −10
	No	−10, 3	0, 0

increased by 2, but the agent who produces the good pays a cost that reduces his payoffs by 3. Thus, for example, if Joe produces and Irving does not, Joe's payoff is $5 + 2 − 3 = 4$ and Irving's payoff is $5 + 2 = 7$.

a. Express this as a payoff table.
b. Analyze the game in terms of dominant strategies.
c. Is it a social dilemma? Why or why not?

Q3.4. Poison Gas

Mainland and Island are rival nations, often at war, and both can produce and deploy poison gas on the battlefield. In any battle, the payoffs to using gas are as in Table 3.13.

a. Are there any dominant strategies in this game? What?
b. Has this game a dominant strategy equilibrium?
c. What would be the cooperative solution in this game?
d. Is this game a social dilemma? Why?
e. Historically, poison gas was used in World War I, but not in World War II, although Germany opposed France and Britain in the second war as it had in the first. The consensus explanation was that if one side were to use gas in a battle, the other side would retaliate with gas in subsequent battles, with the result that the first user would nevertheless be worse off as a result. Discuss the limitations of the Dominant Strategy Equilibrium in the analysis of social dilemmas in the light of this contrast.

Q3.5. Water Game

Recall from Chapter 2: Eastland and Westria share the valley of Southflowing River, which forms the boundary between them. Each country controls some of the northern tributaries of the river, and could divert water from the tributary streams for their own use. If just one country diverts water from the tributaries, the net benefit to that country will be 3 billion Euros, but the other country will lose 4 billion. However, if both countries divert water from the tributaries, each country will suffer a net loss of 2 billion. Using the normal form of this game, determine whether it has dominant strategies and/or an equilibrium.

Q3.6. Happy Hour

Refer to problem 4 in Chapter 1.

a. Are there any dominant strategies in this game? What?
b. Has this game a dominant strategy equilibrium?
c. What would be the cooperative solution in this game?
d. Is this game a social dilemma? Why?

Q3.7. The Training Game

Two firms hire labor from the same unskilled pool. Each firm can either train its labor force or not. Training increases productivity but there are spillovers in that the rival can hire the trained workers away. Thus, both firms always face the same productivity net of pay, although the net productivity is higher if more workers are trained. The relation between the proportion trained and net productivity is shown in Table 3.14.

(Each firm either trains all of its workers or none, so 0, 50%, and 100% trained are the only possibilities.) Each firm's profit is the net productivity shown in the table minus training cost. Training cost is three if the firm chooses to train its employees and zero otherwise.

Table 3.14. Training.

Proportion trained	Net productivity
0	5
50%	7.5
100%	10

a. Who are the "players" in this game?
b. What are their strategies?
c. Express this as a game in normal form.
d. Are there dominant strategies in this game? What?
e. Is there a dominant strategy equilibrium? What?

CHAPTER 4

Nash Equilibrium

In the last chapter, we saw that some games have dominant strategies and dominant strategy equilibria. When a dominant strategy equilibrium exists, it provides a very powerful analysis of non-cooperative choices of strategy. However, we also saw that some games do not have dominant strategy equilibria. In order to analyze those games, we will need a different equilibrium concept: Nash equilibrium. The Nash equilibrium concept was named after the mathematician who discovered it, John Nash. Nash's life had its sad aspects, which are related in a biography and cinema based on it.[1] Nevertheless, no one had a greater impact on game theory than John Nash.

As usual, we want to begin with an example. In fact, we already have the example, from the end of the last chapter: the game of writing a textbook.

1. A TEXTBOOK-WRITING GAME, CONTINUED

In this game, recall, the players are two professors who are writing rival textbooks of game theory. Each one expects a better payoff if his textbook is longer than the rival textbook. The strategies for each author are write a book of 400, 600, or 800 pages. The payoffs are shown in Table 4.1. This table is the same as Table 3.12 in Chapter 3 but is repeated here for convenience.

---

[1]The book is Sylvia Nasar, *A Beautiful Mind* (New York: Simon and Schuster, 1998). The Cinema (2001) has the same title.

Table 4.1. Writing a Game Theory Textbook (Repeats Table 3.12, Chapter 3).

		Prof. Heffalump		
		400 p.	600 p.	800 p.
Dr. Boingboing	400 p.	45,45	15,50	10,40
	600 p.	50,15	40,40	15,45
	800 p.	40,10	45,15	35,35

We already know that this game does not have a dominant strategy equilibrium because Dr. Boingboing's best response strategy depends on the strategy Prof. Heffalump chooses, and vice versa. Table 4.2 shows each author's best response, depending on which strategy the other author chooses. This is a repeat of Table 3.13 from Chapter 3, except that it does not matter which author is responding to which.

Looking at Table 4.2, we can see that the 800 page strategy has an interesting property. If both authors choose the strategy of writing 800 page textbooks, each one is choosing his best response to the other author's strategy. Professor Heffalump is choosing his best response to Dr. Boingboing's strategy, and Dr. Boingboing is choosing his best response to Professor

> **A Closer Look:** John Forbes Nash, 1928–2015
>
> John Forbes Nash, a mathematician, was born in Bluefield, West Virginia. While a graduate student at Princeton, he proved that the non-cooperative equilibrium concept now called "Nash equilibrium" is applicable to non-constant sum games, for which he shared the Nobel Memorial Prize in 1994. He also developed a mathematical bargaining theory for cooperative games and made many other contributions to mathematics. Nash was troubled for many years by mental illness, portrayed in the 2001 movie, *A Beautiful Mind.*

Table 4.2. Best Responses for Authors of Game Theory Texts.

If one author's strategy is	The best response for the other author is
400	600
600	800
800	800

Heffalump's strategy. And that is **not** true for any other combination of strategies. For example, if Dr. Boingboing writes a 600 page text, and Professor Heffalump writes a 400 page text, then Dr. Boingboing is choosing his best response to Professor Heffalump's strategy; but Professor Heffalump can do better by increasing the length of his text to 800 pages, for a payoff of 50 (as we see at the middle cell of the right column of Table 4.1). However, at that point, Dr. Boingboing can do better. Dr. Boingboing's best response to a strategy of 800 pages is to increase his own length to 800 pages. Thus, neither the middle cell in the left or right column of Table 4.1 has both players choosing their best response. Only the bottom right cell, where each one chooses a length of 800 pages, has that "interesting property."

> **HEADS UP!**
>
> Here are two key concepts we will develop in this chapter:
>
> **Nash Equilibrium:** For any game in normal form: if there is a list of strategies, with one strategy per player, such that each strategy on the list is the best response to the other strategies on the list, that list of strategies is a Nash equilibrium.
>
> **Common Knowledge of Rationality:** Each person is rational (in the sense that they choose the best response) and each knows that the other is rational and knows that the other knows they are rational, and so on.

This "interesting property" means that (800, 800) is the Nash equilibrium of the textbook writing game. A Nash equilibrium is defined as a list of strategies, with one strategy for each player, such that each strategy is the best response to all the strategies played by other players. In this case, since we have only two players, there are only two strategies in the list: one for Professor Heffalump and one for Dr. Boingboing.

Even though it is not a dominant strategy equilibrium, this Nash equilibrium is the predictable result of rational, self-interested, non-cooperative play in the textbook writing game.

2. NASH EQUILIBRIUM

The Textbook-Writing Game gives us an example of a game that has no dominant strategies, but does have a Nash equilibrium. Like the dominant strategy equilibrium, the Nash equilibrium reflects rational action on the part of both authors. Neither player can do better by choosing some other strategy so long as the other player persists in the strategy he has chosen. The strategies are chosen independently, with no coordination. Also like the dominant strategy equilibrium, the Nash equilibrium is non-cooperative. Looking back at Table 4.1, we see that both authors would be better off if they both wrote texts of 400 or even 600 pages. The cooperative solution of this game would be for both to write textbooks 400 pages in length.

One important difference between this example and the examples of dominant strategy equilibrium is in the role of information. For the example of writing a game theory textbook, neither author can make a decision without first making a conjecture[2] as to which strategy the other will take. In a game with dominant strategies, the decision-maker would not need to make any such conjecture, since the dominant strategy is the best response whatever strategy the other chooses. In any interactive decision — any game — the payoff

---

[2]A dictionary definition of the word "conjecture" is "an opinion or judgment of fact formed on the basis of incomplete information."

to a decision-maker depends partly on the decision made by the other. Thus, the best response also depends on the strategy chosen by the other. If the other player has a dominant strategy, that is a little easier — I only need to know that my counterpart is rational, since that means my counterpart will choose the dominant strategy. This applies an assumption that is common in game theory: *common knowledge of rationality*. This assumption says that each person is rational and *knows that the other is rational, and knows that the other knows that both are rational* and so on. For a game with dominant strategies, this assures us that the dominant strategy equilibrium will occur. For some other Nash equilibria, however, it may be a little more complicated. This will be further discussed in the next chapter and in Chapters 16 and 17.

However, the concept of Nash equilibrium is more general than dominant strategy equilibrium. Notice that every dominant strategy equilibrium is also a Nash equilibrium, just as every cocker spaniel is a dog. For example, remember the Prisoner's Dilemma. For each player, "confess" is a dominant strategy, and there is a dominant strategy equilibrium. But it is equally true that when one prisoner chooses "confess," then "confess" is the other prisoner's best response. Each prisoner is choosing his best response to the strategy chosen by the other, so it is indeed a Nash equilibrium. But not every Nash equilibrium is a dominant strategy equilibrium, just as not every dog is a cocker spaniel. Some Nash equilibria are not like dominant strategy equilibria in that the best response strategy for one player depends on the strategy chosen by the other player.

Now let's put these ideas to work with another example, the choice of radio formats by two rival radio stations.

3. A RADIO GAME

We have seen that not all games have dominant strategy equilibria. Here is another example, a game in which two radio stations choose among alternative broadcasting formats. For the example the two players are two radio stations, W*** and K†††, and their strategies are the broadcasting formats they might choose: Top 40, Classic

Rock, and a Blend of the two. Because they have somewhat different, overlapping audiences, different reputations, and different disc jockeys, their payoffs are not symmetrical. The payoffs are proportional to their net advertising revenues. The game is shown in Table 4.3. The first payoff is to W*** and the second is to K†††.

The best responses for the two stations are shown in Tables 4.4 and 4.5.

We see that for each of the two stations, the best response may be different, depending on the strategy chosen by the other station. Neither station has a dominant strategy in this game. Now, suppose that W*** chooses Blend and K††† chooses Top 40. In that case, we

Table 4.3.   A Radio Formats Game.

		K†††		
		Top 40	Classic Rock	Blend
W***	Top 40	30,40	50,45	45,40
	Classic Rock	30,60	35,35	25,65
	Blend	40,50	60,35	40,45

Table 4.4.   Best Responses for W***.

If the strategy of K††† is	The best response for W*** is
Top 40	Blend
Classic Rock	Blend
Blend	Top 40

Table 4.5.   Best Responses for K†††.

If the strategy of W*** is	The best response for K††† is
Top 40	Classic Rock
Classic Rock	Blend
Blend	Top 40

see that each station is choosing its best response to the strategy chosen by the other. Consulting Table 4.4 (or Table 4.3), we see that Blend is W***'s best response if K†††️ chooses Top 40, and conversely, consulting Table 4.5 (or Table 4.3), we see that Top 40 is K†††'s best response if W*** chooses Blend. In short, (Blend, Top 40) is the Nash equilibrium of the Radio Formats Game. And that is NOT true of any other pair of strategies. Consider, for example, the bottom right cell where both choose Blend. Then either can do better by switching to Top 40. Examining each cell in turn, we will find that at least one player can do better by switching to another strategy, except for the lower left where W*** chooses blend and K††† chooses Top 40.

Even though it is not a dominant strategy equilibrium, this Nash equilibrium is the predictable result of rational, self-interested, non-cooperative play in the Radio Formats Game. Neither player can benefit by deviating unilaterally from the Nash equilibrium. In that sense, the Nash equilibrium is self-enforcing.

4. AN HEURISTIC METHOD OF FINDING NASH EQUILIBRIA

Checking each combination of strategies to find a Nash equilibrium can be tedious as the number of strategies gets larger. There are heuristic methods, using visualization, to eliminate non-equilibrium strategies and find the equilibria if there are any. In case the word "heuristic" is new to

Definition: *Heuristic Methods* — Heuristic methods of problem solving are methods that are fast and usually reliable, but are informal and may be inconclusive because they can fail in unusual cases.

you, we can say that heuristic methods of problem solving are methods that are informal or inconclusive, by contrast with mathematical solution methods; but are also fast and usually reliable. In this case, the method is informal, involving visualization and a little drawing.

Table 4.6. Writing a Game Theory Textbook with Underlines (Approximately repeats Table 3.11, Chapter 3).

		Prof. Heffalump		
		400 p.	600 p.	800 p.
Dr. Boingboing	400 p.	45,45	15,<u>50</u>	10,40
	600 p.	<u>50</u>,15	40,40	15,<u>45</u>
	800 p.	40,10	<u>45</u>,15	<u>35,35</u>

Table 4.7. The Radio Formats Game (Approximately repeats Table 4.3, Chapter 4).

		Kttt		
		Top 40	Classic Rock	Blend
W***	Top 40	30,40	50,<u>45</u>	<u>45</u>,40
	Classic Rock	30,60	35,35	25,<u>65</u>
	Blend	<u>40,50</u>	<u>60</u>,35	40,45

The idea behind the Nash equilibrium is that a strategy will be chosen only if it is a best response. One simple way to highlight the best response is to underline the payoff that is the best response for each strategy. This is illustrated for the Textbook Game in Table 4.6. As we see, a Nash equilibrium shows up as a cell in which both payoffs are underlined. Table 4.7 illustrates the underlining method for the Radio Formats Game.

EXERCISE: Revisit the examples and exercises in Chapter 3 and put in the underlines. Don't worry about marking in the book — you're going to want to keep it anyway.

5. DOMINANT STRATEGIES AND NASH EQUILIBRIUM

We have seen that a dominant strategy equilibrium is also a Nash equilibrium. Let's explore the relation of dominant strategies and Nash equilibrium a little further. Here is another example of an

Table 4.8. Breweries.

		Little Colony		
		Amber Bitter	Extra Stout	Belgian Dubbel
	Amber Bitter	3,5	7,3	3,4
Froggy Bottom	Extra Stout	4,4	3,2	6,2
	Belgian Dubbel	1,6	2,4	4,1

Table 4.9. Breweries, with Underlines.

		Little Colony		
		Amber Bitter	Extra Stout	Belgian Dubbel
	Amber Bitter	3,<u>5</u>	<u>7</u>,3	3,4
Froggy Bottom	Extra Stout	<u>4</u>,<u>4</u>	3,2	<u>6</u>,2
	Belgian Dubbel	1,<u>6</u>	2,4	4,1

unsymmetrical game with three strategies for each player. Froggy Bottom Brewery and Little Colony are craft breweries that compete in some markets but not others. Each is considering bringing out a new brew for the coming season. The strategies are the styles: Amber Bitter, a widely popular brew, Extra Stout, which appeals to a specialized taste, or Belgian Dubbel, a fun high alcohol brew. The payoffs are shown in Table 4.8, on a scale of ten.

As a first step to find the Nash equilibrium, we underline the best responses. This is shown in Table 4.9.

Examining the underlined table, we see that the Nash equilibrium occurs when Froggy Bottom brews Extra Stout and Little Colony brews Amber Bitter. We can also see that, for Little Colony, Amber Bitter is a dominant strategy. This means that the decision is a little easier for Little Colony. Their best response is Amber Bitter no matter what Froggy Bottom does. Conversely, though, Froggy Bottom has to make a judgment or conjecture as to which strategy Little Colony will choose in order to discover what their best response will be. That is not very hard in this case: Knowing that

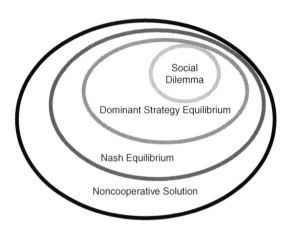

Figure 4.1.   Non-cooperative Solution Concepts.

Little Colony is rational, they can anticipate that Little Colony will choose Amber Bitter. This illustrates a common assumption that underlies the Nash equilibrium: "Common Knowledge of Rationality." That is, we always assume not only that each player is rational, but that each knows that the other is rational, and that each knows the other knows that the first is rational, and so on.

In this example, the game does not have a dominant strategy equilibrium, even though there is a dominant strategy for one brewery. This, again, illustrates that the Nash equilibrium is more general than the dominant strategy equilibrium. This is illustrated by the Venn diagram in Figure 4.1. A social dilemma is a particular kind of dominant strategy equilibrium, and a dominant strategy equilibrium is a particular kind of a Nash equilibrium, and a Nash equilibrium is a particular kind of non-cooperative solution. As we will see in Chapters 16 and 17, there are some non-cooperative solutions that are not Nash equilibria.

6. ANOTHER OLIGOPOLY PRICING GAME

Here is another example to reinforce your understanding. It is similar to the Pricing Dilemma in Chapter 3, Section 4, and like that example it builds on the ideas of Professor Warren Nutter. It is,

Table 4.10. Another Nutter Pricing Game.

		Grossco		
		Low	Medium	High
Magnacorp	Low	20,20	80,10	90,5
	Medium	10,80	60,60	150,15
	High	5,90	15,150	100,000

however, a little more complex. Where the Pricing Dilemma allowed only two prices as strategies, the competitive price and the monopoly price, this example will allow the two firms to choose an intermediate price, so there are three strategies in all.

The players are again Magnacorp and Grossco. Their strategies are to ask a high (monopoly) price, a low (competitive) price, or a medium price somewhere in between. If they charge different prices then the firm with the lower price sells to most of the market and so makes greater profit, although cutting price still further (as from medium to low when the other seller is asking a high price) will reduce profits. Once again, this game is symmetrical. The payoffs are shown in Table 4.10, and the best responses are underlined. We see that this game does not have a dominant strategy. For each company, if the other company asks a medium or low price, then a low price is the best response; but if the other company asks a high price, the medium price is the best response. However, the lowest price is the only Nash equilibrium. In general, we may model oligopoly prices as Nash equilibria. This will be discussed in more detail in Chapter 15, and this game will be further discussed in Chapter 17.

7. A RETAIL LOCATION GAME

The examples so far in this chapter are simple in some important ways, and we will see in later examples that more complex games can have more complex results. Here is a first example to illustrate some of the complexities. It will be a retail location game. The two players are two department stores, Nicestuff Stores and Wotchaneed's. Each is planning a new location in Medium City or its suburbs. Their strategies are the four locations that one or both of them might choose. The strategies and payoffs are shown in Table 4.11. The payoffs might be thought of as annual profit, in millions.

The best responses for Nicestuff and Wotchaneed's are shown as Tables 4.12 and 4.13. As a first step, we might look for a Nash equilibrium in this game, perhaps by underlining the best response

Table 4.11. A Retail Location Game.

		Wotchaneed's			
		Upscale Mall	Center City	Snugburb	Uptown
Nicestuff Stores	Upscale Mall	3,3	10,9	11,6	8,8
	Center City	8,11	5,5	12,5	6,8
	Snugburb	6,9	7,10	4,3	6,12
	Uptown	5,10	6,10	8,11	9,4

Table 4.12. Best Responses for Nicestuff.

Wochaneed's Chooses	Best response for Nicestuff
Upscale	Center City
Center City	Upscale
Snugburb	Center City
Uptown	Uptown

Table 4.13. Best Responses for Wotchaneed's.

Nicestuff Chooses	Best response for Wochaneed's
Upscale	Center City
Center City	Upscale
Snugburb	Uptown
Uptown	Snugburb

payoffs. We find that this game has two Nash equilibria! Whenever one of the two companies locates in Upscale Mall and the other locates in Center City, we have a Nash equilibrium.

For a game like this, with two or more Nash equilibria, we encounter a new problem. Which Nash equilibrium will we observe? This is a problem for the game theorist, but also for the players in the game. Further examples will be discussed in the next chapter.

8. SUMMARY

In a game that does not have a dominant strategy equilibrium, the strategy choices of all the players will nevertheless be stable, predictable and rational if every player is playing his best response to the strategies the other players play. In that case, we call it a Nash equilibrium. Remember that a dominant strategy equilibrium is a kind of Nash equilibrium, but there are other Nash equilibria that are not dominant strategy equilibria. A Nash equilibrium is a non-cooperative equilibrium, and therefore may or may not agree with a cooperative solution for the game. Nash equilibria may be found by elimination, that is, by eliminating all strategy pairs that are not both best responses to one another. This can be visualized by underlining the best responses in each row or column. Nash equilibrium is a very general "rational solution" to a game, but, as we have seen, it may not be unique; and that presents some problems. We will address them in the next chapter.

Q4. EXERCISES AND DISCUSSION QUESTIONS

Q4.1. Solving the Game

Explain the advantages and disadvantages of Nash equilibrium as a solution concept for non-cooperative games.

Q4.2. Location, Location, Location

Not all location problems have similar solutions. Here is another one: For this example, we have two department stores: Gacey's and

Table 4.14.   Payoffs in a New Location Game.

		Gacy's			
		Uptown	Center City	East Side	West Side
Mimbel's	Uptown	30,40	50,95	55,95	55,120
	Center City	115,40	100,100	130,85	120,95
	East Side	125,45	95,65	60,40	115,120
	West Side	105,50	75,75	95,95	35,55

Mimbel's. Each of the two stores has to choose a location for its one store in Gotham City. Each store will choose one of four location strategies: Uptown, Center City, East Side or West Side. The payoffs are shown in Table 4.14.

a.   Does this game have Nash equilibria? What strategies, if so?
b.   Compare and contrast this game with the location game in the chapter.
c.   What would you say about the relative importance of congestion in the location decisions of the firms in the two cases?

Q4.3. Sibling Rivalry

Refer to Chapter 2, Question 1.

a.   Discuss this game, from the point of view of non-cooperative solutions.
b.   Does it have a dominant strategy equilibrium?
c.   Determine all the Nash equilibria in this game.
d.   Do some Nash Equilibria seem likelier to occur than others? Why?

Q4.4. Hairstyle

Shaggmopp, Inc. and Shear Delight are hair-cutting salons in the same strip mall, each groping for a market niche. Each can choose

Table 4.15. Payoffs for Haircutters.

		Shear		
		Punker	Sophisticate	Traditional
Shaggmopp	Punker	35,20	50,40	60,30
	Sophisticate	30,40	25,25	35,55
	Traditional	20,40	40,45	20,20

Table 4.16. A Pet Food Advertising Game.

		Woofstuff		
		Facebook	Radio	Television
Arfyummies	Facebook	6,6	2,4	3,3
	Radio	3,3	3,8	7,2
	Television	4,2	9,2	2,7

one of three styles: punker, contemporary sophisticate, or traditional. Those are their strategies. They already have somewhat different images, based on the personalities of the proprietors, as the names may suggest. The payoff table is shown as Table 4.15.

a. Are there any dominant strategies in this game?
b. Is there a dominant strategy equilibrium?
c. Are there any Nash equilibria?
d. How many? Which? How do you know?

Q4.5. A Dog's Dinner

Arfyummies and Woofstuff are rival pet food companies and are choosing among advertising media. Their strategies are the advertising media that they might choose: Facebook, radio, or television. Their payoffs are shown in Table 4.16.

What Nash equilibria exist, if any?

CHAPTER 5

Games with Two or More Nash Equilibria

In some ways, Chapter 4 is the key chapter in this book. Nash equilibrium, the topic of that chapter, plays a role in all analyses of non-cooperative games, and can enter into cooperative game examples as well. Thus, we will be using those concepts again and again as we continue through the book. However, as we have seen, some games have two Nash equilibria, and it is quite possible for a game to have more than two. We will explore some such examples in this chapter, including some relatively simple games that have often been used in game theory and have become what we might call "classical cases."

1. DRIVE RIGHT!

We have seen that the Retail Location Game has two Nash equilibria, where one of the two companies locates in Upscale Mall and the other locates in Center City. Let's look at another, somewhat simpler example with more than one Nash equilibrium. This will be another two-by-two game — two players, each with two strategies — a bit like the Prisoner's Dilemma. Most of the readers of the book will have experienced the play of this game at one time or another!

In some countries, it is customary to drive on the right side of the road, while in others (Britain, India, and Japan, for example) it is customary to drive on the left side. Let us think of this in terms of game theory and Nash equilibrium. A Mercedes and a Buick are approaching one another on an otherwise deserted two-lane road. Each motorist chooses between two strategies: drive on the left or drive on the right. If both choose the same strategy, then all is well; but otherwise, there is the risk of a crash. (When they are approaching one another, recall, one driver's right is the other's left, so by driving on the same side — each from her own point of view — they pass one another safely. Otherwise, they collide.) This game is shown in Table 5.1, with payoffs in rough proportion to the outcomes.

Here we find a difference from the Prisoner's Dilemma, though. While the Prisoner's Dilemma has a unique dominant strategy equilibrium, the Drive Right Game does not.

HEADS UP!

Here are some concepts we will develop as this chapter goes along:

Coordination Game: A game with two or more Nash equilibria in which the Nash equilibria exist when both choose the same strategy is called a coordination game.

Anticoordination Game: A game with two or more Nash equilibria in which the Nash equilibria exist when two players choose appropriately different strategies is called an anticoordination game.

Focal Equilibrium: In a coordination game, if some clue can lead the participants to believe that one equilibrium is more likely to be realized than the other, the more likely equilibrium is called a focal equilibrium.

Table 5.1.   The Drive Right Game.

		Mercedes	
		Right	Left
Buick	Right	5, 5	−100, −100
	Left	−100, −100	5, 5

In this case, we have an equilibrium whenever both players choose the same strategy. Both are best off if they do play a Nash equilibrium, while they are both worse off if they do not. A game that fits this description is known as a *coordination game*. In a game like this, the decision-makers face a problem of information. The motorists will need some additional information, outside the rules and payoffs of the game, to make a good decision in this coordination game.

> **Definition:** *Focal Equilibrium —* In a game with two or more Nash equilibria, an equilibrium that attracts the attention of all players on the basis of some information outside the game but available to all players is called a *focal equilibrium* or sometimes a *Schelling point.*

But, in this case, the problem is very easy. We know, of course, that the custom is to drive on the left in some countries. In this case, the custom or law can be the source of the information they need. The drivers only need to know what country they are in!

> **Definition:** *Coordination Game —* A game with two or more Nash equilibria that occur when both players choose the same strategy is called a *coordination game.*

Conversely — moreover — the existence of two Nash equilibria provides an explanation for the fact that it is customary to drive on the left in some countries and on the right in other countries. Since either custom is a Nash equilibrium, either is stable.

The custom to drive on one side or the other in this case provides us with the information that enables both drivers to focus their attention on one Nash equilibrium rather than the other. Thus, the customary equilibrium is often called a *focal equilibrium* or a *focal point*. Since this idea originated with Thomas Schelling, it is also sometimes called by his name, as a *Schelling point* or a *Schelling focal equilibrium*.

2. HEAVE-HO

Let's look at another example with more than one Nash equilibrium and think a little further about focal equilibria. This will be another two-by-two game — two players, each with two strategies — again a bit like the Prisoner's Dilemma, and it will also involve driving! We will call it the Heave Ho game. In this game, Jim and Karl are driving down a back road and are stopped by a fallen tree across the road. If they can move the tree out of the way, then they can go ahead. Otherwise, they will have to turn back. In order to move the tree, there will both have to heave as hard as they can to push it out of the road. So, each one has a choice of two strategies: Heave or don't heave.

If they both heave, they can successfully push the tree out of the road. We will call the payoff in that case five to each of them. If one heaves and the other does not, the one who heaves is injured, a minus 10, while the other one has the minor inconvenience of taking the injured party to the hospital: a payoff of zero. If neither man heaves, so that they have to go back, we consider that a payoff of 1 to each. The payoffs are shown in Table 5.2.

Looking at this payoff table, perhaps a drawing a few underlines, we see that there are two Nash equilibria: (Heave, Heave) and (don't, don't). However, these two equilibria are different in one obvious and important way. The first one gives each player a better payoff than the second one — and in fact a better payoff than any another strategy combination. We express this by saying that the

Table 5.2. The Heave-Ho Game.

		Jim	
		Heave	Don't
Karl	Heave	5, 5	−10, 0
	Don't	0, −10	1, 1

"heave, heave" equilibrium is *payoff dominant*.[1] This unique characteristic might seem to make the equilibrium at (heave, heave) a focal equilibrium. It is so obviously superior (it would seem) that each player could assume that the other player would heave and so would himself choose to heave.

And yet experience might still trump this reasonable line of thought. Suppose Jim knows from past experience that Karl is so lazy that he will never make a real effort. And suppose Karl knows Jim's opinion of him. Expecting Karl to shirk and not to heave, Jim will not heave either, to avoid the injury and the payoff of minus 10. And Karl can anticipate that. Even if Karl thinks Jim's opinion is mistaken, Karl will not heave for fear that he will get the injury.

> **A Closer Look:** Thomas C. Schelling, 1921–2016
>
> Born in Oakland, California, in 1921, Thomas Schelling matriculated at the University of California at Berkeley and earned his Ph.D. in Economics at Harvard. His work in government service (1945–1953) and at the Rand Corporation (1958–1959) were important influences on his ideas. Schelling was on the faculty of Harvard and Yale Universities for 31 years, and finally at the University of Maryland. Among his many important works on economic and strategic behavior, The Strategy of Conflict, 1960, stands out as a key contribution to game theory. Schelling shared the 2005 Nobel Memorial Prize with Robert Aumann.

So, we can't be sure that (heave, heave) will be the equilibrium that will occur unless we know something about Karl and Jim's opinions of one another.

---

[1] We see that the word "dominant" is used in different ways in game theory. A "payoff dominant" *equilibrium* and a dominant *strategy* are different things. Don't let that confuse you!

Another possibility in this game is that both might choose to avoid the risk of a minus 10 payoff by choosing (don't, don't.) Because it avoids the risk of a large loss, this is called a *risk dominant* Nash equilibrium. This property, too, could attract attention and make (don't, don't) a focal equilibrium. In experimental studies, both payoff-dominant and risk-dominant strategies can be attractive in different games.

Terminology: *Payoff Dominant and Risk Dominant Equilibria* — If there are more than one Nash equilibria, and (1) one of the equilibria yields a higher payoff to each player than the others do, it is said to be the *payoff dominant* equilibrium, and (2) if one of the equilibria gives the smallest maximum loss to each player, it is said to be the risk dominant equilibrium.

The Heave Ho Game is another example of a coordination game. Both players can get their very best payoff if only they can coordinate their choices of strategies. That seems pretty easy, but as we've seen, they may not succeed if they do not expect to. Success or failure in the coordination game may be partly a self-confirming prophecy

3. ONE MORE DRIVING GAME

Driving has given us some useful examples, and here is just one more. We will call it The Drive On Game. Two cars meet, crossing, at the intersection of Pigtown Pike and Hiccup Lane. Each has two strategies: wait or go. The payoffs are shown in Table 5.3. As we see, if both stop, they simply reproduce the problem, for payoffs of zero; but if both go, they will crash, for payoffs of -100. If one goes and the other waits, the one who goes "wins," getting through the intersection first, for 5, while the other goes through the intersection second (but safely) for a payoff of 1.

Table 5.3. Drive On.

		Mercedes	
		Wait	Go
Buick	Wait	0, 0	1, 5
	Go	5, 1	−100, −100

A Closer Look: Landmarks as Focal Points

Like history or experience, a prominent natural or social landmark can be the basis of a Schelling focal point. In the 1960's, in his classes at Yale University, Schelling would give his students the following mind experiment: you have to meet your friend in New York on a specific date, but you do not know the time or the place, and neither does your friend. Where will you go to meet your friend, and at what time? Most students answered clearly that they would look for the friend under the clock at Grand Central Station at noon. For students in New Haven, in the middle of the 20th century, Grand Central Station under the clock was the conventional meeting place. The convention is enough to break the uncertainty and enable the students to come up with the same meeting place. (Other residents of the New York area, or other regions, might have thought of other places. For example, tourists from other American regions might have thought instead of the Empire state building. Context counts.) And noon, of course, is a prominent landmark on the clock face: both hands straight up. Once again, the landmark is enough to resolve the uncertainty so that the students could come at the same time. It probably would have worked. "Meet me in Manhattan on the first — at the obvious time and place."

For this game, we see that there are two Nash equilibria, each of the strategy pairs at which one car waits and the other goes. In this game, it is necessary for the players to choose *different* strategies in a *coordinated* way, in order to perform a Nash equilibrium, and while they are not equally

Definition: *Anticoordination Game* — A game with two or more Nash equilibria that occur when both players choose appropriately different strategies is called an *anticoordination* game.

well off at the equilibrium, both are better off than they will be at a non-equilibrium strategy pair. In some recent writing on game theory, games such as this are called *anticoordination games*. Once again, the decision-makers will need a bit of information from outside the game in order to appropriately coordinate their strategies; but for an anticoordination game, it is necessary that each gets a different bit of information, that can signal one to go and the other to stop. One possible source of this information could be a stoplight at the intersection of Pigtown Pike and Hiccup Lane.

The traffic light is an important game-theoretic invention. There were a number of independent inventors of the traffic light or similar signals, including Lester Wire (1887–1958), William Potts (1883–1947) and Garrett Morgan (1877–1963). The first two were policemen and Morgan was an African-American inventor. Of course, these inventors were not thinking in terms of game theory — it had not yet been developed — but this provides a good example of a technology that solves a problem that is best understood in terms of game theory.

If we look back at the Retail Location Game, Table 4.11, we see that it too is an anticoordination game. This would probably occur because Center City and the Upscale Mall are the two most profitable locations, but if both stores locate in one of them, their direct competition for customers will reduce the profits for both. We will revisit it and suggest a solution to the problem the two stores face in Chapter 16.

4. PANIC BUYING

In March and April of the year 2020, in the early stages of the global pandemic of the COVID virus, problems of "panic buying" emerged. That is, expecting that stores might be closed and merchandise unavailable, people bought more than they currently needed to obtain reserves against their future need. The result, though, was that the inventories of stores and wholesalers were exhausted, so that the merchandise was unavailable. It seemed as if shortages were a self-fulfilling prophesy. As a first-stage analysis, let us treat this as a two-person game in normal form. This is shown in Table 5.4. The payoffs, as in the driving examples, are arbitrary but correspond to the relative advantages and disadvantages of the decision. In particular, we are assuming that if one player chooses "panic buy" and the other chooses "routine buy," there is no shortage and thus the panic buyer faces storage costs that are avoidable, but if both choose "panic buy" there is a shortage and the storage cost is offset by the advantage of not running short.

We see that this game has two Nash equilibria, where both players choose the same strategy. This is a coordination game, but as in the Heave-Ho game, one of the two Nash equilibria is payoff dominant and the other is risk-dominant. How would the two decision-makers determine which equilibrium is likely to occur? As in the game of choosing which side of the road to drive on, information outside the game might make the difference. If, for example,

Table 5.4.   Panic Buying.

		B	
		Panic buy	Routine buy
A	Panic buy	2, 2	−1, 0
	Routine buy	0, −1	5, 5

some new information were to warn of coming shortages, this could cause the buyers to suddenly switch from one equilibrium to the other. This seems to be what happened in the spring of 2020. Conversely, when we observe sudden, widespread and parallel shifts in the decisions of many people, this might be explained as a shift from one equilibrium to another in a coordination game. Of course, panic buying is not really a two-person game, but we will rethink it as a game with many players in Chapter 15.

5. CLASSICAL CASES: STAG HUNT

Two-by-two games have been studied very extensively in game theoretic research, beginning with the Prisoner's Dilemma. Examples of two-by-two games with more than one Nash equilibrium contrast with the Prisoner's Dilemma, and some are very important in ongoing research. They are among the "classical cases" we shall consider. The Stag Hunt is a classical game with two players and two strategies, one that has been of interest to philosophers as well as other game theorists. It originated in some of the writings of Jean Jacques Rousseau.

We suppose that a number of agents can choose between hunting a stag or hunting a rabbit. They can catch the stag only if all of them choose the strategy of stag hunting. Rousseau writes in Part II of the *Discourse on Inequality* (G. D. H. Cole translation), "If a deer was to be taken, every one saw that, in order to succeed, he must abide faithfully by his post: but if a hare happened to come within the reach of any one of them, it is not to be doubted that he pursued it without scruple, and, having seized his prey, cared very little, if by so doing he caused his companions to miss theirs."

It is conventional to think of this (for simplicity) as a two-person game in normal form. There are two hunters, each of whom can be pretty sure of catching a rabbit for dinner if he hunts alone. However, the rabbits in their country are pretty skinny, and they can both feed their families better if they work together and hunt a stag. The problem is that it will take both of them to catch the stag, and if one of them hunts a stag alone he will fail and go hungry. Game theorists often express this along the lines of payoff Table 5.5.

Table 5.5.   Stag Hunt.

		Hunter 2	
		Stag	Rabbit
Hunter 1	Stag	10, 10	0, 7
	Rabbit	7, 0	7, 7

As we examine this game, we see that it has two Nash equilibria: (Stag, Stag) and (Rabbit, Rabbit). This is another instance of a coordination game. The equilibrium at (Stag, Stag) is payoff-dominant, but as we have seen before (Rabbit, Rabbit), is risk-dominant — it avoids the risk of going home completely hungry.

If the two hunters have, in their past experience, always hunted rabbits, then this experience can create a Schelling focal point that could make (Rabbit, Rabbit) very stable, even though (Stag, Stag) is payoff dominant. The Stag Hunt game may be thought of as a metaphor for modern economic development, in that larger-scale, more productive methods will succeed only if (almost?) all participants are committed to them, and each may be reluctant to take the risk that the others will refuse to change their strategies to the potentially more productive ones.

6. CLASSICAL CASES: THE BATTLE OF THE SEXES

Another "Classical Case" we will consider is the "Battle of the Sexes." Like the Prisoner's Dilemma, it begins with a little story: Marlene and Guillermo would like to go out Saturday night. Guillermo would enjoy a baseball game, while Marlene would prefer a show. (OK, it's stereotypical, but that's what they like.) Mostly, they want to go together. They can't contact one another because the telephone company is on strike, and the e-mail system has crashed, so they are just going to try to meet together at the same place. Each one can choose between two strategies: go to the game or go to the show. The payoffs for this game are shown in normal form in Table 5.6.

Table 5.6. The Battle of the Sexes.

		Marlene	
		Game	Show
Guillermo	Game	5, 3	2, 2
	Show	1, 1	3, 5

This game has two Nash Equilibria: (Game, Game) and (Show, Show). Since both players are better off when they play the same strategy, this is another instance of a coordination game. Once again, there is the problem of determining which equilibrium is more likely to occur. In this case, neither equilibrium is better than the other, from the point of view of both players, so we don't have that sort of Schelling focal point to rely on. In the absence of some other sort of signal or information, there really is no answer. Despite its enigmatic nature — or, actually, because of its enigmatic nature — the battle of the sexes Game has played an ongoing part in game theoretic research.

7. CLASSICAL CASES: CHICKEN

Another widely studied two-by-two game, at once similar to and different from the others we have seen here, is called "Chicken." The chicken game is based on some hot rod movies from the 1950's, or perhaps on the news items that suggested the movies. The players are two hot rodders, whom we will call Mike and Neil. The game is one in which they drive their cars directly at one another, risking a head-on collision. If one of them turns away at the last minute, then the one who turns away is the loser — he is the chicken. However, if neither of them turns away, they both stand to lose a great deal more, since they will be injured or killed in a collision. For the third possibility, if both of them turn away, neither gains or loses anything. The payoffs for this game are shown in Table 5.7.

A little examination shows that this game has two Nash equilibria, one each where one hot rodder turns away and the other one

Table 5.7. Chicken.

		Mike	
		Go	Turn away
Neil	Go	−10, −10	5, −5
	Turn away	−5, 5	0, 0

goes forward. But yet again, with two Nash equilibria, and no signal or clue to define a Schelling focal point, there is no way to say which of the two equilibria is more likely. And this is not only a problem for the game theorist, but also a problem for the hot rodders. There seems to be a real danger that they will fall into the mutual disaster of the collision.

The Chicken Game seems to have had some influence on American nuclear policy during the period of tension between the United States and the Soviet Union from the 50's through the 80's. Certainly the nuclear standoff seems to bear some resemblance to a Chicken Game. It also implies a similar danger. The two hot-rodders might well choose the non-Nash equilibrium strategies "go, go." They will discover their mistakes only when it is too late to correct them. Similarly, in the game of "nuclear standoff," each of the two countries might choose to attack, expecting the other to turn back — only to discover their mistake when it is too late to correct it! The policy of "mutually assured destruction" followed in that period seems to have been designed to make the response so automatic that there was never any real possibility for one to turn back if the other were to attack. This would transform a "Chicken" game into another game, one with a single Nash equilibrium in which neither side attacks, in which peace is the only equilibrium strategy. And it seems to have worked.

The Chicken Game is another instance of an anticoordination game: the two hot-rodders have to choose opposite strategies in order to realize a Nash equilibrium. In any case, the Chicken Game, along with the Battle of the Sexes, continues to be studied as examples of the problems that can arise in game theory because Nash equilibria are not always unique.

8. CLASSICAL CASES: HAWK VS. DOVE

Another example of a two-by-two game with two Nash equilibria comes to us from biologists who study animal behavior and its evolutionary basis, but also from international relations. It is called Hawk vs. Dove. The idea behind this game is that some animals can be quite aggressive in conflicts over resources or toward prey, while others make only a show of aggression, and then run away. Similarly in international relations, a nation can adopt an aggressive or an accommodating strategy. The hawk is symbolic of the first strategy, fighting aggressively, while the dove is symbolic of the second strategy, avoiding a fight.

In population biology, the assumption is that creatures meet one another more or less at random, and dispute over some resource, using the strategies of aggression or running away. The Hawk vs. Dove Game is played out at each meeting, depending on what creatures meet. If two aggressive "hawks" meet, they will fight until both are injured, so both lose even though one of them may come out of the fight in possession of the disputed resource. If two "doves" meet, they will both run away after some show of aggressiveness. Whichever one runs away more slowly will end up in possession of the disputed resource. When a "hawk" meets a "dove," the "dove" runs away, and the "hawk" is left in possession of the disputed resource, at little or no cost.

The payoffs for this game are shown in Table 5.8. These payoffs are derived from assumptions about the benefits of gaining resources and the costs of fighting or pretending to fight, on the average. We won't go into the details of that now, however. (Some implications of these particular numbers will be seen in Chapter 15).

Once again, we see that there are two Nash equilibria: The two combinations in which the birds adopt different strategies. Notice the similarity here to the Chicken Game. Hawk vs. Dove is another example of an anticoordination game.

Table 5.8. Hawk vs. Dove.

		Bird B	
		Hawk	Dove
Bird A	Hawk	−25, −25	14, −9
	Dove	−9, 14	5, 5

Of course, hawks and doves are unlike human game players in some important ways. The rational human being of game theory is a reflective creature who considers the consequences of his actions, aims to maximize his payoffs, and chooses strategies accordingly. Hawks and doves are not like that. What can it mean to say that hawk-like aggressive behavior is a strategy, or that the hawk chooses it? We will have to leave those questions for later, because the Hawk and Dove Game, taken from population biology, is not really a two-person game. It is a game for a population. So, we will return to it, in Chapter 15, after we have begun to explore games with a large number of players.

9. AN ESCAPE-EVASION GAME

Let us consider one further example of Nash equilibrium. Patrolman Pete is pursuing Footpad Fred, a suspected burglar. Fred arrives at a dead-end on Riverfront Road, while Pete is out of sight behind him. Fred can turn either north or south. (These are his strategies). If he turns south, he can take a ferryboat to another jurisdiction and escape. If he turns north, he can hide out in his girl-friend's apartment until she gets a car and drives him to another jurisdiction, where they can escape together. Since he is able to see her, this leaves him with a better payoff. After Fred has made his decision and disappeared, Pete comes to the dead-end. Like Fred he has to decide: north or south. If he turns south and Fred also went south, then Pete can cut Fred off before he boards the ferry and arrest him. If he turns north and catches Fred at the girl-friend's apartment, he

Table 5.9. Payoffs in an Escape-Evasion Game.

		Pete	
		North	South
Fred	North	−1, 3	4, −1
	South	3, −4	−2, 2

can arrest her as an accessory, and bagging them both will get him a modest bonus for initiative. However, if he turns in the wrong direction, he will not catch Fred, and the department will be disappointed in him. Worse, if he turns north and enters Fred's girl-friend's apartment when Fred has gone south, Pete will be reprimanded for harassing the girlfriend.

The payoffs are as shown in Table 5.9.

Careful examination of Table 5.9 will show that no strategy pair in this game is a Nash equilibrium. No matter which cell we begin with, one or the other of the opponents will want to switch. The unpleasant truth is that some games have no Nash equilibrium whatever in *pure strategies*. The lists of strategies we have been dealing with so far are pure strategies. It is possible to combine pure strategies in more complicated kinds of strategies that are not so pure — but we will not get into that until Chapter 8.

This example illustrates yet another shortcoming of Nash equilibrium as a solution to non-cooperative games. The right number of solutions is exactly one. In the case of Nash equilibrium, we can have more than one or (it now seems) none at all. We will see that this new problem has a solution, but that will require some additional tools and a little more mathematics.

10. SUMMARY

This chapter has focused on games with two or more Nash equilibria. Nash equilibrium is a very general "rational solution" to a game, but it has some shortcomings from that point of view.

- The Nash equilibrium may not be unique. Some games have two or more Nash equilibria. This includes coordination and anticoordination games and other, still more complex games. In such a case, the players in the game may find it difficult to determine which Nash equilibrium will occur. This depends on the information they have available. If there is a signal or clue that enables them to see one equilibrium as being much more likely than another, that equilibrium is called a Schelling focal equilibrium. It seems that not all games with multiple equilibria have Schelling focal equilibria, however.
- Not all games (that have only a finite number of strategies) have Nash equilibria. Thus far, we have looked only at games with a finite list of strategies — usually no more than two, three or four. Out of that finite list of strategies, there may not be any Nash equilibrium.

We have considered a number of examples to illustrate these possibilities. Finally, we considered a group of "classical cases," two-person two-strategy (two-by-two) non-constant sum games that have two Nash equilibria. These coordination and anticoordination games include a wide range of possibilities. The Stag Hunt, Battle of the Sexes, Chicken and Hawk vs. Dove Game seem to represent a range of difficulties we often have to cope with in real life.

Q5. EXERCISES AND DISCUSSION QUESTIONS

Q5.1. Sibling Rivalry

Refer to Chapter 2, Question 1.

a. Discuss this game, from the point of view of non-cooperative solutions.
b. Does it have a dominant strategy equilibrium?
c. Determine all the Nash equilibria in this game.
d. Do some Nash Equilibria seem likelier to occur than others? Why?

Q5.2. Location for Complementary Services

Here is yet another location problem. John is planning to build a new movie theater, and Karl's plan is for a brewpub. Note that, instead of competitors, these are complementary services — some customers will have dinner or a drink at the brewpub before or after their movie. Each can choose among several suburban malls for their construction projects. However, Salt-Lick Court already has a brewpub and The Shops at Bitter Springs already has a movie theater. The payoff table is Table 5.10.

a.   What Nash equilibria does this game have?
b.   Could you describe it as a coordination or as an anticoordina-tion game? Explain

Q5.3. Rock, Paper, Scissors

Here is a common school-yard game called Rock, Paper, Scissors or (when I was a child) Paper, Stone and Scissors. Two children (we will call them Susan and Tess) simultaneously choose a symbol for rock, paper or scissors. The rules for winning and losing are:

- Paper covers rock (paper wins over rock)
- Rock breaks scissors (rock wins over scissors)
- Scissors cut paper (scissors win over paper)

Table 5.10. Payoffs with Location Strategies for Complementary Services.

		Karl		
		Sweettown Mall	Sourville Mall	The Shops at Bitter Springs
John	Sweettown Mall	10, 10	6, 5	2, 12
	Sourville Mall	4, 3	12, 10	3, 8
	Salt-Lick Court	11, 4	5, 3	10, 12

Table 5.11. Rock, Paper and Scissors.

		Susan		
		Paper	Rock	Scissors.
Tess	Paper	0, 0	1, −1	−1, 1
	Rock	−1, 1	0, 0	1, −1
	Scissors	1, −1	−1, 1	0, 0

The payoff table is shown as Table 5.11.

a. Discuss this game, from the point of view of non-cooperative solutions.
b. Does it have a dominant strategy equilibrium?
c. Does it have Nash equilibria? What strategies, if so?
d. How do you think the little girls will try to play the game?

Q5.4. The Great Escape

Refer to Chapter 2, Question 2.

a. Discuss this game, from the point of view of non-cooperative solutions.
b. Does it have a dominant strategy equilibrium?
c. Does it have Nash equilibrium? What strategies, if so?
d. How can these two opponents each rationally choose a strategy?

CHAPTER 6

Three-Person Games

In the book, thus far, we have considered only games with just two players. In some ways the two-person games are the purest examples in game theory — since each person has to choose a strategy with attention to the responses of just one rival. But there are also many applications of game theory with three or more players. It is worthwhile to pause (as von Neumann and Morgenstern did) and take a look at three-person games in particular, for two reasons. First, they are simple enough that we can use some of the same techniques that we have used for two-person games, with only a little more complication. Second, on the other hand, many of the complications that come with more than three persons can be found in the three-person games. For example, in a three-person game it is possible for two of the three persons to gang up on the third — a possibility that does not exist in two-person games!

1. AN INTERNATIONAL ALLIANCE

Runnistan, Soggia and Wetland are three countries each of which has a shoreline on Overflowing Bay. They all have military and naval forces stationed on the bay, and depending upon how they deploy their forces, two or more of them may be able to control the Bay effectively, and use the control to increase their own trade and prosperity at the expense of the third. The strategies for the three countries are the positions at which they station their forces. Runnistan can position its forces in the north or in the south; Soggia can position its forces in the east or in the west; and Wetland can

Table 6.1. Payoffs for Three Countries.

		Wetland			
		Onshore		Offshore	
		Soggia		Soggia	
		West	East	West	East
Runnistan	North	6,6,6	7,7,1	7,1,7	0,0,0
	South	0,0,0	4,4,4	4,4,4	1,7,7

position its forces off shore on Swampy Island, which Wetland controls, or on shore.

The payoffs are shown in Table 6.1. Table 6.1 is read in a straightforward way. There are two panels corresponding to Wetland's two strategies. In effect, Wetland chooses the panel, and within each panel, Runnistan chooses the row and Soggia chooses the column. Payoffs are listed with Runnistan first, Soggia second and Wetland third. Thus, if the three countries choose strategies (South, East, Offshore) the payments are 1 to Runnistan, 7 to Soggia and 7 to Wetland, for example.

In game theory, a group of players who coordinate their strategies is called a coalition. Of course, this term came into game theory from politics. In this game, then, there are three possible two-player coalitions — Runnistan and Soggia; and Runnistan and Wetland; and Soggia and Wetland. However, that is not quite the whole story. In addition, it is possible for all three countries to get together, and coordinate their strategies. In another term borrowed from politics, this would be called the grand coalition. Finally, with the single-minded thoroughness that comes into game theory from mathematics, game theorists consider an individual acting independently as a coalition unto himself. The phrase for this sort of coalition doesn't come from politics, though — it comes from card games. A player acting alone is said to be a singleton coalition. In this game, then, there can also be three singleton coalitions.

Naturally, coalitions will fit together. If Runnistan and Soggia form a two-country coalition, then Wetland is left in a singleton coalition. In this way, the group of three players is partitioned into two coalitions: A two-country coalition and a singleton coalition. This is called a coalition structure for the game. (That's why we have to be thorough, and think of a single individual acting alone as a coalition. Otherwise, we would have difficulty in talking about the coalition structure when some players are left out).

Lets list all of the possible coalition structures of this game of international alliances:

{Runnistan, Soggia, and Wetland}.
{Runnistan and Soggia}; {Wetland}.
{Runnistan and Wetland}; {Soggia}.
{Soggia and Wetland}; {Runnistan}.
{Runnistan}; {Soggia}; {Wetland}.

HEADS UP!

Here are some concepts we will develop as this chapter goes along:

Coalition: A coalition is a group of players that coordinate their strategies. A single player who does not coordinate with anyone is called a singleton coalition.

Spoiler: A player who cannot win, but whose play determines which of the other players will win, is called a spoiler.

A Public Good: If a good or service has the properties that everyone in the population enjoys the same level of service, and it does not cost any more to provide one more person with the same level of service, then it is what economists call "a public good."

Definition: *Coalition Structure* — A partition of the players in a game into coalitions, including singleton coalitions, is called the *coalition structure* of the game.

If the grand coalition of all three countries were to be formed, it might choose the strategies (North, West, On shore), since that group of strategies would yield a payoff of six to each of the countries and would yield the maximum total payoff. But that is not a Nash equilibrium. Since this is a non-cooperative game, in that there is no international enforcement mechanism that could force the three countries to deploy their forces in the agreed upon way, this agreement could not be carried out.

> **Definition:** *Grand Coalition* — The *grand coalition* is the coalition of all the players in the game — so when all players coordinate their strategies, we say we have a grand coalition.

However, suppose that Runnistan and Soggia were to form an alliance. They could choose north and east as their coordinated strategies. Given that they choose north and east, Wetland is better off to deploy its forces onshore for a payoff of one rather than off-shore for payoff of zero. This is a Nash equilibrium, so there is no need for enforcement. Neither Runnistan nor Soggia will want to deviate from the agreed upon deployment of their forces.

Similarly, Runnistan and Wetland could form an alliance and deploy their forces north and offshore. With Soggia choosing west, this is a Nash equilibrium. In a similar way, Soggia and Wetland could form an alliance and deploy their forces east and offshore. With a Runnistan choosing south, this is, again, a Nash equilibrium; so no enforcement is needed.

In passing, we have noticed that this is one of those games with multiple Nash equilibria. In principle, there could be some mystery as to which of the equilibria will be observed in practice. In this case, though, any treaty of alliance that may exist between two of the three countries provides the Schelling focal point that resolves the question. Since there is a Schelling focal point Nash equilibrium corresponding to every two-country alliance, we can conclude that any of the two country alliances is a possibility. On the other hand, since there is no Nash equilibrium corresponding to the three-country grand coalition, we would not expect to see a grand coalition in the absence of some system to enforce the deployment strategies corresponding to the grand coalition.

Coalitions can form in non-cooperative games with three or more players, as we have seen. However, in the absence of some enforcement mechanism, we will see only coalitions that correspond to Nash equilibria. If there is some sort of enforcement mechanism, or some other effective way that the players in the game can commit themselves to a coordinated strategy, then the possibilities for coalition formation are richer. We are then dealing with cooperative games, and will take them up in Part IV.

Continuing this chapter, we will explore examples of three-person games from electoral politics, public policy, and stock markets.

A Closer Look: Coalitions and Nash Equilibria

Although coalitions are possible in noncooperative games, they are exceptional. In social dilemmas and many other noncooperative games, no coalitions are stable other than singletons, and in that case, we can ignore coalitions completely — as we usually do in noncooperative game theory. It is only in games with more than one Nash equilibrium, and plenty of opportunity for preplay communication, that we need to allow for coalitions in noncooperative games. But coalitions are very important in cooperative games, as we will see in later chapters, and the terminology is the same.

2. A "SPOILER" IN A POLITICAL GAME

One of the roles that a third party can play is that of a "spoiler." A spoiler is a player who cannot win, but can prevent another player from winning. Some observers of the U.S. presidential election of year 2000 believe that Ralph Nader played the role of spoiler in that election. The same has been said of Ross Perot in 1992 and of George Wallace in 1968. In a closer election, John Anderson might have been a spoiler in 1980. It seems that spoilers are fairly common in American presidential elections.

Table 6.2. Popular Votes for the 2000 Election.

		Nader			
		Run		Don't run	
		Gore		Gore	
		Liberal	Middle	Liberal	Middle
Bush	Conservative	45, 50, 1	45, 49, 3	45, 53, 0	45, 52, 0
	Compassionate	48, 46, 2	46, 47, 3	48, 48, 0	46, 50, 0

Those who think of Nader as a spoiler probably have in mind a game very much like the one shown in Table 6.2. Table 6.2 is read similarly as Table 6.1. Nader, the third party in the game, chooses the panel in which the other two play, and Nader's payoff comes last. Mr. Bush, who was described as a compassionate conservative, chose between the two strategies implicit in that phrase — he could emphasize the conservatism or the compassion. Mr. Gore could run more as a liberal or as a middle-of-the-roader. Nader's strategies are to run or not to run. The payoffs shown are the popular votes. We have to qualify this in two ways. The subjective benefits that motivate the decisions may not have corresponded exactly to the popular vote, in two ways. On the one hand, Nader (and his supporters) might have preferred Gore's election to that of Bush, all other things equal. This does not show up when we record the Nader pay-off as his popular vote. But the Nader supporters had reasons for wanting to maximize their popular vote regardless. A larger popular vote would have given them a better chance of competing in future elections. On the other hand, Gore was running with a handicap. An American president is not elected by popular vote but by the vote in the electoral college.[1] We know in retrospect that he had to win the

---

[1] On three occasions, 1876, 2000, and 2016, the winner in the Electoral College has gotten less popular votes than his main rival.

popular vote by more than 1% to win the election. Here again, though, the more popular votes he won, the better his chance in the Electoral College. In any case, the vote proportions away from equilibrium can only be a best guess, since we only observe the equilibrium.

Nader supporters had a clear dominant strategy: Nader should run. Otherwise, his popular vote can only be zero. Bush, too, had a dominant strategy. That was to emphasize the compassion, not the conservatism, in his message. Gore did not have a dominant strategy, but given that Bush's dominant strategy was to emphasize compassion, Gores best response was to run as a middle-of-the-roader. We see that there is a unique Nash equilibrium.

Thus, the Nash equilibrium strategies are (compassionate, middle, run) and the payoffs are (46, 47, 3) — a win for Bush. Notice that, had Nader not run, Bush and Gore would have chosen the same strategies, but the Gore advantage in the popular vote would have been wider — four points, if he got all the Green vote. No one can be certain that this popular vote margin would have been enough to win the electoral vote, but it seems probable, and many people are sure it would have. That is what it means to call Nader a "spoiler."

3. STOCK ADVISING

When there are three or more players in a game, there may be some advantage for one of them in going along with the majority. Here is an example of that kind.

Luvitania is a small country with an active stock market but only one corporation, General Stuff (GS), and only three market advisors: June, Julia, and Augusta. Whenever at least two of the three recommend "buy" for GS, the stock goes up, and thus the advisors who recommend "buy" gain in reputation, customers, and payoffs. Whenever at least two of the three recommend "sell" for GS, the stock goes down, and thus the advisors who recommend "sell" gain in reputation, customers, and payoffs.

Table 6.3.   Payoffs for Three Stock Advisers.

		June			
		Buy		Sell	
		Julia		Julia	
		Buy	Sell	Buy	Sell
Augusta	Buy	5, 5, 5	6, 0, 6	6, 6, 0	0, 6, 6
	Sell	0, 6, 6	6, 6, 0	6, 0, 6	5, 5, 5

The payoff table for the three advisors is shown in Table 6.3. We see that there are two Nash equilibria: One where everyone recommends "buy" and one where everyone recommends "sell." Whenever one of the three advisors disagrees with the other two, she loses, and so has reason to switch to agreement.

What does this example tell us about the real world? There are, of course, many more than three stock advisers in the real world. The real world is also more complex in other ways. For example, it is possible for the majority of stock advisors to be wrong, and this has happened from time to time. Thus, stock prices must depend on some other things beside the majority opinions of stock advisors: Corporate earnings, for example. However, if stock advisors believe that their advice influences stock price trends, then a stock advisor will not want to be out of step with the majority unless he has a good reason to think the majority is wrong.

Stock advisors do seem to agree with one another most of the time, when they are wrong as well as when they are right. John Maynard Keynes said that stock markets were a good deal like "beauty contests" sponsored by British newspapers early in the 20th century. The newspapers would publish a page with pictures of 100 (or thereabouts) girl's faces, and invite the readers to vote for the prettiest. Those who voted for the prettiest would get a small prize — and the prettiest was the one who got the most votes! The objective (Keynes said) is not to decide which is prettiest but to decide which one would be thought prettiest by the largest number of others — and stock markets, Keynes said, are like that. In the Luvitania example, the objective is not to predict which way stocks will go, but to predict which way the majority of advisors will predict that stocks will go, much as Keynes said.

The fact that there are two Nash equilibria is important, too. There are two ways everyone can be right — but which will be realized in a particular case? Suppose the first bit of news that comes out during the day is good news — then this good news could provide the Schelling focal point by suggesting to each of the advisors that it is likely that the other advisors will say "buy." So they all say "buy." It often seems that the market overreacts to new information — shifting much more than we can explain by the objective content of the news. Perhaps this occurs because it shifts the financial players to a new Schelling focal point.

These conclusions are speculations, not facts. We don't really know why the

A Closer Look: John Maynard Keynes, 1883–1946

John Maynard Keynes was probably the most influential economist of the first half of the twentieth century. With the publication of *The Economic Consequences of the Peace* (1919) he burst (as Joseph Schumpeter wrote in his *History of Economic Analysis*) "into international fame when men of equal insight but less courage and men of equal courage but less insight kept silent." After the disaster of the Great Depression, Keynes was the leading figure in a group of (mostly) younger and very creative economists who attempted to understand and explain the disaster. Borrowing freely from their ideas, Keynes published *The General Theory of Employment, Interest, and Money*, which (again quoting Schumpeter) "was a similar feat of leadership. It taught England, in the form of an apparently general analysis, his own personal view of 'what should be done about it.'" In doing this, Keynes helped to found modern macroeconomics.

stock markets move as they do — and if I knew why stock markets move as they do, I probably wouldn't need to write this book. But the three-person game suggests a possible explanation for some of the things we observe.

4. A CROWDING GAME

We have seen examples of three-person games in which the third person is cut out of a coalition, in which the third person loses unless they conform to the majority, and in which the third person is a spoiler, in that he cannot win but can determine by their strategy who does. Another possibility is "twos company, but three's a crowd." That is, crowding may become a problem, but only after a certain number of people have joined the crowd. The El Farol Game illustrates this.

El Farol is a bar in Santa Fe, New Mexico, where chaos researchers from the Santa Fe Institute often hang out. It is said that the El Farol is at its best when it is crowded, but not too crowded. A key point is that the benefits to bar-goers are nonlinear — as the crowd grows, the benefits increase rapidly up to a point and then decline. Here is a three-person example a little like that. Amy, Barb, and Carole each can choose between two strategies: Go (to the bar) or stay home. If all three go to the bar, it is just too crowded, and each of the three gets a negative payoff. If just two go, those two get the maximum payoff, but if just one goes, she gets a worse payoff than staying home. The payoff to bar-goers is shown in Figure 6.1. Notice the three quite different trends as the bar crowd goes from none to one, then two, then three — a contrast that could not be illustrated with fewer than three players. The payoff to those who stay home is always 1.

The payoffs are shown in normal form in Table 6.4. This game has four Nash equilibria. Each strategy triple in which just two go to the bar is an equilibrium. Thus, we may think of it as a coordination game. The fourth equilibrium, in which everyone stays home, is a bit odd. In fact, though — since it is better to stay at home than to be the only one at the bar — it, too, is an equilibrium. Since it is

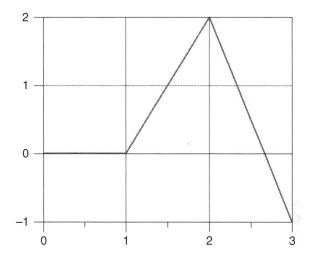

Figure 6.1. Payoffs to Bar-Goers in the Crowding Game.

Table 6.4. Payoffs in a Crowding Game.

		Carole			
		Go		Home	
		Barb		Barb	
		Go	Home	Go	Home
Amy	Go	−1,−1,−1	2,1,2	2,2,1	0,1,1
	Home	1,2,2	1,1,0	1,0,1	1,1,1

inferior to the other Nash equilibria (in that two girls would be better off and none worse off if two were to go to the bar) we might doubt that it would really occur, but no one person can improve on (home, home, home) by changing unilaterally, and in the absence of any other information, each would be uncertain as to which of the other three equilibria could be realized by the shift!

As in the game of international alliances, it is possible that two of the girls might form a coalition and go to the bar, leaving the

A Closer Look: Coalition Proof Equilibria

What about the odd equilibrium at the lower right? Any two girls could form a coalition and shift to one of the other Nash equilibria, and be better off. That would immediately rule the lower right equilibrium out. This Nash equilibrium is not "coalition-proof" since coalitions can form (even in the absence of enforcement) that will shift away from it. It seems that — in games with more than two players and more than one equilibrium — we might "refine" the Nash equilibrium to allow for strategy shifts by coalitions. There are "refinements" in noncooperative game theory that do this. One is **"strong Nash equilibrium."** A Nash equilibrium is strong only if there is no coalition that can benefit, as a whole, by deviating from it to another Nash equilibrium. This game has no strong Nash equilibria at all. A less demanding condition is that the equilibrium be **"coalition-proof."** To determine if it is coalition-proof, we must ask two questions. First, are there coalitions that could deviate to another Nash equilibrium and be better off? Second, are these coalitions stable in the sense that there are no smaller coalitions within them that could benefit by deviating to yet another Nash equilibrium? Clearly, the lower right equilibrium is not coalition-proof, since any two-girl coalition can profit by deviating from it. The two-girl coalitions are stable if no smaller (one-girl singleton) coalition can benefit by a further deviation. We see that the three other Nash equilibria are stable in this sense, so they are themselves coalition-proof. This gives us some reason to think that the lower right equilibrium is a very unlikely Nash equilibrium in reality.

third girl out with a best response strategy of staying home. Of course, these sorts of things do happen — coalitions of this kind are often called "cliques" and are reported in just about every high school.

According to people who have been there, though, the essence of the El Farol game is that the decisions to go or stay home are made at the last minute, without time enough to form a coalition or a clique. There is no communication. The only information people have is from experience, and that can be unreliable. Thus, the problem of uncertainty in coordination games comes to the fore. It often happens that everyone goes to the bar one night, and it is overcrowded, so that they all decide to stay home the following night to avoid the crowd — so that there are too few people for the joint to really rock. Some of the chaos researchers at the institute offer this as an example of mathematical chaos. It is also an excellent example of coordination failures in a coordination game without communication.

5. GLOBAL WARMING

One of the most challenging problems facing the various countries and peoples of the world today is global warming. Many productive processes contribute to the standard of consumption for some people but at the same time contribute to the accumulation of atmospheric gases, such as carbon dioxide and methane, that cause warming. This persistent and cumulative warming in turn contributes to a variety of natural disasters and health problems around the globe. Global warming is another "tragedy of the commons" and we might call it a public bad: In any case reduction of pollution to avoid warming is a global public good.

While our various government and international authorities have discussed and proposed policies to address global warming, limited progress seems to have been made. One reason for this seems to be the different concerns of countries in different parts of the world — as we might say, the north, south and east of the world. For this example, we will try to illustrate how global warming may be influenced by interactive decisions of the three regions of the world. Of course, the global north, south, and east, unlike the two burglars in the Prisoner's Dilemma, are not individual decision makers but large populations. In treating the three global regions as the players in a three-person game, we are making an as-if simplifying assumption. In an earlier section of this chapter, countries were treated as players in the game, and the same might be said — that a country is not an individual decision-maker — in that example. On the other hand, a national state will have some process for making decisions in the interest of its people, and an individual decision-maker, such as a president or king, may play a key part. For a global region such as the north, south and east, each comprising dozens of national states, there truly is no individual decision-maker. But our example will explore the decisions they might make if the regions could each adopt rational decisions in the interests of the people of their regions, and in this way, we might learn something about the persistence of this problem. In this section we suppose that the regions act non-cooperatively and in later chapters the example will be revisited from a cooperative point of view.

We have, then, a game with three players, north, south and east, and each has two strategies "pollute" or "depollute." Here "depollute" means adopting production methods that generate less of the gases that result in warming, but which are less productive and so reduce the standard of consumption in the region. Warming affects all three regions, reducing their payoffs, but the reduction of the standard of consumption affects only the region that chooses the strategy "depollute." The three regions also differ in their productivity. For this example, the gross products of the regions are assumed to be in the proportions shown in Table 6.5, on a scale with 50 as largest. If the region chooses to depollute, there is some reduction in gross products, as shown in the second column. However, there are also costs from pollution — health costs, climatic change and rising sea levels, for example — and, on the one hand, these costs will depend on which regions depollute and which do not; while on the other hand, the pollution costs are deducted equally from the products of the three nations. The pollution deduction is shown in Table 6.6.

The payoff table is calculated by deducting the pollution penalty, in Table 6.6, from each of the gross product amounts in Table 6.5. It is shown in Table 6.7. For example, if the north and south pollute, and east shifts from pollute to depollute, the payoffs

Table 6.5. Production in Three Regions.

Gross product

	Pollute	Depollute
North	50	45
East	25	21
South	10	8

Table 6.6. Global Pollution Deduction.

		South			
		Pollute		Depollute	
		East		East	
		Pollute	Depollute	Pollute	Depollute
North	Pollute	7	5	6	4
	Depollute	4	2	3	0

Table 6.7. Strategies and Payoffs in the Global Warming Game.

		South			
		Pollute		Depollute	
		East		East	
		Pollute	Depollute	Pollute	Depollute
North	Pollute	24,10,2	26,8,4	26,12,1	28,10,3
	Depollute	23,12,4	25,10,6	25,14,3	27,12,5

to north and south are increased by 2 units each, but the payoff to
east is reduced by 2.

Examining this game, perhaps doing a little underlining, we see
that "pollute" is a dominant strategy for each region, so that we have

a dominant strategy equilibrium. Is it a social dilemma? If we understand that the payoffs at the lower right, where all depollute, is the cooperative solution to the game, it is a social dilemma, with a dominant strategy solution that differs from the cooperative solution. And the situation where they all depollute does maximize the total payoff, with 44 as against no more than 42 for any other set of strategies. Everyone is better off at that situation than they are at the dominant strategy solution. Moreover, the situation where all depollute has the property of Pareto-optimality: That is, no-one can be made better off by a shift of strategies without making someone else worse off. Since Pareto-optimality is the definition economists use for efficiency, we may say that the situation where all depollute is efficient, and the dominant strategy equilibrium is inefficient. Moreover, any outcome at which two regions depollute has that same condition of Pareto-optimality. But we see that the payoffs at the efficient situations are very unequal, unlike the symmetrical social dilemmas we have considered. Can that inequality be consistent with a cooperative solution? That is a question we will have to reconsider in Part IV. For now, we may say somewhat tentatively that global warming is a social dilemma.

6. SUMMARY

From some points of view, three-person games are only a little more complicated than two-person games. They can be presented in normal form with tables that are a bit more complicated than those required for two-person games, but still simple enough to get on a single page. From another point of view, three-person games are much more complicated than two-person games. That is, the three-person games bring into game theory many issues that did not arise in two-person games, but do arise in games with more than three players. This combination of simplicity and complication makes three-person games well worth studying.

For example, in a three-person game it becomes possible for two players to form a coalition and cooperate against the third. This is always possible with more than three players, but makes little sense in a two-person game. Of course, in a non-cooperative game, not all coalitions are possible. Only those coalitions that correspond to Nash equilibria are stable. Thus, we can observe three-person social dilemmas, in which any coalition of two or more players could make the participants better off; but none is stable as a Nash equilibrium.

Another complication we find in three-person games is the possibility that one of the three players can be a "spoiler," determining which of the other two can win even though he himself cannot win. This seems to have been a pretty common phenomenon in American presidential elections in the latter half of the 20th century.

In three-person games, too, it becomes possible to model conformism, in which the players each find it advantageous to go along with the majority. We can also find examples in which it is more advantageous to go it alone or in which two's company, but three's a crowd.

The extension to three players does not always make a difference. When we consider a pollution game like the global warming game, for example, we see that it is a three-person social dilemma, with the unique, inefficient dominant strategy equilibrium that we find in social dilemmas in general. All in all, three-person games can be worth studying in themselves, but can also be a source of the ideas to explore in games of more than three persons.

Q6. EXERCISES AND DISCUSSION QUESTIONS

Q6.1. Another Water-Selling Game

Tom, Dick and Harry are all in the bottled-water business. If one or two of them expand their production, they can take business away from those who do not, but if all three expand, they will be back to a break-even situation. The payoffs are shown in Table 6.8.

a. Has this game any Nash equilibria? If so, list them and explain.
b. List all possible coalitions in this game.
c. Which coalitions will be stable as Nash equilibria?
d. Compare and contrast this game with the Global Warming Game and the International Alliance Game in the chapter.

Table 6.8. Payoffs for Bottled Water Vendors.

		Harry			
		Expand		No	
		Dick		Dick	
		Expand	No	Expand	No
Tom	Expand	0,0,0	1,–2,1	1,1,–2	2,–1,–1
	No	–2,1,1	–1,–1,2	–1,2,–1	0,0,0

Table 6.9. Three-Person Effort Dilemma.

		Carl			
		Work		Shirk	
		Bob		Bob	
		Work	Shirk	Work	Shirk
Al	Work	20,20,20	14,25,14	14,14,25	4,20,20
	Shirk	25,14,14	20,20,4	20,4,20	5,5,5

Q6.2. Three-Person Effort Dilemma

Consider the effort dilemma shown in Table 6.9.

a. Are there any dominant strategies in this game?
b. Is there a dominant strategy equilibrium?
c. Are there any Nash equilibria? How many? Which? How do you know?
d. What would happen if two of the guys formed a coalition to increase their effort?

Q6.3. Frog Mating Game

Evolutionary biologists use game theory to understand some aspects of animal behavior, including mating behavior. Here is a Three-Frog

Table 6.10. Payoffs for Eager Frogs.

		\multicolumn{4}{c}{Flip}			
		Call		Sit	
		Michigan J.		Michigan J.	
		Call	Sit	Call	Sit
Kermit	Call	5,5,5	4,6,4	4,4,6	7,2,2
	Sit	6,4,4	2,2,7	2,7,2	1,1,1

Mating Game about three lonely frogs: Kermit, Michigan J, and Flip. The frogs are all males who can choose between two strategies for attracting females: The call strategy (call) or the satellite strategy (sit). Those who call take some risk of being eaten, while those who sit run less risk. On the other hand, the satellites who do not call may nevertheless encounter a female who has been attracted by another call. Female frogs are really, well, you know! So, the payoff to the satellite strategy is better when a larger number of other male frogs are calling, so there are more females around who are confused and, well, you know.

The payoff table for the three frogs is shown in Table 6.10, with the first payoff to Kermit, the second to Michigan J, and the last to Flip.

a. Enumerate all Nash equilibria in this game.
b. Have these equilibria anything in common, so that you could give a general description of equilibria in the game? What is it?
c. Frogs do not form coalitions in mating games. What difference would it make if they did?

Q6.4. Oysterers

Ray, Stan and Tom are Chesapeake Bay oysterers. They all know of an oyster bed off North East that other oysterers are unaware of.

Table 6.11.   A Waiting Game.

		Now		Wait	
		Stan		Stan	
		Now	Wait	Now	Wait
Ray	Now	5,5,5	7,1,7	7,7,1	12,1,1
	Wait	1,7,7	1,1,12	1,12,1	10,10,10

The oysters will be worth more on the market if they are not har-vested until next month, when they are bigger and more mature. However, our three oysterers are considering whether to harvest them now or wait. Those are their two strategies in this waiting game. The payoffs are as shown in Table 6.11.

The payoffs are listed with Ray first, then Stan, then Tom.

a.   What Nash equilibria does this game have, if any?
b.   What kind of a game is it?

Q6.5. Medical Practice[2]

Traditionally, ob/gyn physicians have provided services that included both obstetrics (delivering babies) and gynecology (general wom-en's health care and surgery). Some older physicians stopped performing obstetric care and mainly concentrated on gynecology. Since obstetricians are on call whenever the baby comes, the exclu-sive gynecological practice gives the doctor a better lifestyle. In more recent years, younger doctors have also opted for practices limited to gynecology, both for the better lifestyle and under the pressure of

---

[2]Thanks to Dr. David Toub, a student in my course on game theory for MBA students, for this example.

Table 6.12. Payoffs to Gyn Practitioners.

Number of doctors concentrating on gyn practice	Payoff to gyn only practice
1	20
2	10
3	−5

increased insurance costs associated with medical liability for obstetric care.

Consider three young ob/gyn's: Drs. Yfnif, McCoy and Spock, who are the only doctors qualified for ob/gyn in the isolated city of Enterprise, AK. Each is in sole practice and can choose between two strategies: Limit the practice to gyn or practice ob/gyn. Payoffs to a gyn practice depend on the number of doctors offering gyn services only, as shown in Table 6.12. Payoffs to an ob/gyn practice are always 5.

a. Construct the payoff table for this three-person game and determine what Nash equilibria exist, if any.
b. Realistically, the three doctors probably will not establish their practices at the same time. Suppose there is a sequence in which one chooses first, another second, then the third last. What effect would this sequence have on the actual Nash equilibrium?
c. A partnership can be thought of as a coalition. What possibilities for coalitions exist in this game?

Q6.6. NIMBY

NIMBY stands for "not in my back yard." Table 6.13 shows the payoffs for the NIMBY game. The idea behind the game is that a proposal is made to construct a facility that will provide a public

Table 6.13.   NIMBY.

		c			
		Accept		Reject	
		b		b	
		Accept	Reject	Accept	Reject
a	Accept	2, 2, 2	2, 6, 2	2, 2, 6	2, 6, 6
	Reject	6, 2, 2	6, 6, 2	6, 2, 6	3, 3, 3

good to the three agents in the game. However, the facility will have to be built at the location of one of the agents and will create sufficient local nuisance so that agent will be worse off despite enjoying the public good. Analyze and discuss this as a non-cooperative game.

CHAPTER 7

Probability and Game Theory

Uncertainty plays a role in many games and other human interactions. In games, uncertainty may be created deliberately by throwing dice or shuffling cards. Let's consider the example of throwing a single die. (The word "dice" is the plural of "die," which means a cube with numbers from one to six on its six faces. So, we say: One die, two or more dice, a pair of dice.) When we throw the die, will it show a number greater than three? Yes, no, or maybe? Since the answer is uncertain, it is maybe. But can we do better than that?

1. PROBABILITY

In an absolute sense, we can do no better. The statement is either true, false, or uncertain, and in this case it is uncertain. But in a relative sense, some uncertain statements are less likely than other uncertain statements. When we throw the die, for example, the statement that the number will be greater than five is less likely than the statement that the number will be greater than three. The reason is that there are three different ways we can get a number greater than three — it can be four, five, or six — while there is only one way that we can get a number greater than five. We can often make these comparisons of likelihood, and we put more confidence in the likelier prediction.

Probability is a way of measuring relative likelihood in numerical terms. A probability is a number between zero and one, inclusively. A probability of one corresponds to a certainty, and a probability of zero is assigned to statements that are certainly false. Among uncertain statements, the more likely the statement is, the larger the probability number that is assigned to it. Thus, we might say that when we throw a die, the probability of a number greater than five is one sixth, while the probability of a number greater than three is one half. Thus, the more likely statement has the larger probability.

In some cases, we can do better still. We can tie the probability of an event to the frequency with which we observe that event in a whole series of similar experiments. Suppose, for example, that we throw a die hundred times. We can be reasonably certain that we will observe a number greater than five in approximately one sixth of all the throws. Similarly, we will observe a number greater than three in approximately one half of all the throws. Not only that, but if we throw the die a thousand times, the proportions will be

---

HEADS UP!

Here are some concepts we will develop as this chapter goes along:

Probability: A numerical measure of the likelihood of one of the outcomes of an uncertain event, which corresponds to the relative frequency of that outcome if the event can be repeated an unlimited number of times.

Expected value: Suppose an uncertain event can have several outcomes with numerical values that may be different. The expected value (also known as mathematical expectation) of the event is the weighted average of the numerical values, with the probabilities of the outcomes as the weights.

Risk averse: A person who will choose a safe payment over a risky payment with an expected value greater than that of the safe payment is said to be risk averse.

---

the same, and the approximation will be even better. The proportions remain the same, and the approximations get better, as the number of throws increases. The "limiting frequency" of an event is the number approximated in this way as the number of trials increases without limit. Thus, we identify the *probability* of an event with the *limiting frequency* whenever a limiting frequency makes sense.

In some other important cases, a limiting frequency may not make much sense. Think of the outcome of a research project to find a new technology for automobile engines. Until the research is done, we do not know whether it will be successful are not. As researchers say, that's what makes it research. But once the research is done and the results are in, that same research project will never need to be done again, nor could it be done again. So, a limiting frequency makes no sense in this case. Nevertheless, it seems to me that success in a research project to produce automobile engines using fuel cell technology would be *more probable* than success in a research project to produce automobile engines using nuclear fusion. It would be likelier, so we assign a larger probability number to it. Of course, there is a subjective element in my judgment that success with fuel cell technology is more likely than success with nuclear fusion technology. So, there is a subjective element in the probability attached to it. Nevertheless, we will assume that probabilities for unique events like a research project have the same properties as probabilities identified with relative frequencies, despite the subjective element. This is a basic assumption for probabilities in this book (and a common assumption in many applications of probability).

We can also use the relative frequency approach to *estimate* probabilities, when the relative frequencies are observable. Meteorology gives some good examples of this. What are the chances of a white Christmas next year? Although it is too far in the future to know what cold or warm fronts may be passing through, we can look at the frequency of white Christmases over the past decades and use that information to estimate the probability of a white Christmas next year. The records show that about 20% of Christmases in the Philadelphia area (where I live) have been white Christmases. So, we can say with good confidence that the probability of a white Christmas is about 20%.

Throwing a single die is a pretty simple event, and so is a white Christmas. In applications of probability, we often have to deal with much more complicated events. To compute the probabilities of very complicated events we may use methods from algebra, logic, and calculus. We will not go into that just now. Some of those methods are important in game theory, but we will get to them when we need them for the game theory. However, there is one method of application of probabilities that we will need throughout the book, and it will be explained in the next section.

2. EXPECTED VALUE

Suppose someone were to offer you a bet. You can throw a single die, and he will pay $10 if the die shows six, but nothing otherwise. How much is this a gamble worth to you? If he would ask you to pay a dollar to play the game, would it be worthwhile? What if he asked for two?

One way to approach this is to think what would happen if you played the game a large number of times — let's say that you played it hundred times. Since we know that the probability of a six on each throw is one sixth, you can expect to see approximately 1/6 of 100 sixes in the 100 throws. One sixth of 100 is between 16 and 17. So you could expect to win roughly 16 or 17 times. If you paid $1 for each game, so it you paid a total of $100, and you would win approximately 10 dollars times 16 or 17 wins, that is, approximately $160 or $170, you would be pretty certain to come out ahead. But if you paid $2 per game, for a total of $200, it doesn't look so good.

But we really want the value of a single play. Each play has two possible outcomes — a six, which pays $10, and any other number, which pays nothing. The probability of the $10 payoff is 1/6, and the probability of nothing is 5/6. Multiply each of the payoffs by its probability and add the two together. We have $1/6 * 10 + 5/6 * 0$ for a total of 1.67. Thus, $1.67 is the value of an individual gamble.

Definition: *Expected Value* — Suppose an uncertain event can have several outcomes with numerical values that may be different. The *expected value* (also known as the *mathematical expectation*) of the event is the weighted average of the numerical values, with the probabilities of the outcomes as the weights.

This is an example of the "mathematical expectation" or "expected value" of an uncertain payoff. *The expected value is a weighted average of all the possible payoffs, where the weights are the probabilities of those payoffs.* Thus, in the game of throwing one die, the payoffs of one and zero are weighted by the probabilities of 1/6 and 5/6 and added together to give the weighted average or expected value.

Let's try another example of expected value. Joe Cool is taking a three-credit course in Game Theory, and his grade is uncertain. He is pretty sure that it will be an A, B, or C, with probabilities 0.4, 0.4, and 0.2. His college calculates grade point averages by assigning 4 "quality points" per credit for an A, 3 for a B, 2 for a C and 1 for a D. What is the expected value of the quality points Joe will get from Game Theory? If he gets an A he gets $3 * 4 = 12$ quality points, if a B, 9, and if a C, 6. So we have an expected value of $0.4 * 12 + 0.4 * 9 + 0.2 * 6 = 9.6$ quality points.

EXERCISE: Suppose a gambler offers to throw one die and pay you the number of dollars shown on the die. What is the mathematical expectation of this payment?

3. NATURE AS A PLAYER

In the musical comedy *Guys and Dolls*, a leading character, Sky Masterson, is a notorious gambler. According to a description, he would bet on which of two raindrops dripping down a windowpane would reach the bottom first. This is a good example of natural uncertainty, that is, uncertainty resulting from the complexity and unpredictability of nature. Natural uncertainty plays a part in many games and other human activities.

However, game theory thinks in terms of the interactions of the players in the game. So how do we bring natural uncertainty into game theory? When there is natural uncertainty, the convention in game theory is to think of "nature" as a player in the game. Just as the ancient Romans personified Chance as the Goddess Fortuna, so the game theorist makes Chance a player in the game. But "Chance" or "Nature" is a rather strange player. Unlike the human players in the game, she doesn't care about the outcome and always plays her strategies at random, with some given probabilities.

Let's have an example. Suppose that a company is considering the introduction of a new product. Unknown technological and market conditions may be "good" or "bad" for the innovation. This is "Nature's" play in the game. She decides whether the conditions will be "good" or "bad" and she does so at random with — for the sake of the example — a 50–50 probability rule. The firm's strategies are go or no-go. The payoffs to the firm are shown in Table 7.1, along with the probabilities of Nature's strategies and the expected values¹ of the go and no-go strategies. Since go pays an expected value of 5, while no go pays

> **A Closer Look:** Fortuna
>
> Fortuna, the Roman Goddess of Chance, was the personification of uncertainty in Roman life. She was a popular goddess, worshipped by those seeking success in gambling, war, childbirth, and other uncertain activities. June 11 was her sacred day. She was called Tyche by the Greeks.

> **Definition:** *Natural Uncertainty* — Uncertainty about the outcome of a game that results from some natural cause rather than the actions of the human players might be called *natural uncertainty*. In game theory we introduce natural uncertainty by allowing nature to be a player, and assuming that nature plays according to given probabilities.

nothing, the firm would choose "go" and introduce the product.

We can apply the expected-value idea (along with contingent strategies) to find out how much the businessman's ignorance costs him. Here is the mind experiment: Not knowing whether the conditions are good or bad, the decision-maker can only make his decision based on probabilities. But suppose that our decision-maker can call in a consultant and find out, before making his

Table 7.1. A Game Against Nature.

		Nature		Expected value payoff
		Good	Bad	
Decision-maker	No go	0	0	0
	Go	20	−10	5
Probability		0.5	0.5	

Table 7.2. The Same Game with Contingent Strategies.

		Nature		Expected value payoff
		Good	Bad	
Decision-maker	If "good" then go; if "bad" then go	20	−10	5
	If "good" then go; if "bad" then no go	20	0	10
	If "good" then no go; if "bad" then go	0	−10	−5
	If "good" then no go; if "bad" then no go	0	0	0
Probability		0.5	0.5	

decision, whether the conditions are good or bad. How much will the decision-maker be willing to pay the consultant, i.e., what is the "value of information" that the consultant can supply?

What, in fact, is the advantage of having more information? The advantage is that, with more information, we can make use of contingent strategies. The game of product development with contingent strategies is shown in Table 7.2. We see that the contingent strategy, "if 'good' then go; if 'bad' then no go" gives an expected value of 10. By getting the information, and using it to choose a contingent strategy, the businessman can increase his expected value payoff from 5 to 10. Thus, any consultant's fee less than 5 leaves the decision-maker better off, in an expected-value sense, and the "value of information" in this case is 5.

4. RISK AVERSION

In a sense, the mathematical expectation is the fair value of an uncertain payment. But that may not be the whole story, since it can be a risky value, and people often prefer to avoid risk.

Here is an example. Karen has bought a painting at an antique show. She got it cheap, and she knows she has a good deal. Karen is pretty certain that the painting is by one of two artists. One is a well-known 19th century regional artist. If the painting is by the 19th century artist, then it can be resold for $10,000. The second artist is a 20th century imitator. If the painting is by the imitator, it is only worth $2,000. From her knowledge of the styles of the two painters, Karen knows that the probability that it is by the 19th century artist is 0.25, so that the probability that it is by the twentieth century artist is 0.75.

But another person, who arrived at the antique show just after Karen made her purchase, has offered her $3,500 for the painting. Thus, Karen has to choose between a certain $3,500 and an uncertain resale value of the painting that could be as much as $10,000 or as little as $2,000. Karen's strategies are to accept the $3,500 offer or investigate the identity of the artist and perhaps put the painting on the market when the artist's identity has been documented, if possible. The payoffs and probabilities are shown in Table 7.3.

If Karen accepts the deal, she has $3,500 for her painting, with no uncertainty or risk. However, as we see in Table 7.5, the expected value of the "investigate" strategy is $4,000, which is better. If Karen only cares about the money payoff, she will keep the painting and investigate the author. But Karen may accept the $3,500 offer anyway. The benefits that would motivate Karen to refuse or accept the offer are subjective benefits, and from a subjective point of view, the uncertainty of the resale value of the painting is a disadvantage in itself. Allowing for this disadvantage, the risky resale value isn't really worth as much as its expected value. If this is the way Karen feels,

Table 7.3. Money Payoffs of Art Resale.

		Payoffs		
		19th Century	20th Century	Expected value
Karen	Sell for $3,500	3,500	3,500	3,500
	Investigate artist	10,000	2,000	4,000
Probability		0.25	0.75	

then we would say that Karen is *risk averse.*

Risk aversion seems to be a common fact of human activity. Whenever anyone buys insurance, and whenever a businessman chooses a less profitable investment because it is more secure, these are expressions of risk aversion. Of course, individuals differ, and some may be risk averse while others are not. If a person would choose the risky payment (instead of one that has the same expected value and is certain) we say that that person is *risk loving* rather than risk averse. A person who will always choose the bigger expected value, regardless of the risk, is said to be *risk neutral.*

Definition: *Risk Aversion* — A person who will choose a safe payment over a risky payment with an expected value greater than that of the safe payment is said to be *risk averse.* A person who chooses a risky payment over a safe payment, even though the risky payment has a smaller expected value than the safe payment, is said to be *risk loving.* A person who will always choose the higher money expected value is said to be *risk neutral.*

Economists and finance theorists usually assume that risk aversion is an aspect of an individual's tastes. In any case, they would not expect the individual's risk aversion to change very much from one day to the next, since they are considered to be an aspect of the person's tastes and preferences and tastes and preferences are thought of as being quite stable. But that's an assumption, not a known fact. Most people probably would be less risk averse if they were in a casino than otherwise, so social context does matter. The economist would say that this willingness to take risks in the casino reflects a taste for entertainment, not a change in risk aversion. On the other hand, Mr. Alan Greenspan, the former chairman of the Federal Reserve System and an experienced practical economist, has sometimes spoken about day to day fluctuations in risk aversion on the stock markets. Stable or not, risk aversion is a key factor whenever human beings confront uncertainty. In most of this book, however, for simplicity, we will assume that decision-makers are risk neutral.

5. WILL THERE BE A PANDEMIC?

When the COVID pandemic emerged in the spring of 2020, many decision makers in business and government found that they had made some wrong decisions. Of course, it was known that pandemics are possible — there had been an influenza pandemic a century before, and concern about pandemic diseases had been in the news in the first two decades of the 21st century — but a future pandemic is always an uncertain event — what is its probability? Here is an example of changing information about the pandemic leading to changing decisions.

In May 2020, my friend ordered gardening gloves. The COVID 19 pandemic was then spreading and my friend was one of those who were staying mostly at home, "locked down," so my friend was doing a lot of gardening. But the company apologized, explaining that they had been directed by government under the American Defense Production Act to produce personal protective equipment (PPE) for medical personnel, which were then scarce enough to create a crisis. The PPE was scarce because the pandemic came as a surprise. Everyone knew pandemics were possible, but most people thought that a pandemic was a very *improbable* event, and indeed, considering the 100 years that had passed since the previous one, this was a reasonable judgment. But the emergence of the COVID 19 pandemic changed the game in which personal equipment players were playing.

Table 7.4. No Pandemic.

		Wemake	
		Gardening	Medical
Glover	Gardening	1,1	3,3
	Medical	3,3	1,1

Table 7.5. Pandemic for Sure.

		Wemake	
		Gardening	Medical
Glover	Gardening	0,0	0,5
	Medical	5,0	4,4

For this example, Glovers and Wemake are two companies that can make gardening or medical equipment. If there is no pandemic, demand for both products is small and they are better off to specialize. The game they are playing is shown as Table 7.4.

As the table shows, if it is known for certain that there will be no pandemic, then there are two Nash equilibria, and this is an anti-coordination game. The Nash equilibria take place where the players choose different strategies, that is, where each specializes in one product or the other. On the other hand, if there is a pandemic, the demand for medical PPE is very much increased and the two companies are playing the game shown in Table 7.5.

In this case we see that producing medical equipment is a dominant strategy for both companies. But the companies are not playing either of those games, since a pandemic is a possible but not (in 2019) a certain event. Pandemic and no pandemic are alternative *states of the world,* and Nature decides what will be the state of the world. Thus, the two companies are playing in the 3-person game shown as Table 7.6, with Nature as the third player. The probabilities shown are the consensus subjective probabilities of the two events for this example. (In reality the consensus subjective probability of

Table 7.6. Pandemic Maybe.

		Nature			
		No Pandemic		Pandemic	
		Wemake		Wemake	
		Gardening	Medical	Gardening	Medical
Glover	Gardening	1,1	3,3	0,0	0,5
	Medical	3,3	1,1	5,0	4,4
Probability		0.8		0.2	

Table 7.7. Expected Values.

		Wemake	
		Gardening	Medical
Glover	Gardening	0.8,0.8	2.4,3.4
	Medical	3.4,2.4	1.6,1.6

a pandemic in 2019 was probably much less than 20%, but this will do for a simple example.)

Thus, the two companies will choose their strategies based on the expected values of payoffs. These are shown in Table 7.7.

Once again, we see that there are two Nash equilibria, at each of which the two companies specialize. Suppose, however, that new information makes it seem that a pandemic is more likely (as did happen in early 2020). When there is new information, rational beings revise their subjective probability estimates. (Bayes' rule, a mathematical rule that tells us how best to revise our probability estimates, is discussed in the appendix to this chapter.) Suppose, then, that the probabilities are revised so that the probability of a pandemic is 0.35 and that of no pandemic is 0.65. (By March of 2020, the probability of a pandemic must have been close to one for most rational beings, but these probabilities will illustrate the principle.)

Table 7.8. Revised Expected Values.

		Wemake	
		Gardening	Medical
Glover	Gardening	0.65,0.65	1.95,3.7
	Medical	3.7,1.95	2,05,2.05

With revised probabilities the expected value payoffs are also revised, and so the two companies are playing the game shown in Table 7.8. We see that this relatively small revision of the probabilities quite transforms the game. We see that producing medical equipment is now a dominant strategy.

As the information about COVID 19 emerged, many decision-makers from individuals to companies to government agencies changed their decisions. Without judging how rational or irrational some of those decisions may have been, this example provides a clear explanation of why rational decision-makers would have had good reasons to change their plans, and in many cases to revise those plans radically. This is not to say that every rational decision had to change. We see in this example that the two agents changed their decisions because, in the two states of the world, they were playing different games. If they had played the same game in both states of the world then they would have had no reason to switch. But the widespread changes in demands and supplies that occurred during 2020–2021 may well have been rational, and probably most were. Further, in this example, we have made no distinction between the long and the short run. The decisions in the example are short run decisions. Longer run decisions, such as investment in port facilities, were based on pre-pandemic probabilities and could not be changed and carried out in the few months that followed the new information. That is a major cause of the "supply chain" difficulties that emerged after early 2020. We live in a world of probabilities, not certainties, and will often have to revise our decisions as new information modifies those probabilities — even if, and especially if, we are perfectly rational.

6. BAYESIAN NASH EQUILIBRIUM

In our examples so far, we have assumed that each player in the game knows a good deal about the other player or players — knowing, for example, the strategies available to the other players and the payoff each player would get when all the strategies were chosen. But that may not always be the case. In a conflict, for example, one person may not be sure whether the counterpart will be willing and able to accommodate or will prefer to act aggressively. In a joint enterprise, one may not know whether the other is trustworthy or willing to act opportunistically. Players may be of different *types*, and different types may receive different payoffs for the same strategies, or may be able to play different strategies. If one player does not know what type the other is, then we would not be able to apply the concepts we have used, such as dominant strategy equilibrium and Nash equilibrium without, at least, some further work.

A Closer Look: John Harsanyi 1920–2000

Born in Budapest, Hungary, John Harsanyi was originally educated as a pharmacist, as his parents wished. Harsanyi switched to philosophy in 1946. Scheduled to be sent to a concentration camp in 1944, he escaped from Nazi confinement and was hidden for a year in the basement of a Jesuit monastery. To leave Communist Hungary in 1950, Harsanyi and his fiancée walked through marshes to cross the border illegally.

After moving to Australia, he had to do manual work at first, but later shifted to a new field of study, economics, and obtained his PhD at Stanford in 1959. He joined the faculty of the University of California at Berkeley in 1964. His breadth of interests and experience are reflected in contributions to game theory and other fields that focus on the nature of rationality, especially in circumstances of missing information.

John Harsanyi proposed the following approach, which is now in part of the consensus of game theory. Think of a world inhabited by rational but somewhat uninformed agents. Generally, any one agent does not know what types the others are, but can estimate the probability that agent a is of type α or β or .... These probabilities — the probability that agent a is of type α or β or so forth — are based on public knowledge available to everybody. These are called *prior probabilities,* and it is important that this knowledge is available to everybody. This is called the *common priors* assumption. But one person has unique knowledge about what type agent a is, and that person is agent a. It seems that nobody has better knowledge of the strategies and payoffs to agent a than agent a does! As mentioned above, we have a mathematical approach that tells us how to adjust our estimated probabilities when we have additional knowledge. This mathematical approach is called Bayes' Rule. So, having personal knowledge not available to others, agent a could "update his priors" using Bayes' Rule. Now, Bayes' Rule can be complicated (we will say a little more about it in the appendix to this chapter) but in this case it is simple: Agent a knows that agent a is of type β (or α or ...) with probability one. And in general, we assume that each agent knows their own type with probability one.[1]

Here is an example. Agent B is a schoolyard bully, a new transfer student in the third grade at Blanchard School. He likes to take lunch money from his schoolmates by threatening to beat them up. His strategies are to demand lunch money or don't. His target today is agent A. The strategies of agent A are resist or don't. Schoolboys are of two types: Resisters and non-resisters. Resisters take some satisfaction from fighting back even if it leaves them "bloody but unbowed," so it is in agent B's interest not to demand lunch money from a resister. If agent A is a resister, the payoffs *to agent A* are as shown in Table 7.9.

---

[1] Mystics and some other psychologists have long advised us to "know thyself," which suggests that some of us, at least, do not have such certain knowledge! But we refer only to knowledge of strategies, resources, and payoffs in the game, perhaps not more important dimensions of spirit. We are also thinking of rational individuals.

Table 7.9. The Game for a Resister.

		A	
		Resist	Don't
B	Demand	5	0
	Don't	3	3

Table 7.10. The Game for a Non-resister.

		A	
		Resist	Don't
B	Demand	0	1
	Don't	3	3

Table 7.11. The Game for Agent B.

		A	
		Resist	Don't
B	Demand	−1	5
	Don't	3	3

If Agent A is a non-resister, the payoffs to Agent A are as shown in Table 7.10.

Payoffs to agent B are shown in Table 7.11.

It is known that a proportion p of schoolboys are resisters and $1 - p$ are non-resisters. Since Agent B is a new student at Blanchard school, he does not know which students are which, but knows the "common prior" probabilities, p and $1 - p$. Nature decides whether A is a resister or non-resister. Thus, from Agent B's point of view, the game looks like Table 7.12.

Table 7.12. The Game as Seen by Agent B.

		Nature			
		Resister		Non-resister	
		A		A	
		Resist	Don't	Resist	Don't
B	Demand	–1,5	5,0	–1,0	5,1
	Don't	2,3	2,3	2,3	2,3

Table 7.13. The Game as Seen by a Resister.

		A			
		Strategy 1	Strategy 2	Strategy 3	Strategy 4
B	Demand	–1, 5	–1, 5	5,0	5,0
	Don't	2,3	2,3	2,3	2,3

However, A has private knowledge — he knows what type he is — and thus he can base his choice of strategy on that knowledge. As we have seen in previous chapters, this means that A can choose among contingent strategies:

- Strategy 1. If I am a resister then resist, otherwise resist.
- Strategy 2. If I am a resister then resist, otherwise don't resist.
- Strategy 3. If I am a resister then don't resist, otherwise resist.
- Strategy 4. If I am a resister then don't resist, otherwise don't resist.

If A is a resister, the game is as shown in Table 7.13.

In this case, either Strategy 1 or Strategy 2 are best responses for A. If instead A is a non-resister, the game as he sees it is as shown in Table 7.14.

Table 7.14. The Game as Seen by a Non-resister.

		A			
		Strategy 1	Strategy 2	Strategy 3	Strategy 4
B	Demand	–1, –1	–1, 1	5,0	5,1
	Don't	2,3	2,3	2,3	2,3

Table 7.15. The Game as Seen by Agent B,
in Expected Value Terms.

B	Demand	$-1p + 5(1-p)$
	Don't	2

In this case Strategies 2 and 4 are best responses. In either case, the decisions made by A correspond to Strategy 2: If I am a resister then resist, otherwise don't resist. This is no surprise, of course, and that's the point: Agent B, being rational and knowing that Agent A is rational, will expect Agent A to play according to Strategy 2. But since B does not have the private knowledge that A has, his decision must be based on his expected value payoffs as shown in Table 7.15.

Thus, B will choose "demand" if $5(1 - p) - p > 2$, that is, if $p < \frac{1}{2}$. If $p > \frac{1}{2}$ then B will choose "don't," and if $p = \frac{1}{2}$ then B is indifferent between "demand" and "don't." Since A has used his private knowledge to update the common prior probability, a very simple application of Bayes' rule, this is an instance of a Bayesian Nash equilibrium.

The Bayesian Nash equilibrium thus extends non-cooperative game theory to many interactions that it might not include, that is, to any case in which the players may be of different types but have common prior probabilities. Is the assumption of common priors reasonable?[2] Since the priors are based on all commonly available

---

[2]Our understanding of the common priors assumption owes much to Aumann (1976, 1987, 1998). Concepts such as information, probability, time and space can be fairly clear intuitively, but their mathematical representation can be more

information, and we assume that the agents are rational, it is. If they disagree about the priors, one or both of them must be using less than all commonly available information, and that would be irrational (at least as we understand rationality in game theory and neoclassical economics). If the agents base their probabilities on information that is not commonly available, as agent A does in the example above, this is not a prior probability but a revised probability. Thus, rational agents cannot "agree to disagree" about their prior probabilities.

7. SUMMARY

Probability is a way of expressing the relative likelihood of an uncertain statement in numerical terms. Probabilities range from zero to one, and a larger probability is attached to a "more likely" statement. Probability is a large subject in itself, and we limit ourselves in this chapter to a few applications that are very important in game theory.

   Probabilities can be helpful in assigning values to uncertain payoffs. If we know that the payoff in a decision or game will be one of several numbers, and we can assign probabilities to the numbers, we may compute the expected value of the payoff as the weighted average of the payoffs numbers, using the probabilities as weights.

   In game theory, uncertainty is customarily brought into the picture by making chance or nature one of the players in the game. Nature plays her strategies with fixed probabilities. When a single individual makes a decision with an uncertain outcome, we think of that as a game against nature. One possible solution to a game against nature is to maximize the expected value of the payoff. However, attitudes toward risk may modify this, and risk aversion

---

difficult. At the same time, intuition is limited and may lead us to logical contradictions. Mathematical representation is a way of extending intuition and avoiding its traps. In this case, as Aumann says, the intuition is rather clear, if not compelling. It is the intuition I have tried to outline here. The mathematics begins by representing information and probability in terms of set theory. Once that has been done, as Aumann (1998) says, the proof and the application of Bayes' rule are trivial.

(or risk loving) can be related to the curvature of the utility-of-income function. A player who simply maximizes the money payoff is said to be risk neutral. In most of what follows in this book we will assume, for simplicity, that the decision-makers are risk neutral.

In games with two or more human players, nature may also be a player. Here again, one possible solution is for the players to evaluate their various strategies using the expected value of the payoffs depending on the probabilities of different actions by nature. Using this approach, we may allow for cases in which one player is not sure what "type" of player the other is, but knows that the other player knows what type they are. Thus, the ignorant player makes decisions based on expected value calculations, while the other player can make use of their special information. A Nash equilibrium in such a game is a "Bayesian Nash equilibrium."

Probability and expected value are used very widely in game theory, and will play important roles in many chapters to follow.

APPENDIX A. MEASURING UTILITY

Using the relationship between the expected utility of money and the expected value of utility of payoffs, von Neumann and Morgenstern proposed a way of measuring the utility of payoffs. Suppose we want to know the utility Karen attaches to a riskless payment of \$4,000. We will offer her a lottery, in which the payoff is 10,000, with a probability p, and 0, with a probability $1 - p$. We offer Karen a choice between the lottery and the riskless payment of \$4,000. She will choose the lottery or the riskless payment, depending on whether the lottery ticket gives her more utility than the riskless payment. So, we negotiate with Karen, carefully adjusting the probability p until she doesn't care (is indifferent) which alternative she gets. That means the expected utility of the lottery is the same as the utility of \$4,000.

To measure the utility of \$4,000, or any other amount, we need to agree on units of measurement and on the zero point. We will assign a zero payoff a utility of zero. That's just as arbitrary as any other zero point, but it is easy to remember. Rating utility on a ten-point scale, we will assign 10 to the payoff of $10,000. Therefore, when Karen is indifferent between the two alternatives (doesn't care whether she gets the lottery or the riskless $4,000), the expected utility of the lottery is $10,000p + (1 - p)(0) = 10,000p =$ utility of $4,000. Suppose we try the experiment and find that the probability that does the trick is 0.795. Then we say the utility of $4,000 is $10 * 0.795 = 7.95$.

In general, according to the von Neumann and Morgenstern approach, the first step in measuring utility is to establish payoffs corresponding to an arbitrary zero point and an arbitrary unit of measurement. Then we can measure the utility of any amount X between the lower and upper limits by finding the probability that makes the person indifferent between X and a lottery that pays either the upper or lower limit. The probability is used to measure the utility of X.

APPENDIX B. BAYES' RULE

One of the more advanced methods of probability analysis widely used in advanced game theory is "Bayes' Rule," named after statistician Thomas Bayes. This method will be mentioned in passing in a few of the remaining chapters, but we will not give applications of it in this introductory text. Here is a single example to get the idea.

Suppose that Karen the Collector is examining a painting she might buy. Because of its distinctive style, she thinks it might be by Himmelthal — Karen knows

A Closer Look: Reverend Thomas Bayes 1702–1761

Born in London, England, Bayes was a mathematician and theologian whose work extended from advances in mathematical probability to an attempt to prove the benevolence of God. He is best remembered for his work on probability in "Essay Towards Solving a Problem in the Doctrine of Chances" (1763), published posthumously in the Philosophical Transactions of the Royal Society of London.

that 25% of all the paintings in that style are Himmelthals. Looking more carefully, Karen discovers that the painting is signed "Himmelthal." Himmelthal did not sign all of his paintings — Karen knows from her reference books that he signed 90% of them. But there are forgeries, too, and 35% of all paintings in this style (counting forgeries) are signed Himmelthal. Thus,

P_1 = the probability that a painting in this style is a Himmelthal = 0.25

P_2 = the probability that a Himmelthal is signed "Himmelthal" = 0.90

P_3 = the probability that a painting in this style is signed "Himmelthal" = 0.35

But none of those is what Karen wants. She wants P_4 = the probability that a painting in this style **and** signed Himmelthal **is** a Himmelthal, taking forgeries into account. To get that she applies Bayes' rule, calculating

$$P_4 = P_2\, P_1/P_3 = (0.9)(0.25)/(0.35) = 0.64.$$

The probability that it is a Himmelthal is slightly less than 2/3. In general, Bayes' rule says:

Probability that A is true if X is observed=

$$\frac{\left(\text{Probability of A independently of X}\right)\left(\text{Probability of X if A is true}\right)}{\left(\text{Probability of X independently of A}\right)}$$

For an approachable explanation of Bayes' rule, see Alan S. Caniglia, *Statistics for Economics, an Intuitive Approach* (HarperCollins, 1992) pp. 69–72.

Q7. EXERCISES AND DISCUSSION QUESTIONS

Q7.1. Country Risk

Investors who invest in other countries may be concerned about changes in the countries in which they invest, that have an impact on their profits. This is called "country risk." Of course, country risk is complicated, and the probability of loss is at least partly subjective. Here is a list of countries you might invest in. For each country, estimate the probability that political changes could lead to a serious investment loss. If you do not know anything about the country, take a few minutes with an encyclopedia or on the World Wide Web to learn a little, and then make your best subjective estimate.

(A) Pakistan
(B) Belgium
(C) Ghana
(D) Mexico

Q7.2. Urn Problem

You are to draw one ball at random from an urn containing 50 white balls and 100 black balls. What is the probability that you will draw a white ball? *Hint:* if you were to draw 150 balls, what proportion would be white?

Q7.3. Lottery

You have tickets in a lottery. The lottery will be decided by drawing a ball from an urn. There are a thousand balls in the urn, numbered from 1 to 1,000. The tickets are also numbered from 1 to 1,000, and the person holding the ticket with a number corresponding to that on the ball that is drawn will be the winner. The winner will be paid a thousand dollars. You hold all the tickets numbered from 1 through 20.

a.   What is the probability that you will win?
b.   What is the expected value of your payoff from this lottery?

Q7.4. Investment in Research

You have an opportunity to invest in a company that is working on a research project. If you invest, you will have to put up $1 million. If the project is successful, the result will be a new kind of product that will yield a return of $6 million for your investment. The probability of success is $1/5$.

a. What is the role of nature in this game?
b. Construct a table to represent your choice whether to invest or not as a game, using your answer to the previous question.
c. Compute the expected value of the payoff to your investment.
d. Assuming that you are risk neutral, and you had the million, would you decide to invest, or not to invest?

Q7.5. Risk Aversion

Reconsider your answer to the previous question, assuming that your utility function is SQRT$(Y + 1,000,000)$ where Y is the payoff of either 0 or 6 million, as the case may be, minus the 1 million you have to put up. Now will you make the investment?

Q7.6. Farmer Ramdass

Farmer Ramdass is considering trying a new crop on his farm. If it is successful, it will pay him 80,000 and he will be able to pay off his debts. If it is a failure, however, it will pay only 20,000 and Farmer Ramdass will just be able to get by without paying anything on his debts. The old crop he has planted for years pays 30,000 with certainty, enough for a very small payment on Farmer Ramdass' debts. ASSUMING Farmer Ramdass is risk-neutral, how great does the probability of success need to be in order to persuade him to try the new crop? Is the assumption that Farmer Ramdass is risk-neutral likely to be correct?

Q7.7. Bernie's Umbrella Business

Bernie's umbrella stand sells umbrellas in the Pike Place Market in Seattle. He can consistently sell 30 umbrellas if it does not rain and

100 umbrellas if it does, per day. He sells umbrellas for $12 each. The prediction for tomorrow is that the probability of rain is 50%.

a. Compute the mathematical expectation (expected value) of the number of umbrellas Bernie will sell tomorrow.
b. On an average Seattle day the probability of rain is $1/3$. Compute the mathematical expectation of the number of umbrellas Bernie can sell on an average day.
c. Umbrellas cost $3 at wholesale. How much does Bernie spend to replace the umbrellas he sells per day, on the average?
d. Bernie's costs for operating his stand, including his own income (opportunity cost), amount to $400 plus what he spends to replace the umbrellas he sells. How much does Bernie need to charge per umbrella in order to break even on the average? What is his average profit, if any?
e. Bernie is considering closing the umbrella stand and going to work as a chef for a salary of $300 per day. Treat this decision as a "game against nature" and diagram it in extensive form. What is Bernie's "outside option" as you view the case?

Q7.8. Shame

Mike and Ned are matched to play a game that requires them to choose between making a great effort (working) and making a slight effort (shirking). As in other effort dilemmas in this book, if both make a great effort their work will be highly productive, resulting in high payoffs for both despite the subjective cost of effort, but if just one avoids the subjective cost of effort by shirking, the other agent is made much worse off. However, in this game, agents may be of two types, type A (for ashamed) and type ~A (for shameless). For a type A agent, if the other player works and the type A agent shirks, then the payoff to the type A agent is reduced by intense feelings of shame. A type ~A agent feels no such shame. For a ~A agent, the payoffs are shown in Table 7.16 (as in example 3.2).

Table 7.16. The Game for the Type ~A Agent.

		Other	
		Work	Shirk
~A	Work	10	2
	Shirk	14	5

Table 7.17. The Game for the Type A Agent.

		Other	
		Work	Shirk
~A	Work	10	2
	Shirk	4	5

Table 7.18. Mike and Type ~A.

		~A	
		Work	Shirk
Mike	Work	10,10	2,14
	Shirk	4,2	5,5

For a type A agent, the payoffs are as shown in Table 7.17.

Mike knows that he is type A, but doesn't know whether Ned is A or ~A. Thus, if Ned is ~A, from Mike's point of view, the game is as shown in Table 7.18.

By contrast, if Ned is type A the game is as shown in Table 7.19.

a. Suppose Mike knows that the probability that Ned is type ~A is 0.5. Construct the game table with expected values. What sort of game is it? Is it a social dilemma?

Table 7.19. Mike and Type A.

		A	
		Work	Shirk
Mike	Work	10,10	2,4
	Shirk	4,2	5,5

b. Suppose instead that Mike is ~A. Construct game tables for play between Mike and ~A and A types on that basis, and construct the game table with expected value payoffs in this case. What sort of game is it? Is it a social dilemma? Does it have any efficient Nash equilibria?

c. Suppose we had a large population of agents of both types, divided 50–50, and matches were at random. What proportions would you expect to be profitable?

CHAPTER 8

Mixed Strategy Nash Equilibria

We have seen how game theory uses the concept of expected values to deal with uncertainty from natural sources. But the players may deliberately introduce uncertainty into the game, because it suits their purposes. There are some games in which the best choice of strategies is an unpredictable choice of strategies. The concepts of probability and expected value can help us to understand that. A good example is found in the great American game of baseball.

1. KEEPING THEM HONEST IN BASEBALL

Baseball comes down, on every pitch, to a confrontation between two players: The pitcher and the batter. To keep things simple, we consider a pitcher who has just two pitches: A fastball and a change-up. The change-up is a slower ball that would be easy to hit if the batter knew it was coming. But the change-up may be used to fool a batter who is expecting the fastball. These two pitches will be the pitcher's two strategies. The batter is good — he can hit the fastball and knock it a long way, with a good chance of putting himself in a scoring position, but only if he commits himself to swing before he has seen the ball. If he waits until he can see the ball he will miss the fastball, but he will hit the change-up for a chance at a single — a good result but not quite scoring position. We suppose the count is 3 and 2 — for non-baseball folks, that means this is the last chance for both the pitcher and the batter. The payoffs are shown in Table 8.1.

Table 8.1.   Payoffs for Baseball.

		Pitcher	
		Fastball	Change-up
Batter	Swing early	10,–10	–5,5
	Swing late	–5,5	3,–3

It may seem that the pitcher should always throw his best pitch, especially in a crucial situation such as this. The problem is that if he does, the batter will always swing early, and get a lot of hits and a lot of home runs. Even if the pitcher throws his fastball predictably only in three and two situations, that will make it easier for the batter to decide how to swing. He can just swing early on every three and two situation.

It is very much the same from the point of view of the batter. If he swings predictably — for example, if he just swings early on every pitch — then the pitcher has an easy job. He can just throw a changeup on every pitch. So, the batter, too, needs to be unpredictable.

Assuming, as usual, that they are both rational, each of the two baseball players will choose unpredictably between their two strategies. "Unpredictably" means that

> **HEADS UP!**
>
> Here are some concepts we will develop as this chapter goes along:
>
> **Pure Strategies:** Every game in normal form is defined by a list of strategies with their payoffs. These are the *pure strategies* in the game.
>
> **Mixed or Randomized Strategies:** In a game in normal form, a player who chooses among the list of normal form pure strategies according to given probabilities, two or more of which are positive, is said to choose a *mixed strategy*.
>
> **Mixed Strategy Equilibrium:** A Nash equilibrium in which one or more of the players chooses a mixed strategy is called a *mixed strategy equilibrium*.

there is a random element in the choice of strategies. Each will choose one strategy or the other with some probability. This is called a *mixed strategy*, since it mixes the two *pure strategies* shown in the payoff table. Part of the job of the player who calls the plays — pitcher and catcher in baseball, quarterback in football, point guard in basketball — is to "mix them up," to be unpredictable, so that the opposition cannot guess which strategy is coming and prepare accordingly. And this is one of the toughest jobs in sports — it really isn't easy for a human being to choose at random.

To restate, a pure strategy is one of the strategies shown in the normal form of the game, picked with probability of one (certainty). A mixed strategy is the randomized choice of one pure strategy or another according to given probabilities. The strategies are mixed in the proportions of the probabilities.

But what probabilities will they choose? To figure that out, we have to consider it from both points of view. From my point of view, the key idea is to prevent the other player from exploiting my own predictability. So, I want to choose a probability that will keep him guessing. Let's look at it from the point of view of the batter. Let p be the probability that the pitcher throws a fastball. Table 8.2 shows the payoffs to the batter's two strategies, in terms of the probability p. (Of course, we use the expected value concept, since the strategies are unpredictable.) If p is large enough, swinging early will pay better than swinging late. In that case the batter will choose to swing early. On the other hand, if p is small enough, then swinging late will pay better. In that case the batter will swing late.

Now let's switch to the pitcher's point of view. To limit his losses, and keep the batter guessing, what the pitcher needs to do is to adjust p so that neither of the batter's strategies pays better than the

Table 8.2. Expected Value Payoffs for Batter Strategies.

Swing early pays	$10p - 5(1 - p) = 15p - 5$
Swing late pays	$-5p - 3(1 - p) = 3 - 8p$

other. We can express that with a little algebra. To say that neither payoff is bigger than the other is to say that

$$15p - 5 = 3 - 8p. \tag{1}$$

Solving that for p,

$$p = 8/23. \tag{2}$$

So, the pitcher throws the fastball with probability 8/23. Now, let's look at the batter's strategy choices, from the point of view of the pitcher. Let q be the probability that the batter swings early. Table 8.3 shows the expected value payoffs for a fast ball and a changeup. The batter will try to adjust the probability q so that neither of those strategies gives a better expected value payoff than the other. This is expressed by Equation 3. Solving Equation 3, algebraically, we find a probability of 8/23. That is the probability that the batter will swing early.

$$5 - 15q = 8q - 3. \tag{3}$$

$$q = 8/23. \tag{4}$$

We now have the probabilities: The pitcher will throw a fastball with a probability 8/23 and the changeup with a probability 15/23, and the batter will swing early with a probability of 8/23 and swing late with a probability of 15/23. Each one chooses a "mixed strategy," and in each case the mixture is that players' best response to

Table 8.3.   Expected Value Payoffs for Pitcher Strategies.

Fastball pays	$-10q + 5(1 - q) = 5 - 15q$
Changeup pays	$5q + 3(1 - q) = 8q - 3$

the mixture chosen by the other player. Remember the definition of a Nash equilibrium: Each player chooses a strategy which is his best response to the other players' strategy. It seems that when the two baseball players choose probability mixtures 8/23, we have a Nash equilibrium.

This may be easier to understand if we separate it into two stages. At the first stage, for example, the batter decides whether to randomize or not, and at the second stage chooses the specific probabilities. So long as the pitcher chooses the mixed strategy with probabilities 8/23 and 15/23, the batter is indifferent between his two strategies. Swing early and swing late both give him the same expected payoff. He may as well randomize. But, from the point of view of expected payoffs, the batter could choose any probabilities — 50–50, 90–10, 100,0, any other. So how does he choose the probabilities? He realizes that if he does not choose the probabilities 8/23 and 15/23, the pitcher will not randomize.

Table 8.4 shows the results of three possible probabilities the batter might choose. If he chooses 7/23, then the pitcher can get an expected payoff of 0.43 by throwing a fastball, but only −0.57 for a changeup, so in that situation, the pitcher will not randomize, but will throw a fastball, leaving the batter with an expected value payoff of −0.43. If the batter chooses a probability of 9/23, the pitcher can get a positive 0.13 expected value from a changeup, but −0.87 from a fastball, so again he will not randomize but throw the changeup,

Table 8.4. Expected Value Payoffs with Alternative Probabilities.

Probability of swing early	Expected payoff to fastball	Expected payoff to changeup	Pitcher's best response	Batter's expected payoff
7/23	0.43	−0.57	Fastball	−0.43
8/23	−0.21	−0.21	Either fastball or changeup	0.21
9/23	−0.87	0.13	Changeup	−0.13

leaving the batter with a negative –0.13 expected value. Only the exact probability 8/23 leaves the pitcher indifferent between the fastball and the changeup, and so leads the pitcher to make the best of his bad situation and randomize — leaving the batter with his best expected value possibility, 0.21.

In game theory this kind of Nash equilibrium is known as a "mixed strategy equilibrium." In fact, the mixed strategy equilibrium is the only equilibrium in this game. If you look back at Table 8.1 you can see for yourself that there is no Nash equilibrium in pure strategies. No matter which combination of pure strategies is chosen, one player or the other will want to depart from it. But there is an equilibrium in terms of mixed strategies, and so long as each player chooses an equilibrium mixed strategy, neither player will want to deviate from it independently.

2. PURE AND MIXED STRATEGIES

What we have seen in the previous example is that human beings may make strategic use of uncertainty by choosing mixed strategies rather than pure strategies. The pure strategies are the list of strategies that define the normal form of the game or its extensive form. A mixed strategy is a policy of choosing one of two or more pure strategies according to some probability. We have also seen that a game may have an equilibrium in mixed strategies even though it does not have an equilibrium in pure strategies.

Definition: *Pure and Mixed Strategies* — Every game in normal form is defined by a list of strategies with their payoffs. These are the *pure strategies* in the game. However, the player has other options, namely to choose among the strategies according to given positive probabilities. Every such decision rule is called a *mixed strategy.*

We have seen from a number of examples that not all games have Nash equilibria in pure strategies. This is a well-known shortcoming of the Nash equilibrium concept. However, when we allow for mixed strategies as well as pure strategies, that is no longer true. John Nash discovered that every two-person game in normal form has an equilibrium in mixed strategies, even if it does not have an equilibrium in pure strategies. Von Neumann and Morgenstern showed that this is true for an important subclass of two-person games, the zero sum games, and John Nash extended this by showing that all two-person games have Nash equilibria in mixed strategies if not in pure strategies. Of course, that's why we call it a Nash equilibrium. Nash shared the 1994 Nobel Memorial Prize for that discovery.

There is a sort of intellectual jiu-jitsu involved in the mixed strategy equilibrium idea, and it can be confusing. It may seem odd that the pitcher is balancing the expected value payoffs for the *batter* and equalizing them. But in doing so he his maximizing his own expected value payoff. The pitcher's objective is really to force the batter to randomize *his* hitting strategy.

A person might not want to randomize, reasoning as follows: "I'm not going to randomize, because that's giving up. I'm going to stay a step ahead of the other guy, and choose the right strategy to beat him." But if the other guy is randomizing his strategy, it is no use — he will choose the probabilities so that all of your strategies give you the same expected value, and it doesn't matter which you choose. You just can't out-think the other guy — unless he is trying to out-think you! "You can't con an honest man" and you can't con a randomizer, either.

In game theory we assume "common knowledge of rationality," that is, not only that both players are rational but that each one knows the other is rational. This is especially important for mixed strategies.

So, I reason as follows: "My opponent knows that he cannot out-think me, so he will not try. He will randomize, unless I act irrationally and predictably and so give him an opportunity. He will prevent me from out-thinking him by choosing probabilities that make the expected values of my strategies equal. So, I may as well randomize anyway — but with what probabilities? I want to choose the probabilities that will put him in the worst situation, that will prevent him from out-thinking me — and the way to do that is to choose probabilities that make the expected values of his strategies equal."

In a class studying mixed strategy equilibria, students often ask, "But what *is* the mixed strategy equilibrium?" The simple and direct answer is that the equilibrium is a list of probabilities for each player, with one probability for each of the strategies. But that simple and direct answer sometimes does not seem to resolve the student's doubt.

Some students may have difficulty understanding how it is possible to choose a pure strategy at random. We could think of it as a choice among contingent strategies. We first extend the game to a two-stage game. At the first stage, the only decision is made by nature — or by a random-number generator. This number, which may be called γ, is between zero and one; and for any number g between zero and one, the probability that $\gamma \leq g$ is exactly g. (That is, in statistical jargon, the distribution of γ is uniform.) Then the players choose among contingent strategies such as "If $\gamma < \frac{8}{23}$, then throw a fastball, otherwise throw a change-up" and "if $\gamma < \frac{8}{23}$ then swing early, else swing late." Since γ can take an infinite number of values, there are infinitely many mixed strategies to choose from. It is this infinity that assures us that every two-person game has an equilibrium in mixed strategies if not in pure strategies.

This is not a standard idea in game theory, as mixed strategy equilibria are usually thought of as equilibria of normal form games. If, however, it is easier to think of mixed strategies as strategies contingent on a random variate that becomes known at the beginning of the game, that is no less correct.

3. A BLUE-LIGHT SPECIAL

Let's look at another example of mixed strategies. Some economists have applied mixed strategy reasoning to the scheduling of sales by retail merchants. This would not apply to all sales. Some sales are scheduled quite predictably, for example, on holidays. But some

Table 8.5. Payoffs for a Sale Game.

		Consumer	
		Shop today	Shop tomorrow
Seller	Sale today	5,10	8,4
	Sale tomorrow	10,5	4,8

sales do seem to be unpredictable — for example, there may be a sale without any notice when a blue light is turned on. Why would a retailer want her sales to be unpredictable? This might be a mixed strategy. If the consumer knew when a sale is to be held, she would make a point of coming on those days. But the consumer might also want to be unpredictable. If the retailer knew which days the consumer would be coming, then the retailer would never schedule sales on those days.

Could this kind of reasoning really lead to a mixed strategy? Let's look at a specific example and see. To make things very simple we will treat it as a two-person game. The seller is one player and the consumer is the other player. The seller's strategies are to schedule a sale today or tomorrow. The consumer's strategies are to visit the store today or tomorrow. The payoffs[1] for the two players are shown in Table 8.5.

Now look at the payoffs from the point of view of the seller and let p be the probability that the consumer visits the store today rather than tomorrow. The expected values for the seller's two strategies are shown in Table 8.6.

If one of these numbers is greater, then the consumer is making it too easy for the seller to pick a date for a sale when the consumer

---

[1] Generally, benefits and profits now or in the nearer future are worth more than payoffs further in the future, so both of these players put a higher value on benefits they get today than tomorrow. In economics this is called "time preference." This common tendency has been exaggerated a bit in the example, for simplicity — usually the difference between payoffs only a day apart would be much smaller.

Table 8.6. Expected Value Payoffs for Seller Strategies.

Sale today pays	$5p + 8(1 - p) = 8 - 3p$
Sale tomorrow pays	$10p + (1 - p)4 = 4 + 6p$

will not be there. Therefore, the consumer will adjust p so that the two values in Table 8.6 are equal. That is,

$$8 - 3p = 4 + 6p,$$

solving,

$$p = 4/9.$$

So, we conclude that the consumer will come today with probability $4/9$ and tomorrow with probability $(1 - 4/9) = 5/9$.

Now look at the payoffs from the point of view of the consumer and assume that q is the probability that the sale is today rather than tomorrow. The expected value payoffs to the consumer for her two strategies are shown in Table 8.7. If one of those is greater than the other, the consumer will find it easy to pick a better day with a good chance to get the benefits of the sale.

Accordingly, the seller will adjust q so that the customer's expected values from coming tomorrow and from coming today are the same. That is,

$$5 + 5q = 8 - 4q,$$

solving,

$$q = 3/9 = 1/3.$$

We conclude that this seller will schedule a sale for today with a probability $1/3$ and for tomorrow with a probability $2/3$.

Table 8.7.   Expected Value Payoffs for Consumer
Strategies.

Shop today pays	$10q + 5(1 - q) = 5 + 5q$
Shop tomorrow pays	$4q + 8(1 - q) = 8 - 4q$

Of course, this example is very simplified. In effect, it assumes that there are only two dates on which the people can do business, today and tomorrow, and thereafter the seller will go out of business. The game is also over simple in allowing for only one seller and one consumer. Nevertheless, it shows us how the scheduling of sales could be a mixed strategy equilibrium; and indeed, we can find mixed strategy equilibria in much more complicated and realistic games of scheduling sales, allowing for the reality that there may be many sellers and consumers and that they may continue to do business at any date in the future.

4. EQUILIBRIA WITH MIXED AND PURE STRATEGIES

The games of Baseball and Unpredictable Sales have equilibria in mixed strategies but not in pure strategies. However, some games have both kinds of equilibria. For an example, let us return to Example 5.4. from Chapter 5, the Panic Buying game. The strategies and payoffs are shown in Table 8.8.

For this game, we recall that there are two Nash equilibria in pure strategies, each of the strategy pairs at which the two buyer choose the same strategy. This is a coordination game. But the two buyers each have to guess which strategy the other one will adopt, and if they guess wrongly, one of the non-Nash outcomes might occur — where they choose different strategies, making both worse off. However, there is also a third equilibrium, a mixed strategy equilibrium. Looking at it from the point of view of buyer A, suppose the probability that buyer B chooses panic buying is p. Then the expected values for the buyer A are as shown in Table 8.9. (Since the game is symmetrical, similar results will be obtained from the point of view of buyer B).

Table 8.8. Panic Buying (Repeats Table 5.4).

		B	
		Panic buy	Routine buy
A	Panic buy	2,2	−1,0
	Routine buy	0,−1	5,5

Table 8.9. Expected Value Payoffs for Buyer A Strategies.

Panic buy pays	$2p + (-1)(1 - p) = 3p - 1$
Routine buy pays	$0p - 5(1 - p) = 5 - 5p$

Thus, if buyer B adjusts p so that the expected value payoff of the two strategies for the buyer A are the same, we have

$$3p - 1 = 5 - 5p,$$

solving,

$$p = 6/8 = 3/4.$$

Since the game is symmetrical, both players can reason in the same way and each plays panic buy with probability 3/4 and routine buy with probability 1/4.

So, this game has three Nash equilibria, two in pure strategies and one in mixed strategies. Let us see what the expected value payoff of the mixed strategy equilibrium is. The probability that both choose panic buy is[2] $(0.75)(0.75) = 0.6525 = 9/16$. Thus, they will both panic buy a little over half the time. The probability that they

---

[2]Here we are using a rule from probability theory: The probability that two independent events both occur is the product of the two probabilities, known as the compound probability.

both choose routine buy is $(0.25)(0.25) = 0.0625 = 1/16$. The probability that the strategies are uncoordinated, so that one chooses panic buy and the other routine buy is $(0.75)(.025) = 0.1875 = 3/16$. But this can happen two ways: Where A plays panic buy and where A plays routine buy. Thus, the expected value for each player in this symmetrical game is $0.625(2) + 0.1875(-1) + 0.1875(0) + 0.0625(5) = 1.375$. But notice that both pure strategy Nash equilibria are payoff dominant relative to this expected value payoff from the mixed strategy equilibrium. They are also risk dominant, since both pure strategy equilibria avoid the risk of a negative payoff, and the mixed strategy equilibrium does not. Thus, the mixed strategy equilibrium in this case is a relatively bad one, but given the problem of arriving at a pure strategy Nash equilibrium in a coordination game such as this, we cannot rule it out.

Here is another example of a game with both pure and mixed strategies, an environmental policy example. Two small towns, Littleton and Hamlet, get their drinking water from wells. They can draw from two aquifers. One aquifer is shallow, so a well to reach it is cheap, but will only supply enough water for one town. If both drill for the shallow aquifer, they will deplete it and neither will have enough water. The other aquifer is deeper, so it costs more to drill down to it, but it can provide enough water for both towns, and consequently they can share the costs. Each town has two strategies: Deep or shallow. The payoffs to the two towns are shown in Table 8.10. Notice that there are two Nash equilibria in pure strategies, and notice also that neither is the cooperative solution, which occurs when the two towns share the deep aquifer and payoffs total 20.

Take Littleton's point of view and let p be the probability that Hamlet plays deep. Then the expected values for Littleton's strategies are shown in Table 8.11. Doing the algebra, we find that the equilibrium probability of choosing "deep" is $11/18$. Since the game is symmetrical this is the same for both towns.

The expected value of payoffs from the mixed strategy is 9.17. This is better than the payoffs to a town that independently digs a deep well, but one that digs the only shallow well can do still better.

Table 8.10.   Payoffs for Littleton and Hamlet.

		Hamlet	
		Deep	Shallow
Littleton	Deep	10,10	3,15
	Shallow	15,3	0,0

Table 8.11.   Expected Value Payoffs for Littleton's Strategies.

Deep pays	$10p + 3(1 - p) = 3 + 7p$
Shallow pays	$15p - 0(1 - p) = 15p$

Thus, there are three Nash equilibria once again: Two in pure strategies and one in mixed strategies.

5. GRAPHICS FOR MIXED STRATEGIES

We can get a clearer idea of what this mixed strategy means by doing a bit of graphics. First, let us do a graphic analysis of the baseball example from Section 1. We will take the batter's point of view. Figure 8.1 shows the expected value payoffs for the batter. The probability that the pitcher throws a fastball is on the horizontal axis.

The expected value payoff when the batter chooses "swing early" is shown by the black line, and the payoff when the batter chooses "swing late" is shown by the gray line. Students who have taken a course in the principles of economics know that something important usually happens when two lines cross. That's true in game theory as well. The intersection of the two lines shows the equilibrium probability *for the pitcher*. At any other probability, one or the other of the hitting strategies pays better, so the batter will choose the one that gives the better payoff, and make the pitcher worse off.

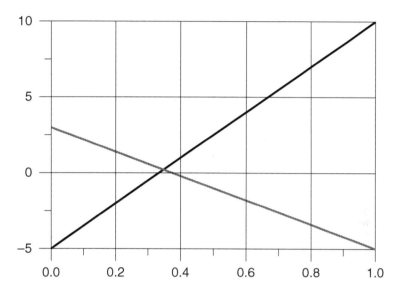

Figure 8.1. Expected Value Payoffs for the Batter in Baseball.

Therefore, the pitcher wants to choose the probability so that neither one is greater than the other — where the lines cross — and as we know, that is 8/23. Since this game is not symmetrical, the diagram for the expected value payoffs to the pitcher will look a bit different, and we will not show it here.

Notice what happens if the pitcher throws a fastball with a probability slightly more than the equilibrium probability of 2/7. Suppose he chooses a probability of 0.4. Then the batter can get an expected value of 1 by swinging early with probability 1, and if he swings early with any probability less than one, he will get less; so, his best response is to swing early with probability 1. But then the pitcher's best response is to pitch the change-up. This pair of strategies isn't stable either, though. As we know, this game has no equilibrium in pure strategies. Thus, the two athletes will keep shifting their strategies and things will not settle down again until they arrive back at the mixed-strategy equilibrium. (*Exercise:* Construct the pitcher's diagram for yourself and use it to follow through the same reasoning about the pitcher's strategies).

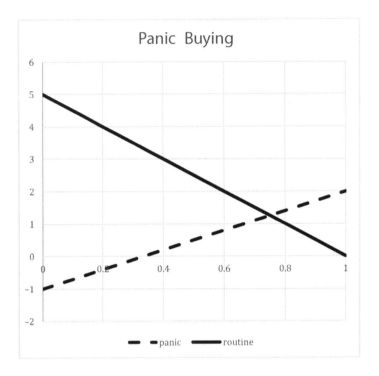

Figure 8.2. Expected Values for Panic Buying Example.

Now let us consider the Panic Buying Game. The expected values for the two strategies for either buyer are shown in Figure 8.2. The horizontal axis shows the probability that one buyer chooses "panic buy," and the vertical axis shows the expected values for playing the two strategies for the other player. The solid line shows the expected value for playing "routine buy" and the dashed line for choosing "panic buy." As before, the intersection of the two lines corresponds to the equilibrium probability of "panic buy," since that is the probability at which the expected values of the two strategies are equal. Since the game is symmetrical, the diagram will look the same from either point of view.

Now, suppose that buyer B chooses "panic buy" with a probability just to the right of the intersection — that is, just a little more

Table 8.12. The Advertising Game (Repeats Chapter 1, Table 1.3).

		Fumco	
		Don't advertise	Advertise
Tabacs	Don't advertise	8,8	2,10
	Advertise	10,2	4,4

than 3/4. Then buyer A's best response is to play "panic buy" with probability one. In turn, player B's best response to that is to play the pure strategy "panic buy" for the pure strategy Nash equilibrium. Similarly, if buyer B plays "panic buy" with a probability just a little less than 3/4, buyer A's best response is to play "routine buy" with probability one, and then B's best response is "routine buy" for the pure strategy equilibrium at "routine buy." We see that any deviation from the equilibrium probabilities, however small, will lead immediately to one of the pure strategy Nash equilibria. In that sense the mixed-strategy equilibrium in this game is unstable. The mixed strategy in the well-drilling game between Littleton and Hamlet is also unstable in the same sense. In general, coordination games with two Nash equilibria in pure strategies will have a third equilibrium in mixed strategies, although the third equilibrium may be unstable.

For another contrast, let us do a graphic analysis of a social dilemma — the Advertising Game from Chapter 1. The two players are two rival companies selling the same product, and their strategies are to advertise or not to advertise. If both advertise, their advertisements largely offset one another. The payoffs for this game are shown in Table 8.12, which repeats Table 1.3 from Chapter 1. In this game, we recall (Advertise, Advertise) is a dominant strategy equilibrium.

This is a symmetrical game, so the analysis will look the same whether we take the viewpoint of Tabacs or Fumco. In Figure 8.3, then, the horizontal axis shows the probability that Tabacs (Fumco) chooses the strategy "don't advertise." Then the black line shows the

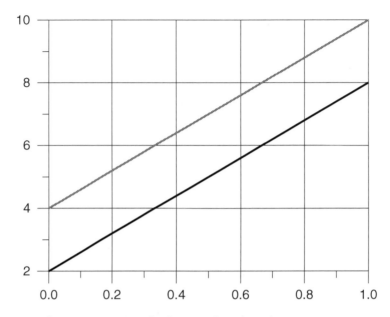

Figure 8.3. Expected Value Payoffs in the Advertising Game.

payoffs to "don't advertise" and the gray line shows the payoffs to "advertise" for Fumco (Tabacs). What we see is that the payoff to "advertise" is always greater than the payoff to "don't advertise." That is to say: "Advertise" is dominant, not only over the pure strategy "don't advertise," but over all possible mixed strategies as well. This applies to social dilemmas in general: The dominant strategy is dominant over mixed as well as pure strategies.

EXERCISE: Show that the mixed-strategy equilibrium for the well-drilling game between Littleton and Hamlet is unstable.

6. SUMMARY

Uncertainty may come from the human players in the game, as well as from nature. A game in normal form, we recall, begins with a list of "pure" strategies for each player. But it may be that playing in the one pure strategy predictably will make the player vulnerable to exploitation by his opponent. In that case, the rational player will attempt to be unpredictable, choosing among the pure strategies with probabilities carefully adjusted to neutralize the opponent's opportunities for exploitation. When strategies are chosen in this way, we call it a mixed strategy.

Since the payoffs to a mixed strategy are uncertain, we evaluate them using the expected value concept. The best response is the strategy (or probabilities for choosing among strategies) that maximizes the expected value of the payoff. For "best responses" defined in this way, all two-person games have Nash equilibria, including those that have no equilibria in pure strategies. Some games may have more than one equilibrium, including equilibria in both pure and mixed strategies. This is true of coordination games, but the mixed strategies in coordination games are unstable.

Q8. EXERCISES AND DISCUSSION QUESTIONS

Q8.1. Matching Pennies

As we noted in Question 3, Chapter 1, matching pennies is a school-yard game. One player is identified as "even" and the other as "odd." The two players each show a penny, with either the head or the tail showing upward. If both show the same side of the coin, then "even" keeps both pennies. If the two show different sides of the coin, then "odd" keeps both pennies.

a. Verify that matching pennies has no Nash Equilibrium in pure strategies.
b. Compute the mixed-strategy equilibrium for matching pennies.

Q8.2. Rock, Paper, Scissors

In Exercise 5.3, Chapter 5 we considered the common school-yard game called Rock, Paper, and Scissors. Two children (we will call them Susan and Tess) simultaneously choose a symbol for rock, paper or scissors. The rules for winning and losing are:

Table 8.13.   Rock, paper and Scissors.

		Susan		
		Paper	Rock	Scissors
	Paper	0,0	1,–1	–1,1
Tess	Rock	–1,1	0,0	1,–1
	Scissors	1,–1	–1,1	0,0

- Paper covers rock (paper wins over rock)
- Rock breaks scissors (rock wins over scissors)
- Scissors cut paper (scissors win over paper)

The payoff table is shown as Table 8.13.

Compute the mixed strategy equilibrium for Rock, Paper and Scissors.

Q8.3. More Mixed Strategies

Compute mixed strategy equilibria for the Heave-Ho, the Battle of the Sexes, Chicken, and Hawk vs. Dove (Chapter 5). Compare.

Q8.4. Escape-Evasion

Recall the escape-evasion game from Chapter 5, Section 9 with the payoff table shown in Table 8.14.

We observed in Chapter 5 that this game has no equilibrium in pure strategies, but we know it must have at least one equilibrium when mixed strategies are included. Compute the mixed strategy equilibrium for this game.

Q8.5. The Great Escape

Refer to Chapter 2, Problem 2.2.

Table 8.14. Payoffs in an Escape-Evasion Game (Repeats Table 5.9, Chapter 5).

		Pete	
		North	South
Fred	North	−1,3	4,−1
	South	3,−4	−2,2

Table 8.15. Payoffs in a War of Delay.

		F	
		North pass	South pass
H	North pass	5,−5	0,0
	South pass	0,0	5,−5

a. Does this game have Nash Equilibria in pure strategies? Why or Why not?
b. Compute the mixed strategy equilibrium for this game.

Q8.6. Punic War

In The Second Punic War, 218–202 BC, the Romans faced a powerful army, with elephants as shock troops, led by one of the great strategic generals: Hannibal. For most of the war the Romans followed the strategy of Fabius Cunctator, that is, "Fabius the delayer:" avoiding combat with Hannibal's army in Italy while undermining his base of support in other parts of the Mediterranean. Here is an escape-evasion game in that spirit. Generals F and H have to choose between two mountain passes. If they choose the same one, there will be a battle and F, whose army is much weaker, will lose. The payoff is shown in Table 8.15.

What Nash equilibria, if any, does the game have?

Q8.7. Football

In American football, the objective for the team on offense is to carry the ball across the goal line, advancing toward the goal line by running with the ball or throwing ("passing") it forward on successive plays. The objective of the defense is to prevent this by "tackling" (physically restraining) the ball carrier or preventing a pass from being caught.

American football, unlike baseball, is a game of coordinated attack by the entire team of 11 players. Crucial as the quarterback may be, he can do little without linemen to block, running backs and pass receivers. The defense is also coordinated. To keep this example simple, we will limit it to two offensive plays. The plays will be the drop back pass and the draw play. Both plays develop from the same initial motion, in which the offensive blocking linemen move backward from the "line of scrimmage" dividing the offense from the defense, and the quarterback moves about 10 yards behind the line of scrimmage.

Defense against the drop back pass often consists of sending the defensive linemen forward rapidly to break down the passing pocket and pulling the faster defensive backfield back toward their own goal to prevent a pass from being caught and to protect against a large gain if it is caught. But this defense is particularly vulnerable to the draw play, since it leaves the line of scrimmage undefended.

Thus, the offense will choose between two strategies: Pass or draw play. The defense will choose between defending against the pass and defending against the run. The payoff for the offense will be the expected value of yards gained and the payoff for the defense will be the negative of that number. The payoff table is shown as Table 8.16.

a. Compute the mixed strategy equilibrium for this game.
b. What is the expected value of the number of yards gained per play in equilibrium?

Table 8.16. Payoffs in American Football.

		Offense	Offense
		Pass	Draw play
Defense	Pass	−1,1	−3,3
	Run	−4,4	1,−1

Table 8.17. Nations in Conflict.

		France	France
		Aggressive	Conciliatory
England	Aggressive	−3, −3	5, −1
	Conciliatory	−1, 5	0, 0

Q8.8. Give War a Chance — or, anyway, a probability

In the 18th century (1700's) England and France fought a series of wars. It seemed that any crisis in their relations could lead to war, with a considerable probability. Let us say that in a crisis, England and France could adopt either of two strategies, aggressive or conciliatory, with a payoff table like Table 8.17.

Determine all Nash equilibria for this problem. Assume that there is a war whenever both countries adopt aggressive strategies, and not otherwise. In any crisis, what is the probability of a war?

PART III

Sequential and Repeated Play

CHAPTER 9

Sequential Games

Thus far, we have dealt mostly with games presented in "normal" or tabular form. While all games can be treated in "normal form," as von Neumann observed, the normal form is particularly useful for games in which all agents choose

> **To best understand this chapter,** you need to have studied and understood the material from Chapters 1–4. Review Chapter 2 particularly.

their strategies more or less simultaneously, such as the Prisoner's Dilemma. Many real human interactions are like that, as we have seen. But there are other important interactions in which the agents have to choose their strategies in some particular order, and in which commitments can only be made under limited circumstances or after some time has passed. These are *sequential games,* and we may say that such games have a *commitment structure.* The extended form of the game can help us to understand the commitment structure and its implications, and so it has become customary to study sequential games in terms of their extended form. Here is an example and a business case.

1. STRATEGIC INVESTMENT TO DETER ENTRY

In the business case in Chapter 2, we saw that the entry of new competition can reduce the profits of established firms. (A great deal of economic theory says the same). Accordingly, we would expect that companies might try to find some way to prevent or deter the entry of new competition into the market, even if it is costly to do so. Here is an example of that kind.

Spizella Corp. produces specialized computer processing chips for workstations. A plant to fabricate these chips (a "fab") costs $1 billion and will produce 3 million chips per year at a cost of $1 billion per year and so at an average cost of $333.33 per chip. Table 9.1 shows the relationship between the number of chips on the market and the prices buyers are willing to pay (or, as economists call it, the demand relationship.)

Spizella's management have learned that Passer, Ltd., are considering building a "fab" to enter this market in competition with Spizella. As things are, selling 3 million chips at $700, Spizella obtains an annual profit of 1.1 billion. But if a second fab comes on line, output will increase to 6 million chips, the price per chip will drop to $400, and revenue per fab will be $1.2 billion, for a profit of 200 million per year for each fab. Worse still, if two new plants come on line, output will be 9 million chips and the price will drop to $200, for a loss of 400 million per plant.

Nevertheless, Spizella is considering investing in a second fab. Their reasoning is as follows:

1. If Spizella builds before Passer makes their decision, Passer will realize that their plant would be the third one, and that if they build it everyone will lose 400 million per plant per year. So, Passer will not build, and Spizella will retain $400 million a year of profit, which is nothing to sneeze at.
2. If Spizella doesn't build the second plant, Passer will, and Spizella will be left with only $200 million of profits on their one present fab.

Table 9.1.   Demand for Chips.

Q	Price per chip
3,000,000	700
6,000,000	400
9,000,000	200

In building the new plant to keep Passer out, Spizella would be engaging in *strategic investment to deter entry*. By building first, Spizella commits itself to retaining dominance of the market and to driving any potential competitor out of the market — since the two plants, plus the third for a new competitor, could not all produce at their capacity without depressing the price below the costs of all competitors.

Let's look at this game in extensive form. The game is shown in Figure 9.1. Each circle represents a decision by one player or the other, a "node" of the game. Each arrow represents one way the player may decide. The node labeled 1 is Spizella's decision whether to build the new fab or not, and the nodes at 2A and 2B are Passer's decision whether to build their new fab, depending on whether Spizella has built its. The payoffs on the far right show the payoff to Spizella first, and that to Passer second.

In analyzing games in extended form, a useful concept is the "subgame." A

HEADS UP!

Here are some concepts we will develop as this chapter goes along:

Extensive form: A game is represented in extensive form when it is shown as a sequence of decisions. The game in extensive form is commonly shown as a tree diagram.

Subgame: A subgame of any game consists of all nodes and payoffs that follow a perfect information node. If the subgame is only part (not the whole) of the game it is called a "proper subgame."

Subgame perfect equilibrium: A Nash equilibrium in a game in extensive form is subgame perfect if it is an equilibrium for every subgame.

Basic subgame: A subgame in a game in extensive form is "basic" if it contains no other proper subgames. Otherwise, it is a complex subgame.

Backward induction is a method of finding subgame perfect equilibria by solving the basic subgames, substituting the payoffs back into the complex ones, solving those, and working back to the beginning of the game.

Behavior strategies: The strategy choices made at a particular stage in a game are sometimes called behavior strategies.

subgame includes all branches that originate from a single, well-defined choice point and also from the branches that originate from all choice points that follow from it. Thus, in Figure 9.1, each of the gray ovals defines a subgame — one, in the lighter gray, originating from choice

Definition: *Extensive Form* — A game is represented *in extensive form* when it is shown as a sequence of decisions. The game in extensive form is commonly shown as a tree diagram.

point 2A, and the other, in the darker gray, originating from choice point 2B. In addition, the whole game is a subgame! After all, it includes the branches from choice point 1 and, since 2A and 2B follow from 1, the subgame beginning from choice point 1 includes those branches too. In general, every game has at least one subgame,

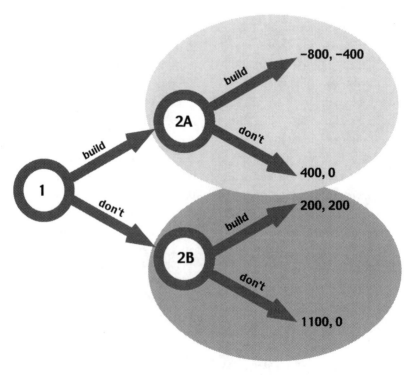

Figure 9.1.   Extensive Form of "Strategic Investment to Deter Entry."

namely itself. The other subgames, the ones that are not identical to the game as a whole, are called *proper subgames*. When I said before that sequential games have a commitment structure, I meant that typically sequential games have one or more proper subgames, and the set of proper subgames is what I mean by the *commitment structure*.

Recall, at the beginning of the example, Spizella "reasoned out" what Passer would do. Spizella's reasoning is based on subgames. In reasoning that "If we don't build the second plant, Passer will," Spizella is observing that the bottom branch with the payoff "1100,0" is not an equilibrium in the lower subgame. Similarly, "If we build before Passer makes their decision, Passer will realize that their plant would be the third one, and that if they build it everyone will lose 400 million per plant per year. So, Passer will not build," says that the top branch, in which both companies build, is not an equilibrium in the top subgame. Anticipating that each of the subgames will be in equilibrium, Spizella can anticipate the results and make their decision accordingly. Thus, Spizella has chosen their contingent strategy on the assumption that every subgame would be in equilibrium. This sort of equilibrium in the game as a whole is called a *subgame perfect equilibrium*. A subgame perfect equilibrium is an equilibrium in which the players always, consistently, anticipate their rivals' decisions in this way.

There may be Nash equilibria that do not have this property. (We will see an example in Section 3). Thus, we may say that subgame perfect equilibrium is a *refinement* of Nash equilibrium. That is, we are selecting among Nash equilibria by using a more demanding concept of rationality. Here is the definition: In order for a Nash equilibrium to be a subgame perfect equilibrium, every subgame must be in a Nash equilibrium.

To find a subgame perfect equilibrium, we may use *backward induction*. We start with the last decision in each sequence, determine the equilibrium for that decision, and then move back, determining the equilibrium at each step, until we arrive at the first decision. Thus, in the game of strategic investment, decision 2A or 2B is last, depending on the decision at 1. At 2A, Passer will decide not to build, since it is better to lose nothing than to lose 400 million dollars! At 2B, Passer will decide to build, since a profit of

200 million dollars is better than nothing. Now we move back to stage 1. Spizella knows that, if they choose "build," their payoff will be 400 million, while if they choose "don't," their payoff will be 200 million. They know this because they know Passer's decision-makers are rational and self-interested and Spizella can figure out what Passer's decisions will be in these cases just as we can. So Spizella chooses "build," and the sequence "build, don't build" is the subgame perfect equilibrium for this game.

Strictly speaking, Passer's subgame perfect strategy is the contingent strategy "*if Spizella builds,* don't build; *otherwise* build." Recall, von Neumann and Morgenstern defined a strategy for a sequential game as a contingent strategy in this sense. However, the strategic choices that the players make within the proper subgames, such as Passer's choice whether to build or not, are called *behavior strategies*, and in more recent game theory they are often just called strategies. In games without a sequence of decisions, there is little reason to make the distinction, since there is no information to base a contingent strategy on. This could lead to some confusion. In what follows, we will take care to make it clear whether "strategies" means contingent or behavior strategies.

In the game of Strategic Investment to Deter Entry, both competitors have reasons for regrets. Passer would rather have the profit of 200 million from splitting the market if they built and Spizella did not. Spizella would rather have the profit of 1.1 billion if nobody built. But these outcomes are not available. Each is doing the best it can, making its best response to the other's choice of behavior strategies at each stage. That's what we mean by "equilibrium" in general. But in this case, the equilibrium must also take into account the fact that Spizella has to commit itself first — the commitment structure. Thus, the subgame perfect equilibrium is the best response equilibrium in terms of behavior strategies.

2. CONCEPTS FOR SEQUENTIAL GAMES

This example illustrates the key concepts for sequential games. First, let's take a little closer look at the concept of a subgame. As a contrast to Figure 9.1, we can take another look at the Prisoner's Dilemma in extended form. For convenience, Figure 9.2 repeats Figure 2.3 from Chapter 2. We suppose that Al makes his decision to confess or not confess at 1, and Bob

Definition: *Subgame* — A *subgame* of any game consists of all nodes and payoffs that follow a perfect information node.

makes his decision at 2. Notice one difference from Figure 9.1. Bob's decision, in the two different cases, is enclosed in a single oval. That tells us that when Bob makes his decision, he doesn't know what decision Al has made. Bob has to make his decision without knowing Al's decision. Graphically, Bob doesn't know whether he is at the top of the oval or the bottom. This uncertainty means that the branches to the right of 2 in this figure are NOT subgames. In Figure 9.1, the fact that 2A and 2B are in separate ovals tells us, in game theory code, that

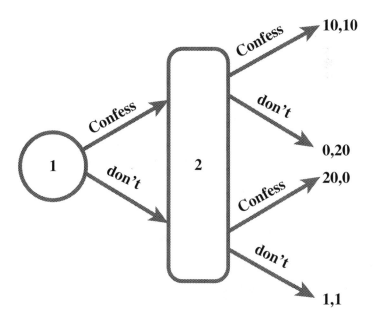

Figure 9.2. The Prisoner's Dilemma in Extensive Form (repeats Figure 2.3, Chapter 2).

Passer knows Spizella's decision before Passer makes its decision. For the purposes of the game, Passer has *perfect information*. (That is, Passer knows all previous decisions).

> **Definition:** *Proper Subgame* — A *proper subgame* is a subgame that includes only part of the complete game.

Recall, the whole game is at the same time one of its own subgames, and a subgame that is only a part of the whole is called a proper subgame. Contrast Figure 9.2 with Figure 9.1. A node like 2A or 2B is a "perfect information node." The decision branches enclosed in the gray ovals are proper subgames in the game of Strategic Investment to Deter Entry, as we have observed. They also illustrate another important property a subgame may have. *Neither of them contains any other proper subgames within them.* Instead, the branches at 2A and 2B lead directly to payoffs. Thus, they are *basic subgames.* By contrast, the subgame beginning at node 1 in Figure 9.1 (which is the entire game) is a *complex subgame.*

That is why the branches to the right of node 2 in Figure 9.2 are not subgames. The Prisoner's Dilemma has no proper subgames. However, node 1 is a perfect information node — there are no previous decisions Al needs to know anything about — so node 1 together with the nodes to its right constitute a subgame. The Prisoner's Dilemma has only one subgame, a trivial subgame consisting of the entire game, and has no "commitment structure."

> **Definition:** *Subgame Perfect Equilibrium* — A game is in *subgame perfect equilibrium* if and only if every subgame is in a Nash equilibrium.

For games like Figure 9.1 that do have proper subgames, however, our equilibrium concept is a refinement of the Nash equilibrium, the "subgame perfect equilibrium" as we have seen. For a subgame perfect equilibrium, we require that every subgame is at a Nash equilibrium. This is illustrated by the sequence "build, don't enter" in the game of Strategic Investment to Deter Entry.

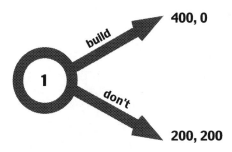

Figure 9.3. Reduced Game of Strategic Investment to Deter Entry.

But we can determine the Nash Equilibrium directly only for the basic subgame, that is, the last subgame in any sequence. In Figure 9.1 at 1, for example, we cannot determine the Nash Equilibrium because the payoffs depend on Passer's decision, and Spizella cannot make a "best response" to a strategy that will be chosen later. Accordingly, we start with the basic proper subgames in each sequence — 2A and 2B. At those nodes, there is no commitment that any later decision can exploit, since there are no later decisions. Accordingly, we can calculate the Nash-equilibrium and payoffs for those two nodes. That tells us the payoffs for node 1. By solving the games at 2A and 2B we have reduced the entry-deterrence game in Figure 9.1 to the smaller game shown in Figure 9.3.

> **Definition:** *Basic and Complex Subgames* — A subgame is *basic* if it contains no other proper subgames within it. If there are other proper subgames within it, it is a *complex* subgame.

This game can immediately be solved. The solution is "build for 400 rather than 200." This illustrates the general method for solving games with commitment structures: Solve the subgames at the last stage in each branch, and in this way "reduce" the game to a smaller one, and then repeat until all subgames have been solved. This gives the subgame perfect equilibrium. In the words of a management consultant's[1] slogan: "Think forward, reason backward." That is, think forward to determine what the proper subgames are, and reason backward to solve them for the best-response strategies and equilibrium.

Backward induction and subgame perfect equilibrium are important principles of rational strategy planning in sequential games. Strategists should indeed "think forward and reason backward." This is no less true in other fields than in business, and applies to military strategy, public policy, and personal life, as we will see in examples to follow in this chapter.

Method: *Backward Induction* —. To find the subgame perfect equilibrium of a sequential game, first determine the Nash Equilibria of all basic proper subgames. Next, reduce the game by substituting the equilibrium payoffs for the basic subgames. Repeat this procedure until there are no proper subgames, and solve the resulting game for its Nash Equilibrium. The sequence of Nash Equilibria for the proper subgames of the original game constitutes the subgame perfect equilibrium of the whole game.

3. NASH AND SUBGAME PERFECT EQUILIBRIUM

We have seen that the subgame perfect equilibrium, like the Nash equilibrium, is a best-response equilibrium. In finding the subgame perfect equilibrium, we first determine the Nash equilibrium in the basic subgames. Clearly the two equilibrium concepts are closely related. To better understand the relationship between the two, let's take another look at another example from Chapter 2, another example of competition from new entries in business — the entry game between Goldfinch and Bluebird. This game in extended form is repeated in Figure 9.4. Table 9.2 shows the game in normal

---

[1]Courtney, Hugh, Games managers should play, *World Economic Affairs*, **2**(1) (Autumn, 1997), p. 48.

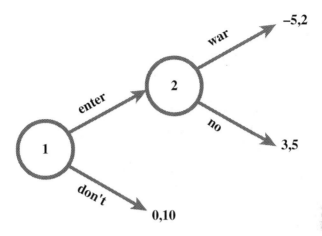

Figure 9.4. The Game of Market Entry in Extended Form (approximately repeats Figure 2.1, Chapter 2).

Table 9.2. The Market Entry Game in Normal Form (Approximately repeats Table 2.1, Chapter 2).

		Goldfinch	
		If *Bluebird enters* then accommodate; *if Bluebird does not enter,* then do business as usual.	If *Bluebird enters* then initiate price war *if Bluebird does not enter,* then do business as usual.
Bluebird	Enter	3,5	–5,2
	Don't	0,10	0,10

form, repeating Table 2.1 in Chapter 2 except that the Nash equilibria of the game are shaded.

The two Nash equilibria of this market entry game are not equal, though. Notice what happens when we apply backward induction to this game. At node 2, we notice that Goldfinch is better off to choose "no war," rather than war, for 5 rather than 2. This puts us in the left column of the normal form table for the game, and shows that the darker shaded Nash equilibrium is the only subgame perfect equilibrium for this game. The lighter shaded Nash equilibrium at the lower right is not subgame perfect. Why is that? It *is* a Nash equilibrium, since no unilateral deviation from it can make either player worse off. But this is less meaningful, since Bluebird chooses first and can commit themselves to the "enter" behavior strategy in such a way that Goldfinch will also shift to the "no war" behavior strategy. To put it in another way, Goldfinch's threat of a price war is not credible, since it is not an equilibrium in the basic subgame.

In this case, there are two Nash equilibria, but only one is subgame perfect. The other corresponds to an "incredible threat." This illustrates the relationship of subgame perfect equilibrium to Nash equilibrium in general. Every subgame perfect equilibrium is a Nash equilibrium, but not every Nash equilibrium is subgame perfect. Game theorists express this by saying that subgame perfect equilibrium is a refinement of Nash equilibrium. A refinement, in general, is a class of Nash equilibria that also fit some more demanding standard for a solution, such as the criterion of subgame perfection.

4. MAKE OR BUY

Here is another business example. Alfa Corp. makes computer chips. Beta, Ltd retails computers and can either make their own chips or buy them from Alfa. Alfa has an option either to mount a costly campaign to convince the public that computers are better with "Alfa inside" or not; and, if Beta elects to buy from them, they can ask a high or a low price. Their game in extensive form is shown in Figure 9.5. The first payoff is to Alfa. How will these companies make their decisions, if they are rational and self-interested?

We will apply backward induction to determine the subgame perfect Nash equilibrium. First, we note that the game has two basic subgames, as shown by the grey ovals in Figure 9.6.

Solving them, we see that Alfa will find the higher price more profitable whether they advertise or not. Thus, we have the reduced game shown in Figure 9.7.

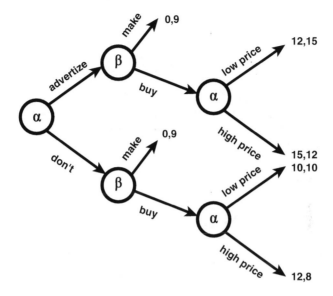

Figure 9.5. The Make or Buy Game.

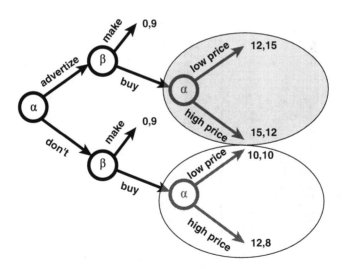

Figure 9.6. Basic Subgames.

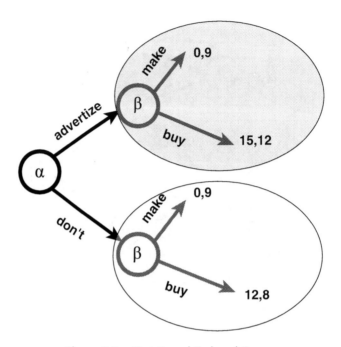

Figure 9.7.   First Round Reduced Game.

Once again, the reduced game has the two basic subgames shown by the gray ovals. Beta's choice in the basic subgames in this reduced game will be buy in the upper branch (since 12 is more than 9) and make in the lower branch (since 9 is more than 8.) The idea here is that "buy" is less profitable in the lower subgame because Alfa's product is not advertised, but nevertheless Alfa will charge the higher price. Thus, solving both basic subgames of the reduced game we have the second-stage reduced game shown as Figure 9.8.

In this case, clearly, Alfa will choose to advertise for 15 rather than not for 0. The subgame perfect equilibrium for this game is the sequence

- Alfa: Advertise
- Beta: If Alfa advertises then buy, otherwise make.

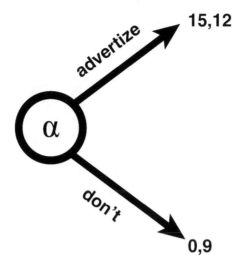

Figure 9.8. Second Round Reduced Game.

Alfa: If Alfa advertises and if Beta buys, then charge high price; if Alfa doesn't advertise and if Beta buys, then charge high price.

Notice that even though the lower basic subgame in Figure 9.6 is never reached, when Alfa plays rationally, nevertheless the decision to charge the high price in that subgame is part of the subgame perfect equilibrium, since the definition of subgame perfect equilibrium requires that *every* subgame be in equilibrium.

5. THE CENTIPEDE GAME

Our next example will be a little more abstract. It will be a four-step game, shown in Figure 9.9. There are two players, A and B, and there is a pot of money. The players alternate with A taking the first turn. At each turn, the player has a choice between two strategies. The first is "grab," represented here by a down arrow. When a player chooses "grab," that player gets the bigger share of the pot of money and the game stops. The second strategy is "pass," represented by the horizontal arrows in the figure. When a player chooses "pass" on

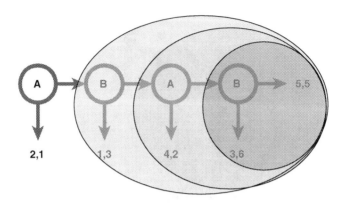

Figure 9.9.   The Centipede Game.

the first three rounds, the pot of money grows larger but passes into the possession of the other player and the game goes on for one more stage. At the fourth stage, the strategy "pass" results in even division of the pot of money, but also its largest amount. The first payoff is to A. As before, we apply backward induction to find the subgame perfect equilibrium.

This four-step Centipede Game has three proper subgames. They are shown by the gray ovals in the figure. However, only one — indicated by the darkest gray oval at the right — is basic. In that subgame, B will choose "grab," because 6 is better than 5. Thus, at the first stage of backward induction, we solve that subgame, resulting in the first-stage reduced game shown as Figure 9.10.

As before, the reduced game has only one basic subgame, the one to the right. In that subgame, A will choose "grab," because 4 is better than 3. Thus, we solve it, obtaining a second round reduced game shown as Figure 9.11.

Yet again, this game has only one basic subgame, B's decision point. In that subgame, B will choose "grab," because 3 is better than 2. Thus, we have the final reduction of the centipede game, shown as Figure 9.12.

Clearly, in this first stage of the game, A will choose "grab," since 2 is better than 1. Thus, the subgame perfect equilibrium of the Centipede Game consists of the following two contingent strategies:

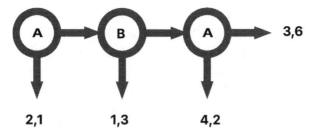

Figure 9.10. First Reduction of the Centipede Game.

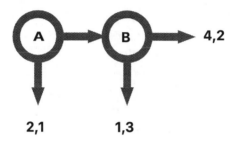

Figure 9.11. Second Reduction of the Centipede Game.

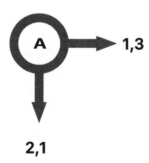

Figure 9.12. Final Reduction of the Centipede Game.

- For A: Grab, and if A does not grab and B does not grab, then grab at the third stage.
- For B: If A does not grab at the first stage, then grab at the second stage, and if B does not grab at the second stage and A does not grab at the third stage, then grab at the fourth stage.

The Centipede Game could have any number of stages, from two without any meaningful upper limit. If you visualize a game with 100 stages, the origin of the name "Centipede Game" may seem clear. In any case, the subgame perfect equilibrium will always be that the pot will be grabbed at the first step — "take the money and run" (to borrow the title of a Woody Allen movie). But this will often be very inferior, for both parties, to the result if they play cooperatively and always pass.

The Centipede Game is a parallel to the Prisoner's Dilemma. Just as the Prisoner's Dilemma has become a standard illustration for the possibility that non-cooperative games in normal form may have inefficient outcomes, outcomes that neither player wants, so likewise the Centipede Game illustrates how that can happen with the subgame perfect equilibrium of a game in extensive form. We next apply the concept of subgame perfect equilibrium in a multi-stage game to an important question of public policy: What is the role of "non-profit" enterprises?

6. THE FUNCTION OF NON-PROFIT ENTERPRISE

Non-profit enterprise is the fastest-growing of the major sectors of American enterprise, and its significance for our economy is increasing. In some ways, though, the term "non-profit" is unfortunate. Non-profit organizations are "non-profit" not because they never make a profit, but because they have other and broader positive objectives. These objectives constitute the mission of the enterprise. For example, a non-profit charity hospital may have its mission defined as provision of medical care to those who cannot pay, or a non-profit museum may have its mission defined as the preservation of certain items of cultural heritage. Thus, we will instead call them "Mission-Driven Enterprises" (MDEs). MDEs are exempt from profits tax in the USA and may be favored in other ways by public policy. But why? What is the advantage of MDE?[2]

---

[2]The model in this section is influenced by the ideas of Henry Hansmann, The role of nonprofit enterprise, *Yale Law Journal*, **89**(5) (1980), pp. 835–901 and Nonprofit

Many MDE's are begun by philanthropic gifts, and the purpose of the gift determines the mission of the enterprise. Once the corporation is formed, the mission becomes the final legal objective of the organization. Such a document-defined mission admits of a much broader range of potential objectives than are possible for a profit-oriented corporation. There are many objectives that seem worthwhile but that cannot be accomplished by market-driven for-profit enterprises. For these purposes, the alternatives are that they might be provided by government or by MDEs. This presumably is a reason for public policy to favor MDE's.

But, when we focus on the mission-driven character of the corporations rather than their non-profit status, a question arises. Non-profit status does not mean only that the firm pays no profit tax: It means that the firm cannot distribute profits if it earns any, and can accumulate profits in reserve and endowment funds only to a legally limited degree. It is not clear that these limitations always advance the missions of the MDEs. For many sorts of objectives, it might be that a very effective way to advance the objective could be to accumulate a profit for a very long time and then undertake a large-scale operation that cannot be undertaken otherwise. But non-profit status will generally rule out these strategies.

Why, then, are MDE's so typically non-profit? Since MDE's are generally targeted on missions that will not return a competitive rate of return, they cannot raise capital from profit-seeking investors, and rely on donors and donation-based endowments for their capital. Conversely, donors will wish to entrust their assets for non-profit objectives to institutional management in order to have them managed by a professional manager and to keep it going after the donor's death. Thus, we envision the formation of an MDE as a game with two (kinds of) participants: A potential donor and a potential institutional manager. The donor's strategies are to donate or not. The manager's strategies are to use the resources to support

---

enterprise in the performing arts, *Bell Journal of Economics*, **12**(1981), pp. 341–361, although those papers do not use game theory.

the donor's objectives or to convert those resources to other purposes, including private profit.

This donation game resembles the Centipede Game from the previous section. The donor's decision comes first, as logically it must. The donor has 10 to donate, and thus if they decide not to donate, they retain that 10 as the donor's payoff. In that case the institutional manager gets nothing. If the donation is made, and is used to support the donor's objectives, the donor gets a subjective benefit equivalent to double the amount (that is her motivation for donating) while the manager gets a payoff of 5, which may comprise subjective satisfactions as well as salary. However, the manager also has the "grab" option — and since a profit-seeking corporation is designed to convert its capital to profit, there are no obstacles to doing this, and the manager can capture the entire donation of 10, leaving the donor with nothing. This is shown as the two-stage centipede game in Figure 9.13. The payoff to the donor is on the left, and the payoff to the manager is on the right.

Now suppose instead that the enterprise is non-profit. This means that obstacles are put in the way of conversion of its resources to private benefit. The obstacles include the prohibition on distribution of profits as well as criminal penalties for conversion, failure of trusteeship, and so on. These obstacles may not be fully effective but mean that the process of conversion will be inefficient, and much of the resources will be wasted if the manager chooses the option to

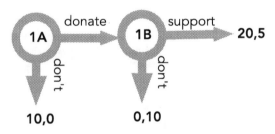

Figure 9.13. A Donation Centipede 1.

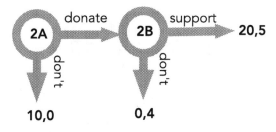

Figure 9.14. A Donation Centipede 2.

"grab" and convert the resources of the MDE to private benefit. The centipede game for a non-profit enterprise is shown in Figure 9.14. The only change is in the case of the "grab" strategy at the second stage. Because the conversion of resources to private benefit is made very inefficient, the manager is able to realize only 40% of the resources, for a payoff of 4, while the donor remains with nothing.

But both for-profit and non-profit organization are possible, so all these decisions are imbedded in a larger game, a three-stage game. At the first stage, the manager chooses whether to organize a profit-seeking enterprise or an MDE. These are the organizer's strategies at the first stage, and they are followed by the decisions to donate or not and to support or grab. The complete game in extensive form looks like Figure 9.15. The node numbered zero is the choice of a corporate form — profit or non-profit. Thus, the game has three stages: Determination of the corporate form, followed by the decision whether or not to contribute, followed by the decision to allocate the contribution to the support of the mission or to distribute it as profits. This game has four proper subgames, highlighted by the light gray ovals. Notice that the subgame that begins at 2A is itself a subgame of the subgame that begins at 1A, and likewise the subgames that begin at 2B and 1B. But only the subgames that begin at 2A and 2B are basic.

We first consider the basic subgames. At decision node 2A, the manager chooses "grab," for 10, rather than "support," for 5,

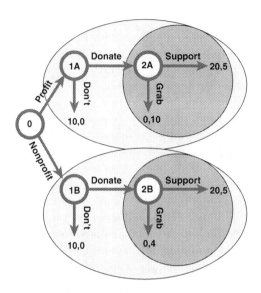

Figure 9.15. The Game of Choosing a Corporate Form.

leaving the donor with nothing. Thus, the payoffs for that subgame are 0,10. At decision node 2B, the manager chooses "support" for 5 rather than "grab" for 4. Thus, the payoffs for this subgame are 20,5.

We now have a reduced game, shown as Figure 9.16. In the reduced game, in the upper branch, the donor will choose between "donate," for a payoff of zero, and "don't," for a payoff of 10. This is, of course, another no-brainer, and she will choose "don't," so the payoffs on the upper branch will be 10,0. On the lower branch of the reduced game, the donor's choice is between "donate" for 20 and "don't" for 10. Again, it is a no-brainer, and the donor chooses "donate," so the payoffs on the lower branch are 20,5.

Using the solutions to those subgames we can "reduce" the game in Figure 9.16 to the one in Figure 9.17. Once again, the "reduced" game is a "no-brainer." Both the donor and the institutional manager are better off if the non-profit form is chosen.

Thus, the method of backward induction (together with the assumption that non-profit status is successful in making the

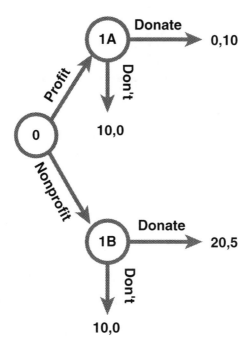

Figure 9.16. The Reduced Game After Solving Basic Subgames.

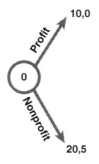

Figure 9.17. The Reduced Game of Choosing a Corporate Form.

conversion of non-profit assets to manager profit very inefficient if not actually impossible) leads us to the conclusion that donation will be a viable basis for a non-profit corporation but not for a for-profit corporation. This could explain the tendency for MDEs to be organized as non-profit corporations. Where-ever capital is supplied by philanthropists, whose benefit is subjective satisfaction with the work done, rather than an efficient distribution of profits, the inefficiency of the non-profit form in converting resources to private benefit is a key advantage for MDEs. This example illustrates, once again, that it may be advantageous to make a commitment in an early stage that restricts one's freedom of choice later on, in a sequential game. It also has lessons for public policy, suggesting that non-profit enterprises may play an important role alongside other kinds of enterprises in an economic system.

Following Hansmann, this model stresses the role of contributions. A contributor has to rely on trust — there is no reason to contribute unless the contributor can trust that the contribution will be used to advance the contributor's objectives. But trust can play an important role in many other decisions. A patient seeking treatment must trust that the treatment they receive will be appropriate, and trust is more difficult and important if the illness and treatment are non-routine. Perhaps, then, it is no coincidence that many hospitals are non-profit and that non-profits are especially likely to offer unusual or experimental treatments. Similarly, a student choosing a college has to trust that the instruction offered will advance the student's capabilities rather than shareholder value, if the two objectives disagree. Here again, non-profits are common, and it may be because they encourage trust. Conversely it seems likely that industries or activities in which trust is important provide a relatively favorable niche for the survival and growth of non-profit organizations.

7. SUMMARY

In this chapter on sequential games, we have studied games in which one or more players make a commitment to which the other player can then respond. In analyzing these games, we make use of the concept of subgame perfect equilibrium. This is expressed intuitively in the slogan "think forward and reason backward." First, the game is analyzed into its subgames, "thinking forward." The condition for a subgame perfect equilibrium is that each subgame of the game is itself in a Nash equilibrium. A subgame is proper if it does not include the entire game, and is basic if it has no proper subgames within it. The basic proper subgames are those at the end of the game tree. To find the subgame perfect equilibrium, we use backward induction, "reasoning backward." We first solve all of the basic proper subgames, and reduce the game by replacing the basic proper subgames with their equilibrium payoffs. We then analyze the reduced game in the same way, continuing step by step until we arrive at a game that has no proper subgames. The solution to that game, followed successively by the solutions of the subgames, gives us the subgame perfect equilibrium.

We have seen that the existence of a sequence of play makes a difference in the outcome of the game. It may be that the one who takes the first move has the advantage, as in the entry game between Spizella and Passer; or as in the Spanish Rebellion, the first mover may be at a disadvantage. As in many non-cooperative games, both players may have reasons to regret the outcome.

Games of threat and retaliation (including price wars) are an important category of sequential games. In such games, threats are credible only if they are subgame perfect. This principle has wide application, extending also to the deployment of military forces, and many other threat and retaliation games. The Centipede games are another important category of sequential games. These games have application in economics and public policy, but are also important for a large volume of recent experimental work on them.

Q9. EXERCISES AND DISCUSSION QUESTIONS

Q9.1. Road Rage

Consider the following simple game, which we may call the "road rage" game. There are two players, Al and Bob. Bob has two choices: to aggress (perhaps by cutting Al off in traffic) or not to aggress. If Bob chooses "do not aggress," then there is no choice of strategies for Al, but if Bob aggresses, Al can choose between

Table 9.2. The Road Rage Game.

		Bob	
		Aggress	Don't
Al	If Bob aggresses then retaliate; if not, do nothing	(−50,−100)	(5,4)
	If Bob aggresses then don't retaliate; if not, do nothing	(4,5)	(5,4)

strategies "retaliate" (perhaps by dangerous driving or by taking a shot at Bob's car with a firearm) or not retaliate. An example in tabular "normal form" is shown in Table 9.2.

As usual, the payoff to the left is the payoff to Al, while the payoff to the right is the payoff to Bob. Draw the tree diagram for the game.

a. What are the subgames of this game?
b. Which subgames are basic?
c. Determine the subgame perfect equilibrium of this game.
d. Does it seem that the subgame perfect equilibrium is what occurs in the real world? Explain your answer.
e. Although many governments have tried to discourage "Road Rage" by penalizing retaliation, the Washington State Police adopted a policy to discourage road rage by increasing the penalty for aggressive driving. Does this make sense in terms of game theory?

Q9.2. Omnicorp

Omnicorp is the established monopoly seller of Omniscanners, which are widely used in business. Newcorp, however, has obtained a monopoly on a process to produce Omniscanners more cheaply. Newcorp has not yet begun selling Omniscanners, and Omnicorp has let it be known that, if the new company does enter the market, Omnicorp is prepared to cut price below their own cost in order to bankrupt the new competitor. If Newcorp enters the market, both companies have the option of pricing at p_1 or p_2, where c_1 is the cost per Omniscanner using the old technology (the technology Omnicorp would have to use) and c_2 is the cost using the new technology, and $p_1 > c_1 > p_2 > c_2$. If the two firms compete and charge the same price, assume that they split the market, each selling $Q/2$ units, while if one charges a lower price, that one firm will sell the entire Q units and the other will sell nothing.

Omnicorp has threatened to retaliate with a price war if Newcorp enters the Omniscanner market. Is this threat credible? Why or why not?

Q9.3. Divorce

Mrs. Jones is seeking a divorce from Mr. Jones. Under the terms of her prenuptial agreement, her settlement will be $100,000 if she can prove that Mr. Jones has had an affair, but $50,000 otherwise. Her lawyer, acting as her agent, can prove the affair only if he hires a private detective for $10,000, which will come out of the lawyer's fee. Mrs. Jones has the option of paying her lawyer a flat fee of $20,000 regardless of the outcome of the case or $1/3$ of the settlement. The lawyer will hire the private detective only if it is profitable for him to do so.

a. Whose is the first move?
b. Draw a tree diagram for this "game in extended form."
c. Express the strategies in normal form.
d. Which payment method will enable Mrs. Jones to win her case? (Use backward induction.)
e. Can you think of any other way (besides the flat fee and the $1/3$ share) that Mrs. Jones might compensate the lawyer so as to get the best outcome for herself?

Q9.4. Gambling the Night Away

This is a true story, according to my informant. The names are changed to protect the innocent, if any. The game is based on the gambling and entertainment industries in Atlantic City, New Jersey.

A Nevada gambling millionaire, whom we shall call NM, wants to buy one of the largest and most prestigious properties in Atlantic City, a property now owned by the Biggernyou Corporation, BC. BC owns two major properties, one at the middle of the Boardwalk and one on the beach at the south end. The property on the south end is almost a square block, except for the small Dunecreep Casino, which faces the beach and occupies the middle of the block. The property at the middle of the boardwalk is the swank and famous Biltwell Hotel.

NM wants the Biltwell, but BC doesn't want to sell. The Dunecreep, owned by a third party, is available but too small for NM's plans. BC has two strategies — offer to sell the Biltwell or don't offer — and so far, their behavior strategy has been "don't offer." NM has three strategies. He can wait until a better offer comes along (wait), or he can buy the Dunecreep and operate it as a luxury casino (luxury) or he can buy the Dunecreep and operate it as a cheap slot machine hall (slots).

Market research says that the Dunecreep will make more money as a luxury casino, but NM intends to run it as a slots hall. Here is his reasoning: "The slot machines draw a lower class of customers, and the wealthy customers of the Biggernyou Corporation will not want to share the beach with them, so the Biggernyou Corporation will lose customers and money as a result of having a slots hall in the middle of their block. To keep their clientele, they will need to get the Dunecreep away from me, and to do that they will have to give me a chance to buy the Biltwell, which is what I really want."

Figure 9.18 shows this game in extended form. NM has the first decision node and BC the second. The first payoff is to NM and the second to BC. Will NM's strategic plan work? Explain in terms of subgame perfect equilibrium.

Q9.5. War of Attrition

In a war of attrition, there are two or more stages (and there may be an unlimited number of stages) but the game ends when one player drops out. The first player to surrender or withdraw from the game

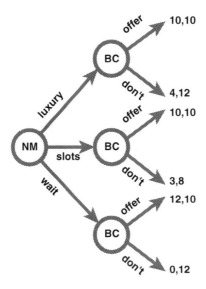

Figure 9.18. Payoffs in a Game of Casinos.

loses, and the other player wins; but the longer the game continues the less is won and the more is lost. The idea of it is that the continuing conflict uses up or destroys resources that the victor might otherwise use to increase her rewards and the loser might otherwise use to repair her position. We will consider a simplified two-stage war of attrition.

At each stage, each player has the choice of fighting or withdrawing from the game. The decisions at each stage are made simultaneously. At the first stage, if both withdraw, they split 150 evenly as 75 each, while if one fights and the other withdraws, the fighter gets 100 and the player who withdraws gets 50. If both fight, then the game goes to a second stage, in which the total payoff is at most 90 because of the resources wasted in the conflict at the first stage. If both fight at the second stage, the further conflict reduces the payoff to 10 each. If one fights and the other withdraws, the fighter gets 55 and the withdrawn player 15.

a. Draw the tree diagram for the game in extended form.
b. Write a list of all strategies in the game for each player, allowing for different responses to the other player's first round play.

c. Using information from part b, write the table for the game in normal form.
d. Has this game any Nash Equilibria? What?
e. Use backward induction to solve the game.

Q9.6. Strike!

A union and the employer expect that there may be a strike. Before the decision to strike, the union can either build up its strike fund, or not, and the employer can either build up their inventories to continue service to customers during a strike, or not. Those decisions are made simultaneously. Thereafter the union decides whether or not to strike.

a. Using the payoff table, Table 9.3 and ignoring the sequence of commitments, would there be a Nash equilibrium in pure strategies? If so, what?
b. Draw the tree diagram for the game.
c. What are the subgames of this game?
d. Which subgames are basic?
e. Reduce the game by solving the basic subgames, write the payoff table for the reduced game, and use it to determine the subgame perfect equilibrium of this game.
f. Compare the answers to Questions a and e. Explain the differences or similarities.

Q9.7. A Drug on the Market

Wellspring Pharmaceuticals Company owns the patent on Grumbledore, a drug for Restless Eye Syndrome. The patent will run out within 2 years. Klever Research Company proposes to do research on an advanced version of the drug, under a contract that would assign the patent to Wellspring if the research is successful. (In effect this would extend the lifetime of the patent considerably). However, this would require Wellspring to release information that would enable Klever to compete with Wellspring in the market for generics if the patent expires. For that reason, it could be profitable

Table 9.3. Strike.

		Union			
		Build		Don't	
		Strike	Don't	Strike	Don't
Employer	Build	–5,–5	–2,–2	–2,–10	–2,0
	Don't	–10,10	0,–2	–5,5	0,0

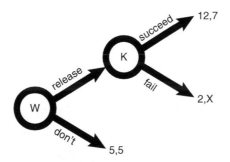

Figure 9.19. Game B.

for Klever to fail, concentrating their resources instead on making themselves ready to compete in the generic market. This is shown in extensive form as Game B, Figure 9.19.

a. Suppose the value of X is 4. Then what is the subgame perfect equilibrium of Game B?
b. Suppose the value of X is 8. Then what is the subgame perfect equilibrium of Game B?

The following exercise uses concepts from this chapter and also from Chapters 7 and 8. If you have not yet studied those chapters, save this one until you do.

Q9.8. Free Samples

Acme Cartoon Equipment (ACE) Corp. relies on other companies to supply the semifinished parts it assembles into cartoon equipment, under contract. There are, however, two types of suppliers: Type R (reliable) and Type U (unreliable). When ACE interviews a potential supplier, ACE has two strategies: Buy or not buy. Some suppliers give free samples and others do not. With a type R supplier, the game in extensive form is as shown in Figure 9.20, in which the first payoff is to ACE and the second is to the supplier. Node S is the decision node for the supplier, to give free samples or not, while node B is the node for the buyer, to buy or not. When the supplier is Type U, however, the payoffs are as shown in Figure 9.21.

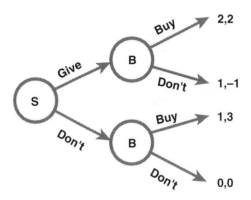

Figure 9.20. Acquisition Game with Type R Supplier.

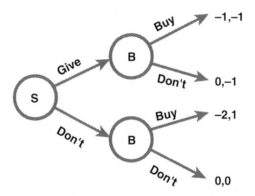

Figure 9.21. Acquisition Game with Type U Supplier.

ACE does not know which suppliers are R and U. Suppose that the probability that the supplier is type R is 0.5. (This is a common prior probability as explained in Chapter 7, Section 6).

a. What are the subgame perfect equilibria in the two games taken separately?
b. Given the probabilities of 0.5 that the supplier is R or U, what is the optimum strategy for the buyer and what is the payoff from it?
c. Suppose that ACE were to adopt the rule of buying only from suppliers who give free samples, and suppliers know this. What would the results be?

CHAPTER 10

Repeated Play

All of the games we have investigated so far in this book are played just once, as if the players would never interact again in the future. In some applications that seems clearly appropriate. In The Spanish

To best understand this chapter, you need to have studied and understood the material from Chapters 1–4 and 9.

Rebellion, with which the book began, there can be no second or subsequent play of the game. When two cars meet at an intersection, matched at random to play the Drive On Game, the chances that they will meet again are slight enough to overlook. But if the two drivers live on the same block, they are not matched at random, and are likely to meet one another again and again, so that they can become accustomed to one another's habits. When we look at a social dilemma like the Advertising Game, the supposition that the game is played just once with no future interaction seems quite wrong. The same firms will continue to compete in the same market year after year.

Game theorists suspected very early in the development of the field that repeated play could make a difference, especially in social dilemmas. The suspicion was so strong that it seemed as if someone must have already proven the case. Today, game theorists speak of the "folk theorem" — and we are using the word folk as it is used in the phrase "folk tale." The "folk theorem" was supposed to say that when social dilemmas were played repeatedly, cooperative outcomes would be rather common. In fact, there was no theorem and no proof: Only a folk tale. And repeated play in social dilemmas has proved to be more complex than the "folk theorem" suggests. Nevertheless, there is an element of truth in the "folk theorem." This chapter will explore the complexities of repeated play in simple two-person games. The next chapter will try to discover the nugget of truth. Here is an example of repeated play in a social dilemma of public goods provision.

1. THE CAMPERS' DILEMMA

Amanda and Buffy are camp counselors for the summer and they are sharing a room with a TV and DVD player. DVD's can be rented from the camp store for the week-end for $5. Amanda and Buffy would each get $4.00 worth of enjoyment from a week-end movie DVD, so if each of them rents a DVD on a particular week-end they can each get $8.00 worth of enjoyment at a cost of a $5 rental. Their strategies are rent or don't rent, and the payoffs for the DVD rental game are shown in Table 10.1.

We see that this is yet another social dilemma, very much like the Public Goods Contribution Game. In fact, the DVD movies are indeed a public good to the two campers. The game in extensive form is shown in Figure 10.1.

In a social dilemma there are basically two kinds of strategies: Cooperate (in this case, rent) and defect (in this case don't rent). The idea behind the term "defect" is that the player who chooses a non-cooperative strategy is "defecting" from a (potential) agreement to cooperate.

Table 10.1. The Campers' Dilemma.

		Buffy	
		Rent	Don't
Amanda	Rent	3,3	−1,4
	Don't	4,−1	0,0

A social dilemma is a problem! But perhaps it is not so much of a problem as all that. After all, the summer is just beginning, and there are 10 weekends ahead of them before they return to their colleges in different states for the fall term. It seems likely that they will choose "cooperate," at least for the first few weeks. After all, if (for example) Amanda chooses to "rent" this week, Buffy can reward her by continuing to choose "rent" next week, whereas if Amanda chooses the non-cooperative "don't rent" strategy, Buffy can penalize or "sanction" her by turning non-cooperative with a "don't rent" strategy next week and perhaps for several future weeks.

But are those rational "best response" strategies? To answer that question, we have to use the theory of games in extensive form, and subgame perfect equilibrium. Surprisingly, they are not best response strategies. To get the flavor of the

> ## HEADS UP!
>
> Here are some concepts we will develop as this chapter goes along:
>
> **Repeated games:** When a "game" is played repeatedly, we must analyze the sequence as a whole, and focus on the subgame perfect equilibrium of the sequence.
>
> **Folk theorem:** The widely held intuition that non-cooperative games played repeatedly may often have cooperative equilibria is called the "folk theorem" of game theory.
>
> **Games played a limited number of times:** If a game with a Nash equilibrium in pure strategies is played repeatedly, the repeated play of the Nash equilibrium is always subgame perfect. If the game is a social dilemma and is repeated a definite number of times, repeated play of the dominant strategy equilibrium is the only subgame perfect equilibrium — contrary to the "folk theorem."

reasoning, let's suppose that the DVD game is played for just two stages. The two-stage game is shown in extensive form in Figure 10.2.

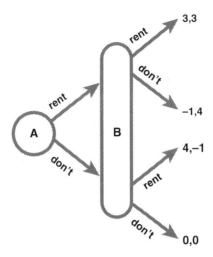

Figure 10.1.   The Campers' Dilemma.

The two-stage Camper's Dilemma has four basic proper subgames, shown by the gray ovals. Applying backward induction, we first solve these subgames. If we express them in normal form, we find that each of them is a social dilemma with a dominant strategy equilibrium at (don't, don't). When we substitute the equilibrium payoffs for the four subgames, we have the game shown in Figure 10.1 and Table 10.1. Since the second round equilibrium payoffs are (0, 0) it simply reproduces the original social dilemma! The conclusion is that the repeated play in this game makes no difference: The subgame perfect equilibrium is to choose the non-cooperative "don't rent" strategy at both stages.

We can extend this reasoning to the whole 10 weeks of camp. No matter how many weeks camp continues, the method to use is backward induction. As we have seen,

Rule: *Repeated Social Dilemmas* — When a social dilemma is repeated for a definite number of times, the subgame perfect play is always for both players to defect, just as in the original social dilemma.

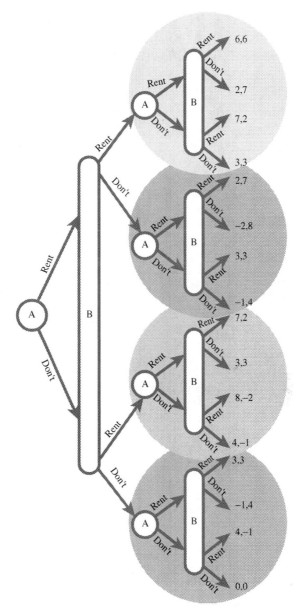

Figure 10.2. The Repeated Camper's Dilemma.

whatever may happen in the earlier weeks, the best response strategy in the last week is the non-cooperative "don't rent" strategy. There cannot be any rewards or sanctions on the next round because there will not be a next round. Now we proceed to the 9th week.

We already know that there will be no rewards or sanctions in the 10th week, since only non-cooperative strategies will be played in the tenth. That being so, there is no reason to play anything other than the non-cooperative strategy in the ninth. Now we proceed to the 8th week. Since we already

> **Definition:** *One-off Game* — When a game is played just once — and not repeated — we describe it as a **one-off game**.

know that there will be no rewards nor sanctions in the last 2 weeks there is no reason to play anything other than a non-cooperative strategy. And so on .... Reasoning in this way we induce the result all the way to the first round of play, when there is no reason to choose the non-cooperative strategy. The conclusion is that no cooperative strategy will ever be played.

It seems that the "folk theorem," however true it may be in some other cases, is not applicable to this particular game. Here is another example.

2. AN EFFORT DILEMMA

For one further example, let us consider an effort dilemma. Andy and Bill are contract workers on a project that will continue for the next 20 weeks. Each week their payoffs depend on how effectively they work together. As usual, each week, Andy and Bill can each choose between strategies "work" (with unpleasant effort) or "shirk" (with less unpleasant effort but also less productivity.) Their payoffs each week are as shown in Table 10.2.

Once again, we have a social dilemma. For each player, shirk is the dominant strategy but work, work is the cooperative solution. But perhaps these two workers could bring about a cooperative

Table 10.2. An Effort Dilemma.

		Bill	
		Work	Shirk
Andy	Work	10,10	2,14
	Shirk	14,2	5,5

outcome by threatening retaliation. If Bill were to shirk on one round, Andy might respond by shirking on the next round.

This policy is called a "Tit-for-Tat" strategy rule. In general, when one player defects from the cooperative solution on one round, and the other player responds by defecting from the cooperative solution on the next round, the second player is said to be playing according to a Tit-for-Tat strategy rule. The Tit-for-Tat rule says: Always play cooperatively

> **Definition:** *Tit-for-Tat* — In game theory, *Tit-for-Tat* refers to a rule for choosing strategies of play in a repeated social dilemma or similar game. The rule is to play "cooperate" until the opponent plays "defect," and then to retaliate by playing "defect" on the following round.

unless the other player has played non-cooperatively on the previous round, but in that case play non-cooperatively. Strategies for a repeated game, like this one, are highly complicated contingent strategies, and there can be a large number of them. For the 20th play, there are 4^{19} patterns of play on the prior 19 stages that would be distinct contingencies in a contingency plan for that last play. Given a tit-for-tat strategy, only the other player's play on the previous round of play needs to be considered. But will it work?

It will not. Once again, we use backward induction. On the twentieth round, there can be no threat of retaliation on the following round, so the workers will shirk on the last round. But on the 19th round, it is known that there will be no cooperative play on the 20th, so there is no incentive to cooperate on the 19th round. The promise that cooperative play on the 19th round will be rewarded on the 20th round (as the tit-for-tat rule requires) is not a credible promise. Similarly on the 18th round. We see again that the only subgame perfect equilibrium of the sequential game of repeated play of this social dilemma is always to choose the non-cooperative dominant strategy.

3. PRESSING THE SHIRTS

Nicholas Neatnik likes to have his shirts pressed every week. He can do them himself or take them to the Neighborhood Cleaners, just down the street. However, Nicholas is concerned that Neighborhood Cleaners will not take enough care, and will spoil some of his shirts. Nicholas' strategies are to have his shirts pressed by the cleaners ("press") or not ("no"), and the cleaner's strategy is to take care ("care") or not ("no"). This game is shown in extended form, with its payoffs, in Figure 10.3. Nicholas is A, Neighborhood Cleaners is B, and the first payoff is to Nicholas, the second to Neighborhood Cleaners.

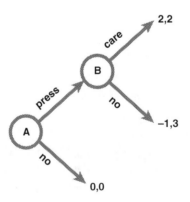

Figure 10.3.   Pressing the Shirts.

This game has a basic proper subgame in which the cleaner decides whether or not to take care, and the rational decision is not to take care, for a payoff of 3 rather than 2. That means Nicholas can anticipate a payoff of minus one if he takes his shirts to the cleaner. The payoff of zero (doing his own pressing) is better than minus one, so the subgame perfect equilibrium is that Nicholas does his own pressing. (The subgame perfect equilibrium is the only Nash equilibrium in this game). Unlike our other examples, this is not a social dilemma. There is no dominant strategy — Nicholas' best response is "press" if he can anticipate that neighborhood cleaners will take care, but "no" otherwise — but nevertheless the equilibrium is inefficient and differs from the cooperative solution, "press" and "take care."

However, this is a repeated game. Nicholas will want to get his shirts done every week for the next 5 years. At the end of 5 years, he plans to retire and never wear a pressed shirt again. Thus, the Pressing Game will be played for 5 × 52 = 260 times in succession. Intuition suggests that Nicholas might be able to persuade the cleaners to take care by punishing the non-cooperative strategy "no." Nicholas might — for example — continue to bring his shirts only if they are carefully pressed this week, and respond to a "no care" strategy by never bringing his shirts to the neighborhood cleaners again. Then Nicholas would be playing a "grim trigger" strategy rule. The Grim Trigger rule says: Play cooperatively unless the other player plays non-cooperatively, and if so play non-cooperatively from that time on. Could this policy bring about a cooperative solution — "press," "care" — in the pressing game?

Definition: *Grim Trigger* — In game theory, the *Grim Trigger* refers to a rule for choosing strategies of play in a repeated social dilemma or similar game. The rule is to play "cooperate" until the opponent plays "defect," and then to retaliate by playing "defect" on all following rounds.

Once again, a rule such as this simplifies what would otherwise be an enormous family of contingent strategies. Consider Nicholas' choice to take his shirts in to be pressed, or not, on the 201st week of his 5 years. This choice may depend on the choices made in each of the previous 200 rounds of play. On the first round there are three possible patterns of play — (press, take care), (press, take no care) and (don't press, do nothing). Thus, Nicholas' choice on the second round is based on those three contingencies.[1] In the third round, his choice is based on $3 \times 3 = 3^2 = 9$ contingencies: Three for the first, each of them followed by three for the second. In the fourth round his choice is based on $3 \times 9 = 3^3 = 27$ contingencies. For the 201st round, his choice is based on 3^{200} possible sequences of play that may have gone before. If he plays according to a Grim Trigger rule on round 201, however, the first 199 rounds do not matter. The rule groups together 3^{199} sequences of play that may have taken place before and says "*if you played press and Neighborhood Cleaners took no care on the last play*, then play no on this and all subsequent rounds; *otherwise*, play press."[2] This rule gives the same result for all of the 3^{199} sequences of play on the first 199 rounds. But is it rational to play according to a Grim Trigger rule?

To explore that, Figure 10.4 shows a tree diagram for two successive rounds of the game. The payoffs shown are the sum[3] of the payoffs in two rounds, with the payoff to A first and B second as

---

[1]Neighborhood Cleaners' strategy choice is based on six contingencies — those three if Nicholas does bring his shirts on round two and the same three if he does not — though only four of those six leave Neighborhood Cleaners with a choice to make.

[2]This means Nicholas plays press even if he has played "no" on the previous round, so that Neighborhood Cleaners had no opportunity to ruin his shirts.

[3]Since the payoffs are only 1 week apart, we are ignoring the fact that the future payment is worth less than a payoff today. An economist would not approve — it is important to get this right and to discount future values to the corresponding present value! We will revisit that point in the next chapter and find that we gain more than just the satisfactions of getting the economics right: But for now, we will leave it inexact for the sake of simplicity.

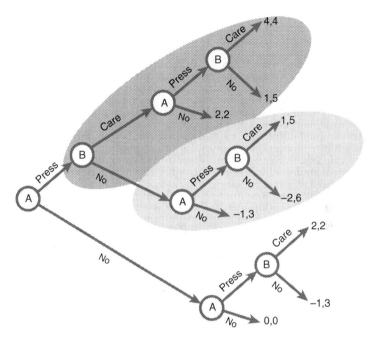

Figure 10.4. Two Stages in the Pressing Game.

before. If Nicholas plays according to a Grim Trigger rule, there is an implicit threat and an implicit promise. The implicit threat is: If Neighborhood Cleaners takes no care, Nicholas will retaliate by not bringing them any further business. The promise is: If Neighborhood Cleaners takes care, he will reward them by bringing his business next week. What we must ask is: Assuming Neighborhood Cleaners are rational, will they find either the threat or the promise credible?

In game theory, remember, credibility is associated with subgame perfection. So, we rephrase the question, asking whether the Grim Trigger rule with its threat and promise leads to subgame perfect play. As we have already noticed, a play of "no" by Nicholas (Player A) is a subgame perfect Nash equilibrium in this subgame. So, it *is* subgame perfect to carry out the threat — the threat is credible.

If the promise is credible then the threat and the promise together *are* enough to persuade Neighborhood Cleaners (Player B) to use care in stage 1. If he does not use care, he ends up at the bottom of the light gray oval with a payoff of 3. If he uses care then he is in the darker gray, upper oval. Within the dark gray oval, if Nicholas carries out the "promise" and brings in his shirts, then Neighborhood Cleaners can do no worse than a payoff of 4, better than the three they get in the lower, light gray oval. However, Nicholas will bring in his shirts only if he can get the (4, 4) payoff at the upper right. If Neighborhood Cleaners plays "no care" at this step, he will be better off to renege on the promise and not bring his shirts to be pressed. Will Neighborhood Cleaners repeat their choice of "take care" on this last round? Not on the last round. Nicholas cannot rationally play the Grim Trigger on the 260th round because there is no 261st on which he can reward good performance by Neighborhood Cleaners. Nicholas' promise of reward is only credible if Nicholas is continuing his unbroken sequence of Grim Trigger plays. Instead, on that last round, Neighborhood Cleaners will choose no care (upper right in dark gray oval) and since Nicholas can anticipate that, he will choose "no pressing" on round 260. However, since we now know that Nicholas will choose "no pressing" on round 260, regardless of what happens on round 259, Nicholas cannot play a Grim Trigger rule on round 259. His promise to reward good behavior on round 259 with his business on round 260 is no longer credible to Neighborhood Cleaners, so they will take no care on round 259; and because he can anticipate that, Nicholas will definitely choose no pressing on round 259. But since Nicholas will definitely choose no pressing on round 259, that means he cannot credibly play a Grim Trigger rule on round 258, either, for the same reasons — and we can continue this sequence all the way back to round one. The answer to the question we started with — can the threat and the promise be credible? — is no, because there can be no unbroken sequence of play into the future. The sequence breaks at round 260, and unravels back from round 260 to round 1. Once again, we see that the "folk theorem" is not applicable to games that are repeated a definite, limited number of times.

If only the sequence of repeated play could go on forever, the grim trigger could work (and so could tit-for-tat) — but, of course, nothing goes on forever, right? I'm going to break one of the textbook-writer's taboos, here, and speak for myself, not even putting it in a footnote. I use something approximating a Grim Trigger or Tit-for-Tat strategy in dealing with merchants myself. It's a little more complicated and probably less predictable, but if I get service, I am not happy with, I will skip doing business with that merchant. I may give them another trial later, as with Tit-for-Tat, or depending on the problem I might not go back, as with the Grim Trigger. As with those strategy rules, I'm relying on the implicit threat and promise. Am I irrational? Well, maybe, but ....

> **A Closer Look:** Ravel or Unravel
>
> It is often said that a game such as the Campers' Dilemma or Pressing the Shirts, a game with a definite last round of play, the possibility of cooperation "unravels" from the last round of play back to the first. We might visualize a cooperative arrangement as a braid that is not fastened at the end and so comes loose. Actually, "unravel" is an odd word in that both "unravel" and "ravel" mean the same thing. Shakespeare writes of "sleep that knits up the ravell'd sleeve of care." But, for a braid or a cooperative arrangement, to come apart is a negative thing, and the negative prefix "un" intensifies that; so "unravel" is the term usually used.

We could think of this result as a puzzle. Strategies like Tit-for-Tat and the Grim Trigger seem very promising in repeated games, but in the examples we focus on in this chapter, in which the repetition comes to an end, those strategy rules won't work. Is there something wrong with the rules, or something missing from the examples? That is a question we will return to in the next chapter.

Threat of retaliation plays a part in other important games, and the next example played an important part in the development of the concept of subgame perfect Nash equilibrium — and in Reinhard Selten's share of the Nobel Memorial Prize.

4. THE CHAIN STORE PARADOX

Here is another example of repeated game reasoning. A large chain store we shall call Chainco has stores in twenty American communities. There are local companies preparing to enter those twenty markets one after another in the future. Thus, Chainco expects to play twenty market-entry games over the next few years. Intuition suggests that Chainco should retaliate in the early games, even if it takes losses doing so, in order to create a reputation as a retaliator and thus deter future entrants. But is this a subgame perfect strategy? It is not!

A single play in the Chain Store game is shown in Figure 10.5. At node 1 the new company decides whether to enter or not. If they do not, then the Chain Store earns profits of 10 (on a scale of 1 to 10) in that market. If none of the local companies enters their markets, Chainco will earn 10 in each market and $10 * 20 = 200$ in all. If the local company does enter, then at 2 Chainco decides whether to

A Closer Look: Reinhard Selten 1930–2016

Born in 1930 in Breslau, Germany, which is now the Polish city of Wroclaw, Selten was partly of Jewish descent and thus suffered hardship in his youth. While a high school student, he read a Fortune magazine article about game theory and studied *The Theory of Games and Economic Behavior* by von Neumann and Morgenstern. While a research assistant at the University of Frankfurt am Main, he earned a PhD in mathematics. Though a mathematician, he has worked on experimental studies and applications in economics; and by working on games in extensive form, then an unpopular field, he was able to do path-breaking work that led to his sharing the Nobel Memorial Prize in 1994. He founded the Laboratorium für experimentelle Wirschaftsforschung at the University of Bonn and was its chairman for several years.

retaliate by engaging in a price war. If it does, Chainco will make

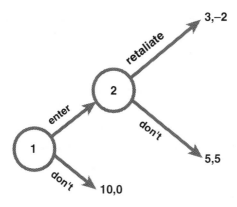

Figure 10.5. One Play in an Entry Game.

only 3 of profits in this market, but the new entrant will take a loss of 2, indicated by –2. If Chainco decides not to retaliate, the market is divided and both firms earn profits of 5.

Consider the last in the series of 20 repetitions of the game. Clearly there is nothing to be gained in this case by retaliating: There are no further threats of entry. We notice then that (enter, don't retaliate) is the subgame perfect solution for the single play. Now consider the 19th repetition. Since it is known that there will be no retaliation on the last round, retaliation on the 19th round will not prevent entry on the 20th.

Therefore, there is no purpose in retaliating in the 19th repetition either. The same reasoning applies to the 18th, the 17th, and so on. We conclude that retaliation to create a reputation is not a "best response" strategy in this game, and a rational chain store will never retaliate.

5. SUMMARY

When social dilemmas are played "one-off," non-cooperative play leads to bad results all around. However, a strong intuition suggests that repeated play changes all this, since cooperative behavior can be rewarded, and non-cooperative play sanctioned, in future rounds of play. This intuition has been expressed as a "folk theorem."

Such sequences of repeated play can be analyzed by methods from the theory of sequential games, such as subgame perfect equilibrium. This allows us to test the intuition behind the "folk theorem." However, if the repetition continues only for a definite number of rounds, repeated play does not in fact lead to cooperation, since the only subgame perfect Nash equilibrium is a sequence of non-cooperative plays. This paradoxical result extends beyond social dilemmas to other kinds of games, such as market entry games, in which the subgame perfect equilibrium is one in which an established monopolist is unable to prevent entry, and loses profits as a result.

This is a surprising result of backward induction. In every case, we begin the analysis with the last repetition of play. In this case, the play is exactly like play in a one-off game, and so the equilibrium must be the same. Any cooperative rule for retaliation or reward, such as the Tit-for-Tat strategy rule, is inapplicable on the last round. Reasoning backward, we extend that result to each repetition of the play. The strategy rule unravels from the end right back to the beginning of the game. Thus, so long as there is an end point, cooperative play cannot be an equilibrium in a repeated social dilemma. Something similar happens in a threat game like the Chain Store Game. Retaliation is not subgame perfect in the last round, and any reputation the Chain Store might have gained does not change that. Once again, the motivation to build a reputation as a retaliator unravels back to the first round. Intuition can be very misleading when repeated play has an end point.

Q10. EXERCISES AND DISCUSSION QUESTIONS

Q10.1. Repeated Battle of the Sexes

Sylvester and his Tweetie Pie want to get together after work either for a baseball game or an opera, but they can't get in touch with one another to decide which, because Sylvester's e-mail isn't working

Table 10.3. Battle of the Sexes.

		Tweetie	
		Game	Opera
Sylvester	Game	5,3	2,2
	Opera	1,1	3,5

and Tweetie's cell-phone battery is dead. (They have season's tickets together at both the opera and the baseball park.) Their strategies are "game" and "opera" and they payoffs are as shown in Table 10.3.

Sylvester and Tweetie expect to be out together for two dates in succession. Thus, they will be playing a repeated game.

For this repeated game in normal form, enumerate the strategies of each person. How many are there? Draw the game tree. Find four different subgame perfect solutions to this repeated game. Is there any reason to think that one of these solutions might be more likely than another? Noted Unitarian-Universalist minister Robert Fulgham wrote that "All I Really Need to Know I Learned in Kindergarten" and said that one of the most important things he learned in kindergarten was "taking turns." From a commonsense point of view, does this example tell us anything about the advantages or disadvantages of taking turns? What?

Q10.2. Getcha Hot Dog for Five More Weeks

Frank and Ernest sell hot-dogs from neighboring vending trucks. They compete against one another every week, but after five more weeks Frank is retiring and selling his truck. Each week they choose between the strategies of pricing their dogs at $1.50 or $2.00, and the payoffs are as shown in Table 10.4.

What is the unique subgame perfect equilibrium of this repeated game? Why?

Table 10.4. Payoffs to Hot Dog Vendors.

		Frank	
		$1.50	$2.00
Ernest	$1.50	10,10	18,5
	$2.00	5,18	12,12

Q10.3. Congressional Gridlock

"Gridlock" has been a common problem in American Congressional politics in recent years. Gridlock occurs when neither party has a strong majority in Congress so that each is able to block initiatives by the other. If they do not act cooperatively, and in fact do block initiatives by the other party, nothing gets done — nothing moves, and we have "gridlock." Analyze this in the terms of this chapter. *Hint*: remember the old saying: "There is no tomorrow after the next election."

Q10.4. Whiskey and Gin

Recall the advertising game from Chapter 1. In that example, we observed that advertising could be a social dilemma, in that when both firms advertise the profits for both are less than they would be if neither were to advertise. The distilled alcoholic beverage industry faced a similar dilemma in the period of the 1960's–1990's. In the 1990's, an unspoken agreement not to advertise on television broke down when one of the major distilled beverage companies was in danger of bankruptcy.

Suppose that Table 10.5 gives the payoffs (on a ten point scale) for Ginco and Whisco. Assume also that Whisco is in financial trouble and it is clear that Whisco will be bankrupt after three more years. What will happen in the meantime? Will one or the other of the companies advertise this year, next year and the year after? Which? Why?

Table 10.5. Another Advertising Game.

		Whisco	
		Don't advertise	Advertise
Ginco	Don't advertise	8,8	2,10
	Advertise	10,2	4,4

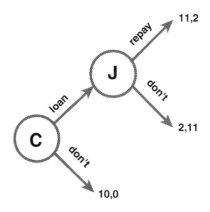

Figure 10.6. Loaning Money to Joe Cool.

Q10.5. Joe Deadbeat

Joe Cool is a student at Nearby College. Joe runs short of money from time to time, and would like to borrow from his fellow dorm residents to get by, and would pay a reasonable rate of interest. However, his classmates are reluctant to lend to him, since they don't know whether he is a good risk to pay back the loan or not. This is a repeated game of moneylending, and each loan is a sub-game like the one shown in extended form in Figure 10.6. In the figure Joe's classmate, C, makes the first decision and the first payoff is to the classmate, and Joe makes the second decision and gets the second payoff.

Joe may not have much money, but he is an excellent student and sure to graduate on schedule in 4 years. Joe would like to create a reputation as a good risk by paying promptly if someone will lend him money this term, but no-one will take the chance. Using the theory of repeated play, explain why Joe cannot get a loan.

CHAPTER 11

Indefinitely Repeated Play

As we think back over the examples in the previous chapter, there is a common factor that causes all the trouble, a common factor that causes both the difficulty in reaching cooperative arrangements and the counterintuitive

To best understand this chapter, you need to have studied and understood the material from Chapters 1–4, 7 and 9–10.

results. The common factor is that the interactions have an end point. In the Camper's Dilemma, camp is over after 10 weeks — so the girls will not be seeing one another again, and there is no reward for cooperative behavior in that last week. Similarly in the effort dilemma. The motive for cooperative behavior unravels from the last week to the first. Similarly, in the Pressing the Shirts example, the relationship ends at Nicholas' retirement after 260 weeks, and in the Chain Store example, there is a last entry threat when all other markets have been entered. There is no motive for retaliation in that last market, and the motivation for retaliation unravels from there to the first entry threat. If there were no end point — if the game were to continue to infinity — this particular problem would not arise. We now consider an example of that kind.

1. A REPEATED EFFORT DILEMMA

Let us look again at the effort dilemma from Section 2 of the previous chapter. Effort dilemmas are social dilemmas based on the contribution of effort to a common enterprise. Like some other social dilemmas they are likely to be played repeatedly. Table 11.1 repeats Table 10.2 for convenience. It shows an effort dilemma played by two working teammates, Andy and Bill. We recall that it is a social dilemma, since the dominant strategy is "shirk." Andy and Bill will play this game once, and possibly more. They don't know how many times they will play the game, but they will play for some time and then quit. How does this work?

The idea is that there is always a 10% probability that this play will be the last one — and conversely, a 90% probability that there will be

HEADS UP!

Here are some concepts we will develop as this chapter goes along:

Indefinitely Repeated Games: When a "game" is played repeatedly, but with no definite end point, we treat the game as if it were repeated infinitely many times.

Discount factor: In future plays of the game, payoffs are "discounted" both for the passage of time and the probability that the game will not be played so many times. The discount factor for the next play is

$$\delta = p\left(\frac{1}{1+r}\right)$$

where p is the probability that the game will be played again and r the rate of discount for time passed between plays.

Trigger Strategy: A rule for choosing strategies in individual repetitions in an indefinitely repeated game is called a "Trigger Strategy" if the rule is that non-cooperative play triggers one or more rounds of non-cooperative play by the victim in retaliation.

Tit-for-Tat: A trigger strategy that responds to a defection by a single defection on the next round.

Grim Trigger: A trigger strategy that responds to a defection by defection from that time on. Other trigger strategies are "forgiving triggers."

Table 11.1. An Effort Dilemma
(Repeats Table 10.2).

		Bill	
		Work	Shirk
Andy	Work	10,10	2,14
	Shirk	14,2	5,5

another round of play. And that will be true in the same way on every future round of play: There is always a 10% probability that there will be no further round of play. How likely is it that there will be at least two further rounds of play? To find that out we have to compound the probability of two events: With a 90% probability that there will be one more round of play and a 90% probability that (if there is a new round) there will be another new round after that. The question is, what is the probability that both of these things will happen, so that there will be two new rounds of play? The probability that both things will happen is the product of the probabilities, 90% times 90% = 81%. The general rule is that the probability that two things in succession will both happen is the product of their probabilities. The probability that n things happen in succession is the product of their n probabilities. Thus, the probability that there will be at least n more rounds of play is the power $(0.9)^n$. Table 11.2 shows the probabilities of 1, 2,…,20 further rounds of play for this example.

Suppose that Andy and Bill always play the dominant "shirk" strategy, so that payoffs are always 5 on every round, so long as they play at all. If they do not play, of course, payoffs are zero. Since it is uncertain what (if anything) the payoffs will be after the present round, the total payoff has to be calculated as an expected value. Thus, the expected value payoff is the sum[1] $5 + 0.9 * 5 + 0.9^2 * 5 + 0.9^3 * 5 + 0.9^4 * 5 + \cdots = 5(1 + 0.9 + 0.9^2 + 0.9^3 + 0.9^4 + \cdots)$.

---

[1] This may seem to apply only if the non-cooperation takes place on the first round, but that is not correct since an infinite series always has infinitely many rounds remaining to be played, and every play is the first round of such an infinite series. As the saying has it, "this is the first day of the rest of your life."

Table 11.2. Probabilities of More Rounds of Play.

Round	Probability
Present	1
1	0.90
2	0.81
3	0.73
4	0.66
5	0.59
6	0.53
7	0.48
8	0.43
9	0.39
10	0.35
11	0.31
12	0.28
13	0.25
14	0.23
15	0.21
16	0.19
17	0.17
18	0.15
19	0.14
20	0.12

We can simplify this by relying on a useful fact from algebra. The useful fact[2] is that for any sequence y, y^2, y^3,..., provided $0 < y < 1$, $1 + y + y^2 + y^3 + \cdots = \frac{1}{1-y}$. In particular, since $0 < 0.9 < 1$ ($1 + 0.9 + 0.9^2 +$

---

[2]If you have taken macroeconomic principles, this may seem somewhat familiar — in that it resembles the multiplier formula in Keynesian economics. This is because the multiplier formula uses the same useful fact from algebra.

$0.9^3 + 0.9^4 + \cdots) = \frac{1}{1-0.9} = \frac{1}{0.1} = 10$. Therefore, $5 + 0.9 * 5 + 0.9^2 * 5 + 0.9^3 * 5 + 0.9^4 * 5 + \cdots = 5 * 10 = 50$. If Andy and Bill always shirk, the expected value of their payoffs on all future play is 50.

The "Folk Theorem" suggests that the two team-mates might instead cooperate on the following basis: If one of them shirks, the other could retaliate on a subsequent round of play by playing "shirk." This threat is credible, since "shirk" is a dominant strategy equilibrium and is thus always subgame perfect. This, of course, is the Tit-for Tat strategy rule, as discussed in the previous chapter.

So, we suppose that Bill chooses his behavior strategy on each repetition according to the Tit-for-Tat rule, and Andy knows that he does. We want to know whether that will deter Andy from *defecting* by playing "shirk" *even once*. If Andy defects once, his payoffs will be

$$14, 2, 10, 10, 10,....$$

and the expected value is

$$Y_1 = 14 + (0.9)2 + (0.9)^2 10 + (0.9)^3 10 + \cdots,$$

while if he plays "work" on every round of play, Bill never deviates from "work" either, and Andy's payoffs are

$$10,10,10,10,....$$

and the expected value is

$$Y_2 = 10 + (0.9)10 + (0.9)^2 10 + (0.9)^3 10 + \cdots$$

Notice that, since Y_1 and Y_2 differ only in the first two terms, we can ignore every term after the second, and Andy will be deterred from defecting if

$$Y_1 < Y_2,$$

that is,

$$14 + (0.9)2 < 10 + (0.9)10$$
$$15.8 < 19.$$

Since that clearly is true, Andy will lose 4.2 whenever he defects. Moreover, the more he defects, the more he loses — so Andy will not defect *even once.*

If Andy also chooses his behavior strategies according to the Tit-for-Tat rule, then they will always choose the cooperative "work" behavior

Indefinitely Repeated Games: When a "game" is played repeatedly, but with no definite end point, we treat the game as if it were repeated infinitely many times. Such games may also be said to be "infinitely repeated."

strategies; and this is a self-enforcing arrangement, since each one is acting in his own self-interest by cooperating. Since there is always a next round of play (with 90% probability) there is always an incentive to cooperate, and the cooperative arrangement does not unravel as it would if there were a last round of play.

How large does p have to be to assure this result? Andy will be deterred from shirking if

$$14 + 2p < 10 + 10p$$

$$4 < 8p$$

$$p > \frac{1}{2}$$

If the probability of another round of play were less than $1/2$, then Tit-for-Tat would definitely not deter defection.

It seems we have found the kernel of truth in the "folk theorem." When a social dilemma is repeated with some definite probability, but without any predictable end, there is at least the possibility that

A Closer Look: Piano Assembly

Port Townsend, Washington, is a small port city on Puget Sound that is host to a summer jazz festival. My wife and I were having lunch in a pub that would be a venue for the jazz festival in the evening, and we watched a two-man team assembling a piano for the evening's performance. Their moves were perfectly coordinated, choreographed as if a dance. It was clear that they had often done this job together before. One result of that frequent repetition was skill: They knew from experience just what needed to be done at every stage. But they also made exactly the needed effort — confident, no doubt, that they would be doing the same job over and over in the future, so that they would have plenty of future opportunities to benefit from their cooperation and skill.

the cooperative outcome may be an equilibrium. In the real world, people work together over unpredictable but extended periods, so it is likely that many effort dilemmas are "indefinitely repeated" with some high probability.

2. THE DISCOUNT FACTOR

When we talk about infinitely repeated games, we seem to be talking about a world in which nobody ever dies, retires or moves away! But not necessarily. The point is that there is no *definite* end to the repetition. If there is a definite end, then retaliation and reward strategies will unravel. But suppose, as in the repeated effort dilemma, that there is no definite ending time, but there is some probability that the play will end on any particular round. Then we have a different situation. As we have seen, the expected value of future payoffs is finite.

The role of the probability is very much like discounting future values of payoffs that only come after some time has passed. This is a familiar idea from finance and economics. Taking it for granted that people prefer payoffs now to payoffs in the future, financial economists define a discount factor, δ, which answers the question "How much would a person pay today to get a payoff of one dollar a year from today?" The answer is, the person would pay δ today for a dollar 1 year from today. In general, since payoffs now are preferred to payoffs in the future, $\delta < 1$. If we then ask "How much would a person pay today to get a payoff of 1 2 years from today?" the answer is δ^2; for 3 years δ^3, and so on.

The discount factor is, of course, related to the interest rate ("discount rate") on loans and investments. Let's say that Y is the payoff next year, and V the discounted value of the payoff on the next round. If I lend money at a rate of r, I get $1 + r$ dollars at the end of the year for every dollar I lend at the beginning of the year. Therefore, a dollar now is worth $1 + r$ dollars at the end of the year, and conversely, a dollar at the end of the year is $1/(1 + r)$ now. Therefore, $V = \frac{1}{1+r}Y$ and $\delta = \frac{1}{1+r}$. In general, a high rate of discount (low discount factor) is associated with *impatience,* and a low rate of discount (high discount factor) correspondingly associated with *patience.*

Instead, as in the effort dilemma, we might assume that the next payoff comes so soon the discounting to present value can be ignored, but the probability of a new round of play is $p < 1$, Y the payoff if there is a next round, and V the expected value of the payoff on the next round. Then $V = p(Y) + (1 - p)(0) = pY$. Therefore, $\delta = p$.

Putting both ideas together, let $p < 1$ be the probability of another round of play, but the next round will not come until next year, and let r be the rate of time discount for 1 year. Then the expected value of the payoff is

$$V = p\left(\frac{1}{1+r}\right)Y + (1-p)\left(\frac{1}{1+r}\right)(0) = p\left(\frac{1}{1+r}\right)Y$$

$$\text{and } \delta = p\left(\frac{1}{1+r}\right).$$

A Closer Look: The Discount Rate

Suppose, for example, that the discount rate is 5%. We want to know the value of a $10,000 payment 3 years from today. Booting up the trusty spreadsheet, we find that every dollar we invest today at a 5% interest rate will be worth $1.05 at the end of 1 year and worth $1.158 at the end of 3 years. We divide the $10,000 by 1.158 to find that $8,638.38, invested at 5% for 3 years, would give us the same $10,000 after 3 years. Accordingly, $8,638.38 is the discounted present value of $10,000 3 years in the future, at a discount rate of 5%.

In the example of the effort dilemma, the time elapsed between repetitions was short enough that we could ignore this, and the time might be different from one job to the next, but the 90% probability of another round leaves plenty of room for errors that result from ignoring the time discount. But we will apply it in the next section, which is an important business application.

3. COLLUSIVE PRICING

The most important applications of indefinitely repeated play in economics are probably to collusive pricing. This has been discussed in other chapters so we may be brief. An oligopoly is an industry with few sellers. Such an industry faces an opportunity and a problem. The opportunity is that it might be able to maintain a monopoly price. The problem is that this is not a Nash equilibrium, and in fact, the competitive price is the only Nash equilibrium strategy. Our example is shown again in Table 11.3, where the oligopolistic companies are Magnacorp and Grossco. From their point of view, "maintain price" is the cooperative solution but "cut" is the non-cooperative dominant strategy solution.

Table 11.3. Payoffs in the Pricing Dilemma
(Repeats Table 3.4, Chapter 3).

		Grossco	
		Maintain price	Cut
Magnacorp	Maintain price	5,5	0,8
	Cut	8,0	1,1

However, pricing games are played repeatedly. Suppose, then, that the two companies set their prices annually and discount future profits at 10%; and that the probability that they will compete again next year is 0.95. (That is, one company or both might go bankrupt or the government might impose price regulations, or their situation might change in some other way, but the probability of such changes is no more than 5%.) Then the discount factor is

$$\delta = p\left(\frac{1}{1+r}\right) = 0.95\left(\frac{1}{1.1}\right) = 0.95(0.909) = 0.864.$$

Is this enough to support cooperative play? In particular, if Magnacorp knows that Grossco is playing according to the Tit-for-Tat rule, will that deter Magnacorp from defecting from the cooperative solution and asking the competitive price for a payoff of 8 rather than 5? Magnacorp knows that if they defect once, then return to cooperative pay, Grossco will retaliate once, their expected value of payoffs will be

$$EV(cheat) = 8 + 0 * \delta + 5 * \delta^2 + 5 * \delta^3 + \cdots,$$

while if they play cooperatively on every round, the expected value payoff will be

$$EV(cooperate) = 5 + 5 * \delta + 5 * \delta^2 + 5 * \delta^3 + \cdots$$

Notice that the two sequences differ only in the first two terms, $8 + 0 * \delta$ and $5 + 5 * \delta$. From the third period on they are the same. Thus, Magnacorp will be better off to cooperate if

$$5 + 5 * \delta > 8$$

That is, after a little algebra, they are better off to cooperate if

$$\delta > \frac{3}{5}$$

Now, $\delta = 0.864 > 0.6$, so Tit-for-Tat play by Grossco will motivate Magnacorp to cooperate, and similarly, if Magnacorp plays according to the Tit-for-Tat rule, Grossco will be motivated to cooperate.

We have seen that in this example, Magnacorp will be worse off if they deviate just once and return to cooperative play. But they will be still worse off if they deviate more than once, so Tit-for-Tat play will deter all non-cooperative play in this example: Magnacorp is worse off *even* if they deviate only once.

Thus, it should not be difficult for the members of a stable duopoly to collude to keep prices up. What is modestly good news in many other situations is modestly bad news for price competition and antitrust policy, since cooperative behavior by duopolists in pricing dilemmas corresponds to prices that exploit monopoly power visavis consumers. When the members of an oligopoly keep their price high without making an explicit agreement to do so, this is called "tacit collusion." What the example suggests is that in duopolies, tacit collusion may be common, and difficult for public policy to prevent.

However, oligopolies may have more than two firms — three, certainly, perhaps four or more — the upper limit for oligopoly is vague and probably depends on the circumstances. All of our examples on repeated games have been two-person games, including this oligopoly example. There has been some research on trigger strategies with more than two players, but this remains an open area of research.

4. OTHER TRIGGER STRATEGY RULES

However, this is still not the whole story. There are other rules for choice of behavior strategies. Tit-for-Tat is one instance of what are called trigger strategies. (A non-cooperative strategy on one play *triggers* a sanction on the next. Strictly speaking, we should say "trigger strategy rule," since they are rules for choosing behavior strategies in particular stages of the repeated game, but the term "trigger strategies" is often used for brevity.) We could also consider a number of other trigger strategies, including (as we recall from the last chapter) the Grim Trigger, which means that a single non-cooperative strategy from one player triggers a switch to "never cooperate" so the retaliator chooses non-cooperative strategies from that time on. By contrast, Tit-for-Tat is known as a "forgiving trigger." The "Grim Trigger" is grim because it means no further cooperation whatever.

Return to the pricing dilemma, but suppose that the probability of another round of price competition is not 0.95 but 0.6. (Perhaps one of the firms is in serious danger of bankruptcy and liquidation.) That would mean that

$$\delta = 0.6\left(\frac{1}{1.1}\right) = 0.545.$$

As we have seen, Tit-for-Tat will motivate cooperation only if $\delta > 0.6$, so with these assumptions Tit-for-Tat would not do the trick. What about the Grim Trigger? If both players play cooperatively from now on, the value of payoffs is

$$EV(\text{cooperate}) = 5 + 5\delta + 5\delta^2 + 5\delta^3 + \cdots.$$

A little algebra will help, here. Factoring out 5, we have

$$EV(\text{cooperate}) = 5\,(1 + \delta + \delta^2 + \delta^3 + \cdots)$$

Here we apply the "useful fact" from Section 1 above:

$$EV(\text{cooperate}) = 5\left(\frac{1}{1-\delta}\right).$$

If δ is 0.545, this is

$$5(2.2) = 11.$$

Now, if Magnacorp defects on this round, Magnacorp knows that Grossco will never play cooperatively in any future round. To play cooperatively on a future round when Grossco does not will only leave Magnacorp even worse off. Thus, Magnacorp will play the non-cooperative dominant strategy "cut" in every future round. It follows that the expected value of payoffs in this case is

$$EV(\text{cheat}) = 8 + \delta + \delta^2 + \delta^3 + \cdots$$

We can again apply the "useful fact" from Section 1, but it is a little more complex in this case. We can factor δ in this case, so we have

$$EV(\text{cheat}) = 8 + \delta(1 + \delta + \delta^2 + \cdots) = 8 + \delta\left(\frac{1}{1-\delta}\right)$$

Therefore, Magnacorp will be better off to cooperate if

$$5\left(\frac{1}{1-\delta}\right) > 8 + \delta\left(\frac{1}{1-\delta}\right)$$

To simplify this inequality, we may multiply both sides by $(1 - \delta)$, obtaining

$$5 > 8(1 - \delta) + \delta = 8 - 7\delta$$

That is

Game Theory (Fourth Edition)

$$\delta > \frac{3}{7} = 0.429$$

Since in this case, $\delta = 0.545 > 0.429$, it seems that when Grossco plays according to the Grim Trigger rule, Magnacorp will be motivated to play cooperatively. Double-checking we see that with $\delta = 0.545$, the expected value of defecting is

$$\text{EV(cheat)} = 8 + \frac{0.545}{0.455} = 8 + 1.2 = 9.2 < 11,$$

so, again, Magnacorp is better off to cooperate, since 11 is the payoff from cooperation.

What we see here is that the Grim Trigger is a more powerful threat than Tit-for-Tat. This is true in general: The Grim Trigger can enforce cooperation where forgiving triggers do not.

Other forgiving trigger strategies include:

- A Tit-for-Two-Tats, meaning that the player retaliates with one round of non-cooperation only after two rounds of non-cooperation. Notice that this strategy rule can be "beaten" by a counterstrategy of alternating cooperation and non-cooperation, since in that case the aggressor gets the benefits of acting non-cooperatively on alternative rounds without ever suffering retaliation. Thus, A Tit-for-Two-Tats is likely to be weakly dominated by Tit-for-Tat.

- Two-Tits-for-a-Tat, which means that the player retaliates with two rounds of non-cooperation for every non-cooperative play by the other player. In the Camper's Dilemma, Two Tits for a Tat will lead to cooperation whenever δ is greater than 0.2637. Notice that this is slightly more than the 0.25, which is the limit for the Grim Trigger, but less than the 0.33 needed for Tit-for-Tat. This makes sense, since the threat of two rounds of non-cooperation is intermediate between one round and every future round. In the pricing dilemma, if a defection is followed by two rounds of non-cooperative play, after which cooperative play resumes, then an oligopolist is better off to cooperate if $\delta > 0.5$, and as we see, in this case this is between 0.429, the break-point for the Grim Trigger and 0.6. the breakpoint for Tit-for-Tat.

It must be clear that there could be an infinite family of variations on these themes for an infinite game, and many of them will be equilibria. Not all of the equilibria will be efficient cooperative play, either. In fact, in many games there are equilibria with average payoffs at all levels between purely non-cooperative and purely cooperative play.

From a mathematical point of view, this range of possible equilibria is an embarrassment of riches. The mathematician wants one solution, and wants the solution to have given properties, such as efficiency. Two points might support the optimistic view. First, if a trigger strategy is effective, the trigger strategy equilibrium corresponding to cooperative play will be payoff dominant. Second, it is only the payoff dominant equilibrium that is strong. That is, given any equilibrium with cooperative play, the agents could form a coalition and agree to play the trigger strategies, leading to self-enforcing cooperation. We owe this insight — along with much else — to Robert Aumann.[3]

5. POISON GAS

Poison gas was used as a weapon by both sides in World War I, but was not used in World War II. Of course, World War II was not a picnic in the park. Weapons even more terrible than gas were used, and horrors that some would find more terrible than any weapon took place. Yet gas was not used.

Table 11.4 is an example based on the experience of the use of poison gas. The two countries are Mainland and Island. If one uses gas and the other does not, then the country that uses gas gains a

---

[3]See Aumann, R. J., Acceptable points in general cooperative n-Person games, Contributions to the Theory of Games, Volume IV, *Annals of Mathematics Studies*, 4(40) (1959), pp. 287–324.

Table 11.4. Gas (Repeats Table 4.16, Chapter 3, Exercise 4).

		Island	
		Gas	No
Mainland	Gas	–8,–8	3,–10
	No	–10,3	0,0

slight advantage, indicated by a payoff of 3 rather than zero. The country that is the victim of gas suffers great loss, indicated by a payoff of –10. If both use gas, both suffer losses nearly as great, indicated by payoffs of –8 for both. Looking at the table, we see that this is a social dilemma. Table 11.3 repeats the Table from Exercise 3.4 in Chapter 3, where it was used as an example of a social dilemma.

Played as a one-off game, the Gas game has a dominant strategy in which both sides use gas — as did happen in World War I, and also in the Iran–Iraq war of 1980–1988. But it is not necessarily a one-off game. In a long war, there may be many occasions on which gas may be used by one side or another or both, and there is also some probability that the war will continue until a next following occasion for the possible use of gas. Thus, we can apply the theory of indefinitely repeated games.

If neither side uses gas, the payoffs in this repeated game are a string of zeros. Suppose then that Mainland plays Tit-for-Tat and Island "cheats" just once on this round by using gas. Then the payoffs for Island are

Play "gas" $3, -10, X_3, X_4,\ldots$

Play "no" $0, 0, X_3, X_4,\ldots$

The cooperative strategy "no" is better if

$$3 + \delta(-10) < 0 + \delta 0 = 0,$$

that is

$$\delta > \frac{3}{10}.$$

The conclusion is that the Tit-for-Tat strategy can lead to a cooperative outcome in which nobody uses gas provided the chances of the war continuing until an occasion of retaliation are no less than 3 in 10.

What about the Grim Trigger? The Grim Trigger may be realistic in this case, as the issue is often put this way: "If we use gas, that will open the door, and gas will be used from that point on." So, suppose that Mainland plays the Grim Trigger and Island uses gas on this round. Then the payoffs for Island are

Play "gas" 3, −8, −8, −8,...

Play "no" 0, 0, 0, 0,...

The cooperative strategy is better if

$$3 + \delta \frac{1}{1-\delta}(-8) < \frac{1}{1-\delta} 0$$

that is

$$\delta > \frac{3}{11}$$

So, the Grim Trigger can maintain the non-use of gas so long as the probability of the war continuing to at least one more occasion for retaliatory use of gas is at least 3/11. We see, as always, that the Grim Trigger, with its more complete retaliation, can maintain cooperation when Tit-for-Tat cannot, that is when $\frac{3}{10} > \delta > \frac{3}{11}$.

So, we may be able to understand how it is that gas was not used in World War II. But gas was used in World War I and in the

Iran–Iraq war in the 1980's, and nuclear weapons were used in World War II. How are these contrasts explained?

The payoffs to the use of gas may have been greater in World War I than in World War II. In World War I, gas was used in trench warfare, as one part of a coordinated attack to overcome the enemy's trenches. Iran and Iraq were also involved in trench warfare. Trench warfare was not as important in World War II, so the payoffs to using gas could have been even less. But, in fact, even if the

---

A Closer Look: Gas in World War II

There is evidence that thinking in terms of trigger strategies played at least some role in the decision not to use gas in World War II. According to Robert Harris and Jeremy Paxman,[*] an early British proposal to use gas was rejected on the grounds that "British use of gas would immediately invite retaliation against our industry and civil population" (p. 112). Winston Churchill, the British prime minister, advocated use of gas in some circumstances but the circumstances never arose, and Churchill attributed concern about British retaliation to the Germans. He said, "But the only reason they have not used it against us is that they fear retaliation" (p. 130). Under interrogation, German government figure Hermann Göring confirmed this (p. 138). American President Franklin Roosevelt explicitly threatened retaliation amounting to a Grim Trigger if the Japanese were to use gas against the Chinese (p. 120). Other motives, including treaty obligations and moral restraint, clearly played some role too, but underlying it all was the fear of retaliation.

---

[*]*A Higher Form of Killing* (New York, Random House Trade Paperback Edition, 2002). I am indebted to my student Ying Zhang for bringing this book to my attention.

payoffs were the same, the different outcomes in different cases should not be surprising. Remember, the repeated game has more than one equilibrium, and "always defect" is always one of the equilibria in a repeated social dilemma. It may be that we simply observed one equilibrium ("always defect") in World War I and another equilibrium (symmetrical Grim Trigger) in World War II. Moreover, the use of gas in the Iran–Iraq war of the 1980's once again realized the "always defect" equilibrium.

As to nuclear weapons, nuclear weapons were used in World War II in circumstances that, (1) the enemy did not possess nuclear weapons with which to retaliate; and (2) it was expected that the use of the weapon would end the war, so that there would be no further occasions for the use of nuclear weapons between the United States and Japan. Thus, it was not a repeated game.

All in all, it seems that gas in World War II was nearly an ideal case for cooperative strategies via a trigger strategy. There are many other instances of self-restraint in war that fit the same reasoning. All the same, we should not lose sight that there is nothing inevitable about the cooperative equilibria, and unrestrained use of all weapons and tactics, however dreadful, is a possibility whenever nations go to war.

6. ERRORS

The Grim Trigger strategy seems rather, well, grim, since it shuts off all possibility of cooperation in the future, to the disadvantage of both parties. But it is not really that bad, in theory. The Grim Trigger will never be adopted unless it is sufficient to induce both parties to cooperate, so when it is adopted, we will see cooperative behavior, and the sanction of cutting off cooperation will never be invoked. But this runs into some difficulties if we allow for any errors or irrationality at all. Suppose that the players sometimes have "trembling hands:" They always know the best-response strategy but sometimes, by accident, choose the wrong one. Then a single error could lead to a Grim Trigger sanction, and no further cooperation between the players. That really is pretty grim.

Forgiving Trigger Strategies may be more tolerant of error, but even so, errors could have serious consequences. Suppose that both players are playing Tit-for-Tat. One chooses non-cooperation in error, the other retaliates on the following round, and the first retaliates for that on the next round, and so on, with no further mutual cooperation. However, application of the Tit-for-Tat strategy needs not be quite that mechanical. Expected play on rounds $n + 2$ and so on does not influence play on round n, with the Tit-for-Tat strategy, whereas the Grim Trigger gets its power from the expectation of non-cooperation on all future rounds. Thus, a Tit-for-Tat strategy could be modified by some error-correction routines. We might have a "Tit-for-Tat-but-Pardon-Me" strategy: "Retaliate on round $n + 2$ for a non-cooperative play on round $n + 1$ except when I myself have played non-cooperatively by error on round n." (An apology might help, too.) The victim might apply a "Tit-for-Tat-with-Pacific-Overtures," along the lines of "retaliate on round $n + 1$ for a non-cooperative play on round n, but skip the retaliation after four rounds of alternating non-cooperative play." These more complex routines could allow recovery from errors, but might be open to exploitation by a shrewd opponent.

Trigger strategies with errors have not been much studied, however, so there remains a good deal to learn.

7. SUMMARY

As we have seen, repeated play in itself does not solve the problem of social dilemmas. The answer to the paradox seems to be in the fact that people often don't know just how long their relationships will last. When we allow for the fact that there is some probability of continuing the relationship for another play — but the probability is usually less than one — we can introduce "trigger strategies," such as Tit-for-Tat and the Grim Trigger. These trigger strategies can often support cooperative equilibria. (As a dividend, we also give a correct treatment of discounting payoffs to present value in games that continue over time.) When we see cooperation in repeated social dilemmas, it is probably because the game does not have a definite end, but rather there is always some probability of another round of play.

Thus, repeated play does make a difference, but does so primarily when the game continues for an indefinite, not a definite period. Even then, there may be many equilibria. After all, there are infinitely many trigger strategies, and generally there will be equilibria with average payoffs at the cooperative level, the non-cooperative level, and every possibility in between. So even in indefinitely repeated games, cooperative play is not a certainty. But it is a possibility.

Q11. EXERCISES AND DISCUSSION QUESTIONS

Q11.1. Pressing the Shirts

Reanalyze the "Pressing the Shirts" example from Chapter 10 (Section 3) on the assumptions that the probability of Nicholas retiring and not having any more shirts pressed is 0.02 on any round of play and that a dollar today is worth $1.00094 a week from today (which corresponds to a 5% rate of discount for a year). Does the Tit-for-Tat rule work in this case? Explain.

Q11.2. Tourist Trap

Stores in tourist resorts have a reputation for being untrustworthy, offering poor-quality merchandise at excessive prices, which leads to the stereotype of the "tourist trap." Assume that there is some truth to the stereotype — that stores in tourist resorts are at least somewhat more likely to victimize their customers than other stores — and explain why this might be, using concepts from this chapter.

Q11.3. Getcha Hot Dog From Now On

Exercise 2, Chapter 10. Frank and Ernest sell hot-dogs from neighboring vending trucks. Each week they choose between the strategies of pricing their dogs at $1.50 or $2.00, and the payoffs are as shown

Table 11.5. Payoffs to Hot Dog Vendors (Repeats Table 10.4, Exercise 2, Chapter 10).

		Frank	
		$1.50	$2.00
Ernest	$1.50	10,10	18,5
	$2.00	5,18	12,12

in Table 11.5. Suppose Frank changes his mind and decides to continue in the business as long as his health permits. What difference would this make?

Q11.4. The CEO Game

The CEO of Enrob Corp. is very powerful. He administers the employees' pension fund and could convert it for his own profit if he wished. This strategy is "grab" and the CEO's other strategy is "don't grab." The employees' strategies are "work" and "shirk." The payoffs are as shown in Table 11.6.

a. Treat this game as a one-off game and determine its Nash equilibrium or equilibria. Contrast the Nash Equilibrium in this analysis with the cooperative solution.
b. Observers note that the employees of Enrob are very loyal and hardworking and the employees' pension fund is very generous and stable. (It is all invested in Enrob stock, but that stock has been rising consistently.) Explain this, using concepts from this chapter and assuming that $\delta = 1/2$.
c. Finally, the CEO is notified by his CFO that some innovative investment strategies have failed, losing a great deal of money, so that the CEO is certain to be fired at the next meeting of the board of directors. How is this likely to change the equilibrium of the CEO game?

Table 11.6. The CEO Game.

		Employees	
		Work	Shirk
CEO	Grab	7,0	1,1
	Don't grab	5,3	0,4

Table 11.7. A Dog's Dinner.

Payoffs to PG, DD		DD	
		Adulterate	Don't
PG	Adulterate	5,5	1,6
	Don't	6,1	2,2

Q11.5. A Dog's Dinner

Poochgrub, LLC, and Doggydins Partners produce dogfood and are the only competitors in the market of East Dogritania, where the price is controlled so that they cannot compete on price. However, they can adulterate their product with cheap grain. If both do so then their profits are increased, but if just one adulterates, that firm will lose market share and profits. The game in normal form is in Table 11.7. (Payoffs are profits in millions). They make product decisions annually, discount profits at a rate of 15%, and estimate the probability that they will compete again next year at 0.9. Is it likely that they will cooperate in offering adulterated dog food? Explain.

Q11.6. Soda Pop

Fizzypop and Sploosh are competing brands of flavored, colored water. Each can choose between a high, monopoly price and a lower, competitive price. Table 11.8 shows their profits in billions, depending on the price strategies.

Table 11.8. Payoffs for Fizzypop and Sploosh.

		Sploosh	
		High	Low
Fizzypop	High	5,5	1,8
	Low	8,1	2,2

They make pricing decisions annually, discount profits at rate of 15%, and estimate the probability that they will compete again next year at 0.75. Is it likely that they will collude and charge a high price? Explain.

Q11.7. Don't Chicken Out

Buck Backaw is an independent chicken farmer. Verloren Farms is a major packager and wholesaler of chicken meat. Verloren may (or may not) choose to advance some scratch to Buck to cover the cost of hatching, feeding, and raising a load of chickens, expecting Buck to sell them his market-ready chickens at a sub-wholesale price in return. However, when he has his chickens grown, Buck might instead wholesale them himself. (Contract enforcement is a little vague where Buck lives, especially since the local judge is his brother-in-law). Since Buck does not have a wholesaling operation supported by infrastructure and contacts, he will not get as much as Verloren Farms does, but perhaps more than they will pay him.

Here are some data on Buck's operation:

Cost of hatching, raising, and feeding a load of chickens	$8,000
Wholesale value to Verloren Farms	$30,000
Wholesale value to Buck	$12,000
Contractual price to Buck from Verloren	$7,500

a. Analyze this as a game in extensive form, for a single play, using the concept of subgame perfect Nash equilibrium. *Hint*: if Verloren does not advance anything to Buck, then Buck's only choice is to pay his own costs and wholesale the chickens himself.

b. Suppose that the game is played repeatedly with a fixed probability of repetition and Verloren adopts a Tit-for-Tat strategy rule. Is it likely that this will assure a cooperative outcome? Explain.

c. Suppose that the game is played repeatedly with a fixed probability of repetition and Verloren adopts a Grim Trigger strategy rule. Is it likely that this will assure a cooperative outcome? Explain.

d. What do you suggest that Verloren Farms do? What do you suggest that Buck do?

PART IV

Cooperation

CHAPTER 12

Cooperative Games in Coalition Function Form

All of the examples so far have focused on non-cooperative solutions to "games." We recall that there is, in general, no unique answer to the question "what is the rational choice of strategies?" Instead there are at least two possible answers, two possible kinds of "rational" strategies, in non-constant sum games. Often there are more than two "rational solutions," based on different definitions of a "rational solution" to the game. But there are at least two: A "non-cooperative" solution in which each person maximizes his or her own rewards regardless of the results for others, and a "cooperative" solution in which the strategies of the participants are coordinated so as to attain the best result

To best understand this chapter, you need to have studied and understood the material from Chapters 1–4 and 6.

Definition: *Cooperative and Non-cooperative Games and Solutions* — If the participants in a game can make binding commitments to coordinate their strategies then the game is *cooperative,* and otherwise it is *non-cooperative.* The solution with coordinated strategies is a cooperative solution, and the solution without coordination of strategies is a non-cooperative solution.

for the whole group. Of course, "best for the whole group" is a tricky concept — that's one reason why there can be more than two solutions, corresponding to more than one concept of "best for the whole group."

Games in which the participants cannot make commitments to coordinate their strategies are "non-cooperative games." The solution to a "non-cooperative game" is a "non-cooperative solution." In a non-cooperative game, the rational person's problem is to answer the question "What is the rational choice of a strategy when other players will try to choose their best responses to my strategy?"

Conversely, games in which the participants can make commitments to coordinate their strategies are "cooperative games," and the solution to a "cooperative game" is a "cooperative solution." In a cooperative game, the rational person's problem is to answer the question, "What strategy choice will lead to the greatest mutual benefits if we all choose a common, coordinated

HEADS UP!

Here are some concepts we will develop as this chapter goes along:

Cooperative games and solutions: If the participants in a game can make binding commitments to coordinate their strategies then the game is cooperative. The solution with coordinated strategies is a cooperative solution.

Coalition: A group of players who coordinate their strategies is called a coalition.

Side Payment: When part of the payoff is transferred from one member of a coalition to another, so that no member of the coalition needs to be worse off as a result of adopting the coordinated strategy of the coalition, this transfer is called a side payment.

Solution concepts: There are a number of solution concepts for cooperative games. Two are important for this book: The core, which includes only those arrangements that cannot be destabilized by some coalition dropping out and separately coordinating its strategies for the advantage of its members, and the Shapley value, a method of imputing a net payoff to each player based on the player's marginal contribution to each coalition.

strategy?" Let's begin with an example in which this makes a very great difference.

1. A DIVISION OF LABOR GAME

Adam Smith wrote, "The greatest improvement in the productive powers of labour, and the greater part of the skill, dexterity, and judgment with which it is anywhere directed, or applied, seem to have been the effects of the division of labour." Here is a game of division of labor between two workers, Adam and John. They produce a divisible good that we will call "output." There are two methods of production of output, a one-stage technique and a two-stage technique with division of labor. For the two-stage technique, output is produced only if one of the two workers takes the first stage and the other takes the second stage. This is an instance of division of labor. If output is produced by the two-stage method the overall productivity of labor is doubled. However, with the two-stage method, the entire product is possessed by the worker who takes the second stage, and so finishes the work. There are three strategies available to each worker:

1. Use the one-stage technique.
2. Take Step 1 in the two-stage technique.
3. Take Step 2 in the two-stage technique.

Payoffs are the money revenues from selling the output that the worker possesses at the end of the game. The game in normal form is shown in Table 12.1.

It can quickly be verified (for example, by underlining) that the only Nash equilibrium in this game is for each to choose Strategy 1. It also seems clear that this is not a cooperative solution. If the two can arrive at a common agreement, there is the potential to double their output and revenue from 10 to 20. But such an agreement would leave at least one of the two worse off, with zero.

There is of course a solution to this problem, and it is a familiar one, One of the two might say to the other, "I will take Stage 2, you take Stage 1, and I will pay you $w > 5$ from the revenue from the

Table 12.1. The Division of Labor Game in Normal Form.

		John		
		1	2	3
Adam	1	5,5	5,0	5,0
	2	0,5	0,0	0,20
	3	0,5	20,0	0,0

production of 20 units of output." The payment of w is what we commonly call a wage, but it is an instance of a more general kind of payment, known in game theory as a "side payment." A side payment is a money transfer that redistributes the payoffs from a game after the play is complete. Clearly side payments are important in the real world: As we have seen, wages are an instance, as are salaries, interest, dividends, taxes, bribes, and most sales revenues.

So how will this game be played, supposing the contract for a cooperative solution is enforced and so trustworthy? To investigate this, we might construct a diagram shown as Figure 12.1. The diagram shows the net payoff to Adam on the horizontal axis and the payoff to John on the vertical axis. The downward-sloping line shows the payoff to John as a function of the payoff to Adam. In the jargon of economics, it is a payoff-possibility frontier: Payoffs above or to the right of the line would more than exhaust the total payoff to the coalition of Adam and John, and so would be impossible. However, we know that neither Adam nor John will accept less than 5, since each can assure themselves of 5 by choosing strategy 1 instead of agreeing to join the coalition. These constraints are expressed by the dotted lines at payoffs of 5. The segment of the downward sloping line between α and β, highlighted in gray, shows all payoffs that are mutually beneficial, which is the minimum condition for a cooperative solution. That is, it shows all payoffs that assure each player of more than the player can be sure of getting by individual action. In that sense, all are potential cooperative solutions.

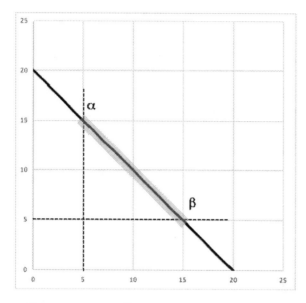

Figure 12.1.   Payoffs in a Division of Labor Game.

This is not very helpful! While we see that side payments will make it possible to arrive at a cooperative solution in this game, there are infinitely many possible solutions. This is essentially what von Neumann and Morgenstern found in their book at the beginning of game theory. While their definition of a solution was more general (and thus more complex) the set of solutions given here is an instance for this particular game. The set of such solutions is called the "solution set." Much of the early research in game theory was devoted to narrowing that wide range of possible solutions.

Much of that research, and much research in cooperative game theory, made use of some strong simplifying assumptions:

1. Members of a coalition may make side payments to redistribute payoffs in any way without costs or other changes that would change the sum total of the payoffs.

2. Increasing money payoffs is the motive for rational decisions.

Since von Neumann and Morgenstern assumed, as most economists do, that people would maximize their utility, the second assumption tells us that transferring money is equivalent to trans-

> **Terminology:**   *Singleton Coalition* — A single player who goes it alone in a cooperative game is called a *singleton coalition.*
>
> *Grand Coalition* — A coalition of all the players in the game is called the *grand coalition.*

ferring utility. Thus, games in this analysis are "transferable utility" (TU) games. Moreover, a TU game can be thought of as having two stages: Play of strategies to produce the largest total payoffs, and distribution of the payoffs. But, from the point of view of game theory, the first stage is trivial: Within a coalition, it is in everybody's interest to choose the strategies that maximize total payoffs. We may leave that for operations research. Thus, we may throw away all the information about strategy choices and identify each coalition with the maximum total value that it can generate. When the game is represented in this way, it is called "a game in coalition function form." With just two players, there are just three coalitions with one or more member: The grand coalition {Adam, John} and the singleton coalitions {Adam} and {John}, the coalition function for the Division of Labor game is rather simple and is shown in Table 12.2. This chapter will discuss games in coalition function form using two widely researched concepts of solution: The core of the game and the Shapley value.

The division of labor game shown here is a two-person game, but division of labor will usually involve many more than two players, and in cooperative game theory we are usually concerned with games of more than two players. In games with more than two players, there are many more possibilities for coalitions. Coalitions can also form in non-cooperative games with three or more players, as we saw in Chapter 6. However, in the absence of some credible commitment, we will see only coalitions that correspond to Nash

Table 12.2. The Division of Labor Game in Coalition Function Form.

Adam, John	20
Adam	5
John	5

equilibria in non-cooperative games. In the rest of this chapter, we will assume that credible commitments can be made, and will see that there is a wider range of possibilities. Here is an example with three players and a fourth party, not a player in the game but an entrepreneur who wants to get some of the players together in a coalition.

2. THE CORE

Jay, the real-estate developer, wants to put together two or more parcels of property in order to develop them jointly. He is considering properties owned by Kaye, Laura, and Mark. Jay wants to propose a deal that will be stable in the sense that none of the three property owners will want to renegotiate with some other property owners, or go it alone. Having studied game theory, Jay recognizes that the property consolidations are coalitions in a cooperative game, and that his offer needs to be a solution of the game. But what does that mean in this case? To find out, Jay will look in detail at the various ways the property owners might combine (or be combined) in coalitions and what the payoffs will be. All of the possible coalitions, and the payoffs to each coalition, on a scale of zero to 10, are shown in Table 12.3.

The first line of Table 12.3 shows a coalition of all three property owners. In game theory terms, this is called the *grand coalition*. On the second, third and fourth lines we see the coalitions of two property owners. On the fifth, sixth and seventh lines of Table 12.3, we see each of the three property owners going it alone as a *singleton coalition* — in other words there is no consolidation of the property

Table 12.3. Payoffs in a Real Estate Coalition (Version 1).

K,L,M	10
K,L	7
L,M	7
K,M	6
K	3
L	3
M	3

at all. A player who goes it alone is called a singleton coalition. The payoffs to the singleton coalitions on this line are what an economist would call the opportunity costs of the property owners when they enter into any coalition. They could also be called the property owner's *outside option*.

We see that the grand coalition offers the three property owners a chance to do better than they can do separately. In fact, there is a surplus of 1 to divide among the three property owners. But how should it be distributed? Jay understands that the only payoffs that will work are for L to get 4, and the other two 3. Here is the reason. First, the payoffs to K and L together have to be at least 7, since otherwise they would be able to do better forming the coalition on the second line and leaving M out. Thus, if y_K and y_L are the payoffs to K and L we must have

$$y_K + y_L \geq 7$$

By similar reasoning,

$$y_K + y_M \geq 6$$

$$y_L + y_M \geq 7$$

If we add these three inequalities, we observe that

$$2(y_K + y_L + y_M) \geq 20$$

that is,

$$y_K + y_L + y_M \geq 10$$

So, we see that the grand coalition can generate just enough payoffs to keep all of the two-person coalitions happy. This is one condition for the stability of a solution to a three-person game. We must also have

$$y_K \geq 3$$

$$y_L \geq 3$$

$$y_M \geq 3$$

$$y_K + y_L + y_M \geq 9$$

and we already know that that condition is fulfilled. Suppose that $y_M > 3$. Then $y_K + y_L < 7$, and so K and L will drop out and form a separate coalition for 7. Thus, $y_M = 3$ is the only payoff to M that will satisfy all the conditions. By similar reasoning, $y_L = 3$. Thus, y_K must be 4. Thus $y_K = 3$, $y_L = 4$, $y_M = 3$ is said to be in *the core* of this cooperative game.

A payment schedule like $y_K = 3$, $y_L = 4$, $y_M = 3$ is called an *imputation* or an *allocation* of the value of the grand coalition. In general, the core of a cooperative game includes any and all imputations that will reward each player sufficiently that no group of players will want to drop out of the grand coalition and go it alone. This set of possible solutions is called *the core* of this cooperative game. In this case, there is only one imputation that meets this standard.

Let us see how the example changes if we make a small change in the payoffs. In his next deal, Jay is dealing with properties owned

Table 12.4. Payoffs in a Real Estate Coalition (Version 2).

G,H,I	10
G,H	7
G,I	6
H,I	6
G	3
H	3
I	3

by Gina, Harry and Inez. The coalition values are as shown in Table 12.4.

Once again, each of the (potential) two-person and singleton coalitions will have to be paid enough to keep them content with the grand coalition. The inequalities expressing this will be

(1^*) $y_G + y_{II} \geq 7$
(2^*) $y_G + y_I \geq 6$
(3^*) $y_{II} + y_I \geq 6$

and, as before, each player must get at least three, and otherwise may choose to remain as a singleton coalition. As before, combining $1^*, 2^*, 3^*$, we have

(4^*) $2(y_G + y_{II} + y_I) \geq 19$

that is,

(5^*) $y_G + y_{II} + y_I \geq 9.5$

We see that the grand coalition generates a surplus of 0.5 over the minimum necessary to keep it stable. In fact, there are many ways the payoffs can be distributed that will satisfy these conditions.

For example, if payments are $y_G = 4$, $y_{II} = 3$, $y_I = 3$, the conditions are all satisfied. If payments are $y_G = 3$, $y_{II} = 4$, $y_I = 3$, the conditions are fulfilled, and they are fulfilled by any payments that give G at least 3, H at least 3, and give exactly 3 to I. Notice that, for this game, the solution is not unique. Rather, again, we have a family of solutions, but because each of the two-person groups must have at least

Terminology: *The Core of a Cooperative Game* — The *core* of a cooperative game consists of all imputations (if there are any) that are stable in the sense that there is no individual or group that can improve their payoffs (including side payments) by dropping out or reorganizing to form a new or separate coalition.

as much as they could make separately, the core is more limited than the solution set. In this case the requirement that I gets no more than 3 reflects the fact that G,H must get enough to make it worth their while to remain in the grand coalition.

Now, let us consider one more example along the same lines. In a new real estate consolidation, Jay is now dealing with Noreen, Pete and Quincy {N,P, and Q}. The payoffs to coalitions among these three are shown in Table 12.5.

Table 12.5. Payoffs in a Real Estate Coalition (Version 3).

N,P,Q	10
N,P	7
N,Q	7
P,Q	7
N	3
P	3
Q	3

As before, no-one needs to settle for less than 3, and the grand coalition produces enough value to pay each individual 3 or more. For the two-person coalitions, however, we must have

$(1^{**}.)$ $\quad y_N + y_P \geq 7$

$(2^{**}.)$ $\quad y_N + y_Q \geq 7$

$(3^{**}.)$ $\quad y_P + y_Q \geq 7$

$(4^{**}.)$ $\quad 2(y_N + y_P + y_Q) \geq 21$

$(5^{**}.)$ $\quad y_N + y_P + y_Q \geq 10.5$

But the value of the grand coalition is not enough to pay them all. Therefore, the grand coalition is not stable, since one or another of the two-person coalitions will drop out. But a two-person coalition will not be stable either. Suppose Noreen and Pete form a coalition and divide 7 equally as 3.5 each, leaving Quincy in a singleton coalition with a payoff of 3. Then Quincy could approach either Noreen or Pete and say, "Let's form a two-person coalition and I will settle for 3.25, so that you can have 3.75. Then we will both be better off." In this game, there are no imputations that satisfy the requirements of the core, and therefore we say that the core is *null*, or empty. In this third game, the core is (in other words) a set without any members.

The core is a widely used approach to the solution of cooperative games. However, a recognized shortcoming of the core, as a solution, is the fact that (for some games) there may be many imputations in the core, and on the other hand (for other games) there may be none at all. As we have seen, small differences in coalition values can lead to these different results.

These examples illustrate several simplifying assumptions commonly used in the analysis of cooperative games. First, Tables 12.3, 12.4, and 12.5 are known as the *coalition functions* or *characteristic functions* for the three games. The coalition function is a listing of all possible coalitions with the values of each coalition.

Second, we have not said very much about the specific strategies of the coalitions, once they get together: Whether they are building a mall or new houses or an industrial park. We have limited our

A Closer Look: Who is "in the Game?"

In these examples, we have not treated Jay, the property developer, as a player in the game. Of course, it would be more complete to treat it as a four-person game, with Jay as one of the players — but because Jay's objective is to put together a stable coalition, we wouldn't gain much beside completeness by treating this as a four-person game. Indeed, one way to think about cooperative solution concepts is to think of them as blueprints for arbitrators, facilitators, and deal-makers. There are other solution concepts for cooperative games, with more emphasis on fairness, that lend themselves especially to being thought of as arbitration schemes. The Shapley Value, considered in a later section, is one instance.

analysis in this game to the "coalitional form," which associates a payoff with each coalition without saying just how the coalition coordinates strategies to get the payoffs. This is a very common approach in cooperative game theory.

Third, the coalition function or characteristic function for the game, in these examples, is based on the idea that the values that the various coalitions can produce depend only on the members of that coalition,[1] and not on how the other members sort themselves out into coalitions. As a matter of logic, the value of a singleton in a three-person game could logically be different when the other two players form a coalition than when they play as singletons. For example, if they form a coalition, they might gain market power at the expense of the singleton. But when we expressed the game in

---

[1] This is not an arbitrary assumption. Von Neumann and Morgenstern gave an argument for it and most cooperative game theorists have been persuaded by the argument. For a history and critique of this idea see Roger A. McCain, *Game Theory and Public Policy* (Elgar Publishing Company, 2015, Chapter 3).

coalition function form, we ruled out that possibility.

Fourth, notice that these games are *superadditive*. That is, whenever two of the coalitions merge, the value of the merged coalition is at least as great as (and sometimes greater than) the sum of the values of the coalitions before they merged. That is why we only need to consider the grand coalition: There is nothing to lose by getting everyone together to form the grand coalition, so we can be confident that if any coalitions are formed, the grand coalition will be the one formed.

Definition: *Transferable Utility* — A game is said to have *transferable utility* if the subjective payoffs are closely enough correlated with money payoffs so that transfers of money can be used to adjust the payoffs within a coalition. Side payments will always be possible in a game with TU but may not be possible in a game without TU.

Fifth, in this example and in the previous one, the payoffs are valued in money and transfers of money are used to compensate the players for giving up the alternatives they might obtain in singleton coalitions (or in other multiperson coalitions). These money payments are called "side payments." In a world of non-cooperative games, there could be no buying and selling or hiring — because buying and selling and hiring always means that an enforceable agreement is made and on the basis of the agreement, a payment changes hands. In game theory, the payment is called a "side payment." This terminology comes from gambling games. In a poker game, if one player paid another to bluff or to fold, that would be cheating — the rules of poker do not allow payments outside the game, that is, side payments. But buying and selling is not poker, and side payments are very much part of games of exchange.

3. DOMINANCE

In the previous section, the core has been analyzed as a set of imputations of net payoffs to the members of a coalition that are *stable* in the sense that there is no alternative coalition and imputation that makes the members of the new coalition, as a group, better off. Another way to discuss this makes use of a concept of *dominance*. (Remember, the words "dominant" and "dominance" are used in several different ways in game theory. This is a new sense for these words.)

Think again of the second version of the Real Estate Coalition. Suppose that Jay were to propose imputations of $y_G = 3$, $y_H = 3$, $y_I = 4$. But by forming coalition {G,H}, Gina and Harry can realize total payments of 7, and they can divide 7 between them in such a way as to make both better

> **Notation:** *Brackets* — Following a convention from mathematical set theory, a coalition may be denoted by a list of its members set off with {} brackets, as {G,H,I} or {G,H}, etc. in the text.

off. Then we say that the grand coalition and imputation $y_G = 3$, $y_H = 3$, $y_I = 4$ is *dominated via {G,H}*. Similarly, suppose H and I form the coalition {G,I} and realize a total value of 6. However, they divide it, the grand coalition can offer $y_G = 3.5$, $y_H = 3.5$, $y_I = 3$. This makes Gina better off, so she will support the change, and Inez being no worse off, she will not oppose it. In the jargon of economics, imputation $y_G = 3.5$, $y_H = 3.5$, $y_I = 3$ is a Pareto-improvement over $y_G = 3$, $y_H = 3$, $y_I = 3$. Thus, {G,I} with $y_G = 3$, $y_I = 3$ is *dominated* by the grand coalition with $y_G = 3.5$, $y_H = 3.5$, $y_I = 3$. On the other hand, $y_G = 3.5$, $y_H = 3.5$, $y_I = 3$ is *undominated*. Any movement to another imputation in the grand coalition, or any shift to another coalition will make someone in the coalition worse off, and that person will veto the shift. When we say that imputations in the core are stable, we are saying that they are stable in this specific sense: They are undominated.

In the first version of the real estate game, the grand coalition with $y_K = 3$, $y_L = 4$, $y_M = 3$ is undominated. Any shift to another imputation with the grand coalition will make someone worse off, and no smaller coalition can make either of its members better off without making another worse off. In version 3, there can be no imputation in the grand coalition that is not dominated by some two-person coalition, but each two-person coalition will be dominated by another two-person coalition, so there is no undominated coalition in that game. The core is null.

Let us apply this reasoning to the Division of Labor Game. For this game the only rationality constraints are

(1) Payment to John ≥ 5
(2) Payment to Adam ≥ 5

Since the grand coalition produces a value of 20, the surplus of $20 - 5 - 5 = 10$ can be divided between the two in any way, there are infinitely many imputations in the core of this game.

4. SHAPLEY VALUE

The core is the solution concept most often used in economic applications of cooperative game theory. There are a number of other solution concepts. The most important, apart from the core and especially in applications other than economics, is the Shapley value, and that is the only other solution concept we will consider in this book. The Shapley value is also applied to superadditive games in coalition function form, but within those limits it has the advantage of uniqueness: There is always exactly one Shapley value.

The Shapley value is based on a concept of a marginal contribution. Consider the grand coalition in Table 12.3. Suppose it is formed by adding K, L, and M in that order, and each gets what it adds to the value of the coalition. Thus, since K "merges with an empty coalition to form" a singleton coalition, the value K adds is 3. The value of {K, L} is 7, so L adds 4 to the value of the coalition. Since the value of {K, L, M} is 10, M adds 3 to the value of the coalition. Then the payoffs would be 3, 4, 3 for K, L, M. But the order K, L, M is arbitrary. K could object that he ought to be last rather than first — then he would add 4. Accordingly, the Shapley values are computed by averaging over all possible orders in which the players

Table 12.6. Shapley Values for the
Real Estate Coalition, Version 1.

	K	L	M	Checksum
K,L,M	3	4	3	10
K,M,L	3	4	3	10
L,K,M	4	3	3	10
L,M,K	3	3	4	10
M,K,L	3	4	3	10
M,L,K	3	4	3	10
Average	$3\frac{1}{6}$	$3\frac{2}{3}$	$3\frac{1}{6}$	10

might be added. The Shapley values for this game are computed in
Table 12.6.

To interpret the table, consider, for example, the third line. The
order of joining the coalition is L, K, M. Since L comes first, his
marginal contribution is his singleton value, 3; and this is shown in
the third column. Second in the coalition is K, and they form the
coalition {K, L} with a value of 7, so K's marginal contribution is
$7 - 3 = 4$, shown in the second column. Then M comes last, forming
the grand coalition with a value of 10, so M's marginal contribution
is $10 - 7 = 3$, shown in the fourth column.

For this game, the Shapley values are $3\frac{1}{6}$, $3\frac{2}{3}$, $3\frac{1}{6}$. We see that the
Shapley value disagrees with the unique core allocation in this case.
We saw that the first version of the Real Estate Development Game
has a core consisting of just one imputation: 3, 4, 3. In this game, the
core and the Shapley values disagree. In particular, suppose that the
grand coalition forms with payouts according to the Shapley values.
Then the coalition {K, L} gets $6\frac{5}{6}$, and by dropping out of the grand
coalition and forming their own separate coalition {K, L} can get 7.
Similarly, the coalition {L, M} gets $6\frac{5}{6}$, but can drop out as a separate
coalition and get 7.

When we construct a similar table for the second version of the game, we find that the Shapley values are 3.5, 3.5, and 3. (*Exercise:* Construct the table and demonstrate this.) In this game the Shapley solution is within the core, but as we have seen, that is not always so. For the third version of the real estate game, we find that the Shapley values are 3.33, 3.33, 3.33. (*Exercise:* Construct the table and demonstrate this). There is an easy way to get this result, though. This game is *symmetrical* — if we can interchange two players without changing the coalition values, they get the same payoffs. Shapley showed that if the game is symmetrical, then the Shapley value solution is also symmetrical. Since any of the three players can be interchanged in the third version of the real estate game, the payoffs must be divided equally among them. Symmetry is one of the properties that defines the Shapley value. Another condition is the fact that there is exactly one Shapley value; and there are some technical conditions. Shapley showed that the value is the only solution that has these "nice" properties. However, there are other "nice" properties that it lacks. Notice also that in version 3 of the Real Estate Game, the core is null, so that the Shapley value supplies a solution when the core does not.

Recall the Division of Labor Game from Section 1 of this chapter. This is another symmetrical game. We see that either of the two workers, by "merging with a null coalition," adds 5 to the singleton coalition, and then the other player adds 15 by forming the grand coalition, so the marginal contributions of a worker for either order are 5 and 15, and the Shapley value for this game is even division, 10 each.

There is an algebraic formula for the Shapley value. That's especially helpful in games with more than just a few players. The value for a player i is

$$\phi_i(v) = \sum_{\substack{S \subseteq N \\ i \in S}} \gamma_n(s)\big(v(S) - v(S - \{i\})\big) \qquad (12.4.1)$$

where v is the game in coalition function form, S is any coalition with s members, and

$$\gamma_n(s) = \frac{(s-1)!(n-1)!}{n!}.$$

(12.4.2)

In this formula for the weights, n is the number of players in the game as a whole and ! denotes the factorial operator. These formulas were the way that Shapley expressed his solution: The tables calculating the average marginal contributions are a simplification of the formula readily applicable to games with only a few players. Shapley had made a series of assumptions about the properties a rational solution should have, and deduced that the only values that would have those properties is the one computed by formulae 12.4.1 and 12.4.2. Symmetry was one of those assumptions. Others were that the payoffs exhaust the value of the coalition, that a player who never adds anything to the coalition gets a payoff of zero, and that if two games are played the total payoff would be the sum of the payoffs to the two games taken separately. This assumption enabled Shapley to break a game into its component parts and plays a key role in narrowing the values down to one particular set. For this book, to keep things simple, we will consider only 3-person games and use the simple tabular form to compute Shapley values.

Defenders of the Shapley value would say that the core is not really a cooperative solution — that a cooperative solution is based on a binding commitment, and having made a binding commitment to the three-way merger, landowners K and L, for example, could not secede without violating the commitment. But defenders of the core could point out, in turn, that such a commitment would be shortsighted — that K and L, anticipating that their payoff would be made according to the Shapley value, would refuse to commit themselves to the merger in the first place. What we can say is that a solution in the core has the "nice" property that it is stable against the temptation of groups within the grand coalition to form separate coalitions, and this is a "nice" property that the Shapley value does not have. This illustrates the reason why there are several solution concepts in cooperative game theory and the difficulty of the theory. Different solutions have different "nice" properties, and neither of these solution concepts has all the "nice" properties that we might like.

5. FURTHER EXAMPLES

Let us consider two more examples of three-person cooperative games.

Three small Philadelphia colleges are considering merging. Able Tech, Beta College, and the Charlie School are strong, respectively, in engineering, business, and design arts. The payoffs (in millions) to the various merger possibilities are shown in Table 12.7. They have good reason to consider merging, since two are losing money and the third is barely breaking even! We want to use the theory of the core to determine whether this is a viable merger and how the benefits of the merger are likely to be divided among the three players in this cooperative game.

First, we should verify that this game is superadditive. When we check each possible merger, we find that it is indeed superadditive. Accordingly, we will focus on the grand coalition of all three colleges. For an imputation in the core, it will be necessary that

$$y_A + y_B \geq 8 \qquad (12.5.1)$$

Table 12.7. Payoffs for Merged Colleges.

A, B, C	14
A, B	8
A, C	6
B, C	4
A	0
B	-2
C	-4

$$y_A + y_C \geq 6 \qquad (12.5.2)$$

$$y_B + y_C \geq 4 \qquad (12.5.3)$$

$$y_A \geq 0 \qquad (12.5.4)$$

$$y_B \geq -2 \qquad (12.5.5)$$

$$y_C \geq -4 \qquad (12.5.6)$$

Adding 12.5.1–3,

$$2(y_A + y_B + y_C) \geq 18 \qquad (12.5.7)$$

$$y_A + y_B + y_C \geq 9 \qquad (12.5.8)$$

Since 14 is the value of the grand coalition, there is more than enough to compensate all the two-person coalitions for remaining in the grand coalition. As a result, there are many imputations in the core of this game. For example, value assignments 5,4,5, 6,5,3, and 7,6,1 are all in the core of the college merger game. All three schools can benefit from joining in the merger. Indeed, as we see, after the merger they are quite profitable. Perhaps they will share with their students by granting more scholarships.

Let us explore the Shapley values for the game of merging colleges. Table 12.8 shows the computations. For this game, we see that the Shapley values for colleges *A*, *B*, and *C* are $6\frac{2}{3}$, $4\frac{1}{3}$, and 3. This is one of the imputations in the core of the College Merger Game. (*Exercise*: Confirm this).

In this case, then, the Shapley value can settle the question as to which of the many imputations in the core of the College Merger game might be chosen. But we have seen that this will not always be true, since the Shapley value may not be in the core, depending on the particular game.

Recall Example 6 from Chapter 6, a model of global warming as a game among three great regions of the world, north, south and

Table 12.8. Shapley Values for the Game of Merging Colleges.

	A	B	C
A,B,C	4	6	7
A,C,B	4	3	10
B,A,C	7	3	7
B,C,A	10	3	4
C,A,B	5	10	2
C,B,A	10	5	2
Average	$6\frac{2}{3}$	$4\frac{1}{3}$	3

east. We recall that, when this is treated as a non-cooperative game, it has a dominant strategy solution at which all of the regions pollute and thus contribute to global warming, and the result is inefficient and very unequal. But that inequality raises the question whether the efficient outcome, which is still very unequal, is really a cooperative solution. In this chapters we have seen that cooperative imputations may be unequal, using either the core or the Shapley value as our criterion of cooperative solution. Thus, it is appropriate to revisit this example to see what the concepts of the core and of the Shapley value tells us about it. Table 12.9 reproduces Table 6.6, in Chapter 6.

Using the payoffs for the different cells, we may determine the total payoff to each of the coalitions. Suppose, for example, that north and east form a coalition. By coordinating their strategies — specifically, if they both choose "depollute" or north chooses "depollute," and east chooses "pollute" — they can assure themselves of at least a total payoff of $25 + 10 = 23 + 12 = 35$. They can do better still if south chooses "depollute," but the coalition of north and east has no control over the decision of south, so their coalition payoff is 35, which they can be sure of. Similarly, if east and south

Table 12.9. Strategies and Payoffs in the Global Warming Game (Duplicates Table 6.6).

		South			
		Pollute		Depollute	
		East		East	
		Pollute	Depollute	Pollute	Depollute
North	Pollute	24,10,2	26,8,4	26,12,1	28,10,3
	Depollute	23,12,4	25,10,6	25,14,3	27,12,5

Table 12.10. The Global Warming Game in Coalition Function Form.

North, South, East	44
North, East	35
North, South	28
East, South	13
North	24
East	10
South	2

form a coalition, then they can assure themselves of a total payoff of $10 + 3 = 12 + 1$ provided that either both of them depollute or east depollutes and south pollutes. Again, they would do better if north depollutes, but as north is not part of their coalition, they cannot expect that. The coalition values, calculated in that way, are shown in Table 12.10.

Represent the imputation to north by N, to south by S, and to east by E. To investigate the core, we first list the two-member rationality constraints, and the checksum:

$$N + E \geq 35 \qquad (12.5.9)$$

$$N + S \geq 28 \tag{12.5.10}$$

$$E + S \geq 13 \tag{12.5.11}$$

$$2(N + E + S) \geq 76 \tag{12.5.12}$$

$$N + E + S \geq 38 \tag{12.5.13}$$

First, we notice that since the value of the grand coalition is 44 > 38, the grand coalition can make sufficient payouts that it is rational for each two-person grouping to remain within the coalition. Indeed, there is a continuum of imputations within the core, since the surplus of $6 = 44 - 38$ can be distributed in infinitely many different ways. Could an equal distribution of 14 2/3 be within the core? The answer is no, and for two reasons. First, the total distribution to the north and east, 29 1/3, does not satisfy rationality constraint 12.5.9. Second, the solution must also satisfy individual rationality constraints,

$$N \geq 24 \tag{12.5.14}$$

$$E \geq 10 \tag{12.5.15}$$

$$S \geq 2 \tag{12.5.16}$$

Constraint 12.5.14. is also violated. A distribution of 24, 10, 10 would be within the core, and any more equalized distribution would be outside the core.

Alternatively, we might apply the Shapley value. The calculation is shown in Table 12.11.

First, we can observe that for this game, the Shapley value is within the core, since

$$39\tfrac{1}{6} > 35 \tag{12.5.17}$$

$$31\tfrac{1}{6} > 28 \tag{12.5.18}$$

$$17\tfrac{1}{6} > 13 \tag{12.5.19}$$

Table 12.11. Shapley Values for the Global Warming Game.

	N	E	S	Checksum
N,E,S	24	11	9	44
N,S,E	24	16	4	44
E,N,S	25	10	9	44
E,S,N	31	10	3	44
S,N,E	26	16	2	44
S,E,N	31	11	2	44
total	161	74	2	
Average	$26\frac{5}{6}$	$12\frac{1}{3}$	$4\frac{5}{6}$	44

The individual rationality constraints are also satisfied.

These solutions assume that there might be costless and routine transfers of payoffs among the regions of the world. If there are no transfers, and all choose cooperatively to depollute, then, from the lower right cell of Table 12.9, the imputations would be 27, 12, 5. These payoffs are also within the core. (*Exercise:* check this.) However, they disagree with the Shapley value, which would require the north and the south each to transfer 1/6 to the east. If, then, we understand the core as the criterion for a cooperative solution, then the payoffs 27, 12, 5 are one of very many instances of a cooperative solution. If, on the other hand, we interpret the Shapley value as the criterion for a cooperative solution, then a cooperative payoff cannot be attained without some (modest) transfers from both the rich north and the poor south to the middle-class east.

But negotiations for an international agreement to prevent further global warming have not included any discussion of interregional or international transfers, and indeed, international transfers of even this scale are hardly to be expected on the basis of past experience. Perhaps interregional transfers are not a reasonable expectation. If not, then the core and the Shapley value are not the appropriate concepts of a cooperative solution for this game, based as they are on the "TU" assumption. Accordingly, we will again revisit this game in the next chapter when we explore cooperative games without transfers.

6. SUMMARY

A cooperative game is a game in which the participants can make commitments to coordinate their choices of strategies. There are several concepts of the solution of a cooperative game. Two are considered in this chapter. One, the core, consists of all imputations of the value of the grand coalition that are undominated — meaning that there are no possibilities for people to make themselves better off by leaving the coalitions they are in and associating with some other coalition. This requirement limits both the coalitions and the range of side payments in the core. Thus, the core will be encompassed in the solution set — will be all or part of the solution set or may be empty. But there may be many imputations in the core, or the core may be null. The second solution concept is the Shapley value. The Shapley value assigns a payoff to each individual in the grand coalition. The Shapley value is based on the marginal contribution of each player to the value of each coalition. There is always exactly one Shapley value assignment for a TU game in coalition function form, but there may be some groups that can do better than the Shapley value by forming separate coalitions.

Even though neither the core nor the Shapley value provides a final answer for all cooperative games, they are important additions to our kit of tools for all games in which players can make commitments and coordinate their strategies.

Q12. EXERCISES AND DISCUSSION QUESTIONS

Q12.1. A Business Partnership

Jaye, Kaye and Laura are considering forming a web design company. Jaye is a very skilled programmer, Kaye a designer, and Laura

Table 12.12. Payoffs to Partners.

J,K,L	50
J,K	25
K,L	20
J,L	30
J	15
K	10
L	5

a successful sales person. The value they can create in various coalitions is shown in Table 12.12.

Using the theory of the core of a cooperative game, answer the following questions:

a. What coalitions, if any, are likely to form?
b. Why?
c. It is proposed that the members of a coalition be paid equally. Will this be a stable arrangement?
d. Compute the Shapley values for this problem. Are they in the core?

Q12.2. Choosing Information Systems

Here is a two-person game with a technological twist: Choosing an information system. For this example, the players will be a company considering the choice of a new internal e-mail or intranet system, and a supplier who is considering producing it. The two choices are to install a technically advanced system or a more proven system with less functionality. We'll assume that the more advanced system really does supply a lot more functionality, so that the total payoffs to the two players are as shown in Table 12.13. This is a TU game.

Table 12.13. Payoffs in the IT game.

		User		
		Advanced	Proven	No deal
Supplier	Advanced	−50,90	0,0	0,0
	Proven	0,0	−30,40	0,0
	No deal	0,0	0,0	0,0

Table 12.14. Hospital Costs.

{N,S,C}	225
{N,S}	200
{N,C}	235
{S,C}	260
{N}	100
{S}	125
{C}	150

Supposing that there is a cooperative agreement, side payments will be required. What are the limits on these side payments? Determine the core and the Shapley Values for this game.

Q12.3. Hospital Merger

In West City, three hospitals are considering merger as a "health system." In doing so they hope the reduce their total costs. Table 12.14 shows the total costs for Northern Hospital, Southern Health Facility, and Central Hospital and for all possible coalitions among them in millions. *Hint:* Other things equal, coalition values are inversely related to costs.

Describe the core of this game.

Q12.4. Mallard, Widgeon and Pintail

Mallard, Widgeon and Pintail (M, W, and P) are firms in the widget industry, and are considering prospects for mergers. They are minor players in the industry so there is no danger of government opposition to the merger. Their experts all agree that the profit prospects are as shown in the following coalition function (see Table 12.15).

Use cooperative game theory to analyze this example. On the basis of the theory of the core, what mergers would you expect to see occur? Why? They are considering distributing the shares in a merged corporation and thus the profits on the basis of the Shapley

Table 12.15. Profits.

MWP	50
MW	30
MP	30
WP	20
M	8
W	7
P	6

Table 12.16. Patent Values.

{A, X, G}	10
{A, X}	7
{A, G}	7
{X, G}	5
{A}	2
{X}	2
{G}	2

values. Calculate the Shapley values and comment. Which company will profit most? Why?

Q12.5. A Patent Consortium

A Corp, X Llc and G Partners are technological companies each of which holds some key patents on laser-printing of musical instruments, and they are considering a patent consortium to share their patent rights in further development and production. Table 12.16 is the coalition function for their 3-person cooperative game.

a. Analyze this problem in terms of the concept of the core.
b. Compute the Shapley Value.

CHAPTER 13

Cooperative Games Without Coalition Functions

The previous chapter surveyed the most widely studied models of solution of transferable utility cooperative games in coalition function form: The core and the Shapley Value. But those methods rely on a number of simplifying assumptions, so that the game can be expressed in coalition function form. In this chapter, we consider two generalizations, one in case the value of a coalition may depend on decisions made in other coalitions, and cases in which there are no side payments, that is, non-transferable utility (NTU) games.

1. A PUBLIC GOOD

In economics, provision of public goods is a problem in a competitive (non-cooperative) market system. A *public good* is a good or service with two properties: First, the cost of supplying one more person with the good is zero; that is, it is *non-rivialrous*. Second, there is no practical way to limit the availability of the good to people who pay for it. In short, the benefit of the public good is equally available to everyone; that is, it is *non-exclusive*. Examples are national defense, universal protection of property rights and enforcement of contracts, a lighthouse or similar information service, commercial-free broadcasting, and global satellite positioning service.

Let us explore this kind of problem as a three-person cooperative game with transferable utility. We suppose that each of the three players, A, B, and C again, begin the game with wealth of 5 each.

Each player can choose between two strategies: Producing at most one unit of a public good at a cost of 3 units of wealth, or not producing. Suppose X units of the public good are produced by the other agents. (The variable X could be zero, one, or two.) Then the individual payoff if the player does not produce the public good is $5 + 2X$, and the payoff is $2 + 2(X + 1) = 4 + 2X$ if he

Definition: *Holdout* — When a group of players form a coalition to take action that enhances the payoffs of all players in the game, but some players nevertheless refuse to take part or to share the costs, the players who do not take part are sometimes called *holdouts*, and the resulting inefficiencies are the *holdout problem*.

does produce it. A little reflection will make it clear that not producing the public good is a dominant strategy. Producing a public good is a social dilemma. If all the players operate independently, none will produce the public good, and the payoff to each will be 5.

Suppose, instead, that the grand coalition forms. It can choose to produce 0, 1, 2 or 3 units of the public good, and its best choice is to produce 3 units, since in that case each player gets a payoff of $2 + 2 \times 3 = 8$, for a total payoff of 24, and that is the best payoff that the grand coalition can obtain.

Now, suppose that the two-person coalition {A, B} is formed. This coalition can produce 0, 1, or 2 units of the public good. (Recall, each member can produce no more than one unit of the public good). If the two-person coalition produces 2 units, A and B each get a payoff of $2 + 2 \times 2 = 6$, for a total payoff of 12, which is the best that the two-person coalition can obtain. However, if {A, B} produce 2 units of the public good, and {C} produces nothing, then C's payoff is $5 + 2 \times 2 = 9$. Moreover, this is a Nash equilibrium in play between the two-person coalition and the singleton coalition. In this case, {C} is a *holdout*, and the coalition {A, B}, by producing the public good for its own benefit, is also creating a positive externality to {C}. The simplifying assumptions of the theory of games in coalition function form exclude holdout behavior such as this. One argument for these assumptions is that this example is not *really* or purely cooperative, since the relation between {A, B} and {C} is non-cooperative; Nash equilibrium is a non-cooperative concept. This is right, of course, but holdout behavior and externalities seem to be pretty common in the real world.

Suppose we modify the coalition function to allow for holdout behavior and externalities. As we have seen in this case, the value of the singleton coalition {C} is 5 if the other two players do not form a coalition, but 9 if they do. Now, in terms of mathematical set theory, {{A}, {B}, {C}} and {{A, B}, {C}} are different *partitions* of the grand coalition {A, B, C}, so we can say that the value of a coalition depends on the partition of which it is a part. The partitions {{A}, {B}, {C}} and {{A, B}, {C}} are also called *coalition structures* and another way of saying the same thing is to say that the value of a coalition depends on the coalition structure as a whole. Thus, we could express the coalition values for a case like this as a *partition function*, and the partition function for the public goods game is shown as Table 13.1. The table is read as follows: In the left column, the partition is shown with each coalition listed with each coalition shown, as usual, as lists of members in {} brackets and the values for the coalitions listed in the same order in the right column.

Unfortunately, relatively little is known about solutions to games in partition function form. In principle, we can apply some of the

Table 13.1. A Partition Function for the Public Goods Game.

Partition	Values
{A,B,C}	24
{A,B},{C}	12,9
{A,C},{B}	12,9
{A},{B,C}	9,12
{A},{B},{C}	5,5,5

same criteria for solutions, though they have to be adapted. For example, in Table 13.1, suppose that the grand coalition {A, B, C} forms with equal payouts of 8 to the three players. Any one of them — {C}, for example — might choose to drop out of the grand coalition. What payoff could he expect? It could be 9 if {A, B} continue as a coalition and produce the public good, or 5, if they do not, and if it is 9 then {C} would be better off to secede, but if it is 5, worse off. Now, A

Definition: *Partition Function Form* — When the value of a coalition depends on the other coalitions that form, we say that it depends on the *partition*, or the *coalition structure*, of the game. A function that assigns a value to each coalition depending on the partition is the *partition function*, and a cooperative game represented in this way is represented in *partition function form*.

and B are themselves better off to continue to produce the public good as a coalition, and if {C} anticipates that decision, {C} is better off to secede. In order to discourage that, C would have to be paid 9. That is, the grand coalition with any imputation is dominated via {C} But the same is true of the other two players, and the grand coalition cannot generate enough value to pay each of them 9. So, the grand coalition is not stable, and indeed — on this reasoning — the public goods game has no stable solution. It is an empty "core" game.

But game theorists who adopt the coalition function approach would not reason in that way. Their argument would be that (1) everyone can see that the grand coalition will eventually form, so each would try to enhance their own bargaining power in the grand coalition; (2) the way to do this is for {A, B} to punish {C} by producing zero (at whatever cost to themselves) leaving {C} with 5; and (3) Then {C} does not dominate {A, B, C} and the core includes a continuum of imputations for the grand coalition.

Both of these arguments apply the idea of coalition dominance to a game in partition function form, but they apply it in different ways. Would it indeed be rational for {A, B} to give up a payoff of 12 for a payoff of 5 to punish {C} for dropping out of the coalition? Nash thought it would not, but that lead him to non-cooperative game theory. To punish in this way seems spiteful, but in this case, spitefulness results in the superior cooperative outcome.[1] And it seems that human beings are often spiteful. However, it seems on the face of it that holdout behavior does sometimes create problems, so that the holdout analysis may be the more pragmatically useful approach.

We see that the reasoning about dominance that leads to the theory of the core for games in coalition function form can be applied for games that are not expressed in coalition function form, but that there may be some ambiguity about the result.

2. NON-TRANSFERABLE UTILITY

The theory of games in coalition function form has this great advantage: It puts side payments at the center of the picture. That's good because side payments are important in the actual world. However, side payments may not be possible or practicable in some games. For the Global Warming Game, as we have seen, side payments may be out of the question. Side payments might be costly or rejected as corrupt. For some games, the objectives of the players may simply not be expressed in numerical terms nor in terms of anything transferable like money. Recall the Spanish Rebellion game from Chapter 1, the first example in the book. The outcomes expressed in Figure 1.2 and Table 2.2 are qualitatively described as, for example, "Hirtuleius wins big" or "Good chance for Pius." In Table 2.3, for convenience, arbitrary payoff numbers were supplied, and in many other examples in the book the numerical payoffs have been arbitrary indicators of the preferences of the players in the game.

---

[1] Note Natalie Angier, "*Spite Is Good. Spite Works.*" In the *New York Times*, April 1, 2014, Section D, p. 1, accessible at https://www.nytimes.com/2014/04/01/science/spite-is-good-spite-works.html on December 15, 2021.

There is a modest literature on cooperative solutions for games without side payments and in which there may be no numerical payoffs. It is sometimes called implementation theory and in place of the coalition function it relies on an "effectivity function." We begin with a list of players, a list of strategies, a list of outcomes, and a function from the strategies chosen by each player to the outcome; and, for each player, a preference ranking over the outcomes. A coalition {A, B, C,...} is effective for outcome Q if there are coordinated strategies that the coalition can choose that will result in Q. This function is then applied to determine which outcomes may be dominated much as we did in the theory of the core. An advantage of this approach is that it addresses something that the theory of games in coalition function form does not[2]: How might a coalition coordinate its strategies in the interest of its members?

This is all quite abstract. Let us look at an example.

According to Emmy Lou Harris, if you are going to play in Texas, you have to have a fiddle in the band. The point is that some styles can be played only by a band that can play the right instruments for the style. Bluegrass, for example, traditionally requires a mandolin, banjo, guitar and fiddle, as in the founding band, Bill Monroe's Bluegrass Boys. A band is a coalition like any other group of people who choose their strategies in concert for mutual benefit. We can think of the instruments as the strategies of the individual players and the outcomes are the styles a coalition can play in; to play in a particular style, each must make the right choice of instrument from those they can play.

This example is about a group of four amateur musicians {Abe, Barb, Curt and Deb} who want to get together and play for fun. Since they are amateurs, there is no money for side payments. All can play more than one instrument, so they can play in a number of different styles, but they have different preferences among the styles. Table 13.2 shows the instruments they can play. These are the

---

[2]For a theory that attempts to do both, see McCain, Roger A., Comparing Fairness: Relative Criteria of Economic Fairness with Applications (Cheltenham, U.K.: Elgar, 2021), Chapter 3.

strategies available to each player. The payoff table for this four-person game would be unwieldy, so instead Table 13.3 shows the instruments required for these musicians to play each style. Some of the assumptions behind the table are that Rock and Bluegrass both require at least four instruments,[3] and that you cannot play jazz without a drummer. Using the information from these tables we can construct Table 13.4, which is the effectivity function for this game. The *effectivity function* shows which

Definition: *Effectivity Function* — For a given group of decision-makers who may form coalitions, the *effectivity function* is a function or listing that tells us, for any coalition, the power the coalition has to limit outcomes by a choice of its coordinated strategies. Each coalition is associated with a list of outcomes that has the property that the coalition can coordinate its strategies to assure that the outcome is one from that list.

outcomes each coalition can bring about by an appropriate choice of individual strategies, that is, for which the coalition is *effective*. The preferences of the members of the coalition are shown in Table 13.5, where the numbers are preference rankings so that 1 denotes the *best* outcome from that person's point of view, 2 the *second best,* and so on.

Table 13.2. Musicians and Instruments.

A	Guitar, mandolin, piano
B	Bass, banjo
C	Drums, fiddle
D	Guitar, saxophone

---

[3]Well, Rock and Jazz *could* be played with different instruments, but for the sake of the example *these* four musicians cannot play rock or jazz with fewer instruments.

Table 13.3. Styles and Instruments Required.

Rock	2 Guitars, bass, drums
Bluegrass	Mandolin, banjo, fiddle, guitar
Jazz	Bass, drums, piano or saxophone or both
Country	Solo guitar or guitar with banjo, bass or fiddle
Folk	Guitar or banjo, with or without other instruments

Table 13.4. The Effectivity Function: Coalitions and Styles.

{A, B, C, D}	Rock, Bluegrass, Jazz, Country, Folk
{A, B, C}	Jazz, Country, Folk
{A, B, D}	Country, Folk
{A, C, D}	Jazz, Country, Folk
{B, C, D}	Jazz, Country, Folk
{A,B}	Country, Folk
{A,C}	Country, Folk
{A,D}	Country, Folk
{B,C}	Folk
{B,D}	Country, Folk
{C,D}	Country, Folk
{A}	Country, Folk
{B}	Folk
{C}	None
{D}	Country, Folk

With four players there are fifteen possible coalitions. Suppose, for example, that coalition {A, B, D} is formed. It can play only in a folk or country style, since it has no drummer nor fiddle

Table 13.5. Players' Preferences with Respect to Styles.

	Abe	Barb	Curt	Deb
Rock	2	2	2	2
Bluegrass	4	1	4	1
Jazz	1	5	1	5
Country	5	3	5	4
Folk	3	4	3	3

player — those are Curt's instruments. The preference rankings for those two styles are (3, 4, 3) and (5, 3, 4) respectively. Thus, in the jargon of NTU game theory, coalition {A, B, D} is *effective* for preference profiles (3, 4, 3) and (5, 3, 4). But if the coalition splits up, with Barb as a solo, she can go country while Abe and Deb form a folk duo, for rankings of (3, 3, 3), which improves on both of (3, 4, 3) and (5, 3, 4). Therefore, {A, B, D} is not stable and is not in the core for this game with either of the preference profiles it can put into effect.

Suppose coalition {A, B, C} forms. They can play jazz, country, or folk and so are effective for payoff profiles of (1, 5, 1), (5, 3, 5) and (3, 4, 3). If they choose Jazz or Folk, though, again, Barb will drop out to go country solo. (Barb would play bass in the jazz

Definition: *Effectiveness Form* — When a coalition can choose its joint strategies so as to bring about a particular *outcome* then the coalition is said to be *effective* for that outcome. The coalition's choice of strategies will depend on the *preferences* of the members. When a cooperative game is represented in this way it is said to be in *effectiveness form*. In this example, various kinds of bands are the *outcomes*, and the effectiveness of a coalition for an outcome depends on the instruments the members can play.

combo, but she has more fun playing country banjo.) If they choose country, Abe will drop out and join Deb to form a two-guitar folk group. So, again, {A, B, C} is not in the core with any of the preference profiles for which it is effective. By similar reasoning, we can rule out any coalition other than the grand coalition: Each such coalition is unstable in the sense that some subgroup can drop out and do better, regardless of the style that the group may choose.

What about the grand coalition? By choosing Rock, it can assure each member of at least his or her second preference; that is, the grand coalition is effective for preference profile (2, 2, 2, 2). (If Barb plays rock bass, she can do those rock-guitar moves, and she has a lot of fun with that.) We can rule out any other style, as (for example) if the grand coalition were to choose Bluegrass, Abe and Curt would prefer to drop out and play folk with guitar and fiddle. So, we suppose that the grand coalition would indeed choose to play rock. As we have seen, if {A, B, D} were to drop out, they would all do worse. If {A, B, C} were to drop out, then Barb, at least, will be made worse off, so she will veto any proposal to form that coalition. Similarly, we can show that any group that might drop out of the grand coalition will include at least one person who will then be at a lower preference level than the second preference he or she can get by remaining in the grand coalition and playing rock. Therefore, the core for this game is the grand coalition playing rock with preference profile (2, 2, 2, 2).

We see that coalition dominance reasoning and the concept of the core can be applied to NTU games. However, there are some complications we have skipped over.

3. α AND β EFFECTIVITY

In the previous section we read "A coalition {A, B, C,...} is effective for outcome Q if there are coordinated strategies that the coalition can choose that will result in Q." But this may depend on what non-members of the coalition do. In the example of the band, in the previous section, this issue is avoided since the styles available to any coalition depend only on the instruments they can play, not on what non-members do. In the jargon of the field, the Band Game is a "simple game." For games that are not simple, we may find there are two kinds of effectivity: We say that the coalition is α-effective for Q if there are strategies that will result in Q regardless of the strategies chosen by players who are not members of the coalition, and that the coalition is β-effective for Q if, given the strategies chosen by non-members of the coalition, there are strategies that the coalition may choose that will result in Q. This is a difference in information. We are saying that, for β-effectivity, the coalition needs to know what strategies the non-members will choose, but for α-effectivity, they do not need to know that since they have a strategy combination that will yield Q regardless what strategies non-members choose.

Here is an example to illustrate the distinction of α-effectivity from β-effectivity. Agents A and B would like to get together. They can choose between two destinations for their rendezvous: North or South. To do that they must, at least, choose the same destination. But this is a cooperative game: They can form a coalition and jointly decide on one destination or the other. Thus, the *outcomes* for this game are *rendezvous* and *no rendezvous*. However, Agent C wants to prevent them from getting together. Perhaps C suspects that they will act jointly to do C some harm, if they get together; but we don't need to know the answer to that question — all we need to know is that while A and B prefer rendezvous to no rendezvous, C prefers no rendezvous to rendezvous. To prevent the rendezvous C can block either North or South, but not both. Thus, each agent chooses between strategies "North" and "South," and the relation of outcomes to strategies is shown in Table 13.6.

Now we ask whether the coalition of A and B is α-effective for "rendezvous." The answer is no. Recall, the coalition Ξ is α-effective for outcome Q if there is a strategy that assures Q no matter which strategies are chosen by agents not in Ξ. Whichever destination the coalition of A and B chooses, C can block that destination by choosing the same strategy. But conversely, C is not α-effective for no rendezvous, since whichever destination C blocks leaves it open to {A, B} to choose the other destination and get together. Thus, neither {A, B} nor {C} can assure the outcome they prefer. Neither the

Table 13.6.   Strategies and Outcomes in the Rendezvous Game.

		C			
		North		South	
		B		B	
		North	South	North	South
A	North	No rendezvous	No rendezvous	Rendezvous	No rendezvous
	South	No rendezvous	Rendezvous	No rendezvous	No rendezvous

singletons {A} and {B} can do so either, and two-agent coalitions {A, C} and {B, C} cannot agree on a preferred outcome. The grand coalition {A, B, C} is (always) effective for any outcome. Neither {A, B} nor {C} is effective to veto the outcome they do not like, so neither rendezvous or no rendezvous can be excluded from the α-core of the rendezvous game.

Instead, we might ask whether {A, B} is β-effective for rendezvous. The answer is yes. Recall, a coalition Ξ is β-effective for Q if, given any strategies non-members of Q may choose, there are strategies that Ξ may choose that assure Q. Thus, if {C} chooses North, {A, B} can choose South, and vice versa. But in a similar way, C is β-effective for no rendezvous. Thus, either outcome can be vetoed by a (singleton or two-person) coalition that unanimously prefers the other. On this basis, the β-core of this game is a null set. Here is a qualification: If we interpret β-effectivity as an information condition, the conditions for effectivity of {A, B} is that {A, B} has the advantage of going second, while the condition for effectivity of {C} is that {C} has that advantage, and both of these cannot be true. However, what we can say is that there is no coalition structure *and specific set of strategies* that will be stable in the sense that no coalition or potential new coalition can improve on them by shifting to an outcome that is preferred by all members of the coalition.

In either case, it seems there is no cooperative solution to this game that clearly improves over a non-cooperative solution. If there were numerical payoffs and side payments, and if the game is non-constant sum, then either {A, B} could compensate C or vice versa and thus realize the better outcome. On the other hand, if the game is constant-sum in numerical payoffs, we would expect no cooperation. With data only on preferences, we do not know whether it is constant-sum or not.

4. GLOBAL WARMING YET AGAIN

Once again, recall the model of global warming as a three-person game. Here we consider it as a cooperative game without side payments. Here, as in most of our non-cooperative game examples, the outcome is the vector of payments to the three players. We assume that each player prefers an outcome in which the payoff to that player is larger. (That's the point of numerical payoffs, of course)!

The question is, as we recall, whether the cell at the lower right, where all depollute, can be thought of as the cooperative solution to the game. In particular, we may ask if these strategies and the outcome are in the α-core of this game. The effectivity approach will have to be a little different in this case. Consider a coalition of North and East. By choosing "depollute, depollute," they can assure that the outcome is either {25, 10, 6} or {27, 12, 5}, depending on the strategy chosen by South. Compare this with "pollute, pollute." Either of {25, 10, 6} or {27, 12, 5}, makes North better off, and East either better off or no worse off, so the coalition {North, East} will never choose "pollute, pollute." Thus, even if the effectivity of the coalition is not precise, it gives us some useful information about what a coalition will rationally choose. An effectivity function can show that a coalition has the power to limit the outcome to a set, rather than a specific outcome. More power means that the set is smaller, and in general, the grand coalition will be able to deploy any combination of strategies, and so is effective for any outcome. Smaller coalitions, however, may not be powerful enough to specify the outcome, but at the same time are not powerless. In this case any two-person coalition has power enough to limit the game to just two outcomes. With this in mind, the effectivity function for the Global Warming Game is shown as Table 13.8.

Here are some hints on how to interpret the table. By choosing strategies "Pollute, Pollute" the coalition of {NE} can assure that the outcome will be either (24, 10, 2) or (26, 12, 1), but they don't know which, since it depends on the strategy chosen by {S}. In any case they can do better. To move to the end of this line, by choosing "Depollute, Depollute" they can assure either (25, 10, 6) or (27, 12, 5), again depending on the strategy chosen by {S}. But since these outcomes are Pareto-better for {N, E}, the strategy combinations "Pollute, Pollute, Pollute" and "Pollute, Pollute, Depollute" are unstable and can be ruled out of the core of the game.

What more can we say about the core of this game? First, consider the grand coalition. It is effective for any of the 8 possible outcomes. The grand coalition will not rationally choose any outcome where at least two regions pollute, however, since for any of these, a shift to the lower right cell of Table 13.7 at which all three regions depollute will make some regions better off and none worse off. That is, again, such a shift would be a Pareto-improvement. Thus, the grand coalition would make a shift away from a situation at which fewer than two regions depollute. This excludes half of the eight outcomes from the core. What about the other half, at which at least two regions depollute? All of these outcomes share a kind of stability property: A shift away from the two-region coalition will make one or more of the two regions worse off. They are stable against deviations to other strategy combinations. Thus, the grand

Table 13.7. Strategies and Payoffs in the Global Warming Game (Approximately Reproduces Table 6.8 from Chapter 6).

		South			
		Pollute		Depollute	
		East		East	
		Pollute	Depollute	Pollute	Depollute
North	Pollute	24,10,2	26,8,4	26,12,1	28,10,3
	Depollute	23,12,4	25,10,6	25,14,3	27,12,5

coalition could rationally choose any of them. These four outcomes are shaded in light gray in Table 13.7.

We may also ask whether any of those four outcomes are dominated by a two-person coalition. The answer is no. Consider, for example, the cell at which North and East depollute but south does not, with payoffs (25,10,6). Could this be dominated via {N,E}? Consulting the second line of the effectivity function, Table 13.8, we see that every pair of outcomes for which they are effective other than {(25,10,6), (27,12,5)} contains at least one outcome that makes either North or East worse off. Thus, the coalition {N,E} will not take the risk of a shift from {depollute, depollute), and they do not have any power over the decision of S if S is not in their coalition. So {depollute, depollute, pollute} is not dominated. Similar arguments can be made for any other coalition of two regions. Further, no singleton coalition is effective for a strategy set that improves on {depollute, depollute, pollute}. Thus, none of the shaded outcomes can be excluded — all are in the α-core of this NTU game.

Table 13.8. The Effectivity Function: Coalitions and Outcomes.

{N,E,S}	Any
{N,E}	{(24,10,20), (26,12,1)}, {(26,8,4), (28,10,3)}, {(23,12,4), (25,14,3)}, {(25,10,6), (27,12,5)}
{N,S}	{(24,10,2), (26,8,4)}, {(23,12,4), (25,10,6)}, {(26,12,1), (28,10,3)}, {(25,14,3), (27,12,5)}
{E,S}	{(24,10,2), (23,12,4)}, {(26,8,4), (25,10,6)}, {(26,12,1), (25,14,3)}, {(28,10,3), (27,12,5)}
{N}	{(24,10,2), (26,8,4), (26,12,1), (28,10,3)}, {(23,12,4), (25,10,6), (25,14,3), (27,12,5)}
{E}	{(24,10,2), (23,12,4), (26,12,1), (25,14,3)}, {(26,8,4),(25,10,6),(28,10,3), (27,12,5)}
{S}	{(24,10,2), (23,12,4), (26,8,4), (25,10,6)}, {(26,12,1), (25,14,3), (28,10,3), (27,12,5)}

We recall from the previous chapter that, with side payments, the core and the Shapley value are based on total payoffs of 44, which can only be realized if all three depollute. But side payments among world regions present great difficulties and may well be impossible. Without them, the core is larger, including all outcomes at which only one region pollutes. At the same time, we might see a holdout situation, with each region hoping that the other two will form a coalition and depollute, benefiting the holdout. Perhaps this is the reason there has been so little progress so far.

In chapter 6 the question was posed whether the cooperative solution would be {Depollute, Depollute, Depollute}, despite the inequality of payoffs in that case. What we may say is that, in the absence of side payments, it is a cooperative solution and, though there are others, the dominant strategy solution "pollute, pollute, pollute" is not among them. Perhaps this is reason enough to say that global warming is a social dilemma.

5. SOME POLITICAL COALITIONS

The concept of coalitions came into game theory from parliamentary politics, so here is an example in parliamentary politics. Britain, Canada and Germany are examples of countries with parliamentary governments. In a parliamentary system, a party cannot form a government or continue for long in government unless it has a plurality in the

> **Definitions:** *Majority and Plurality* — A *majority* is more than half of the votes cast. If there are more than two alternatives or candidates, then the alternative or candidate with the largest number of votes is said to have the *plurality*, but this may be less than a majority.

legislature, that is, the parliament. There are often three or more parties with representation in parliament. That means that in order to form a majority government, parties in the parliament who together constitute a majority of the votes have to form a coalition and agree on a common program. A party with a plurality, but less than 50% of the vote, may be able to form a "minority government," but only if the other parties do not get together and form an opposition coalition. Thus, the payoff to a coalition depends on the other coalitions that form, and we may need to use the partition function to represent the game. If there is no governing coalition, no measures can be passed, and a new election has to be called. These are the rules common to countries with parliamentary government.

For example, in 2022, the German Federal Republic has roughly six or seven parties represented in its parliament, the Bundestag. The Christian Democratic Union and the Christian Social Union are parties in different regions in Germany and usually act in the Bundestag as a single party. The other five parties are the Social Democrats, the Greens, the Left Party, Alternative for Germany, and the Free Democratic Party. At various times, both of the bigger parties, the Christian Democrat/Christian Social Union and the Social Democrats, have joined in coalitions, sometimes with one another.[4] In late 2021, the Social Democratic, Green and Free Democratic parties formed a coalition called the "traffic light" coalition after their traditional party colors, red, green and yellow.

Here is a simpler fictional example. The Republic of Mitteleuropa has a parliamentary government, with three parties strong enough to be represented in the parliament. The three parties are as shown in Table 13.9.

The formation of a government in the Parliament of Mitteleuropa is a cooperative game, but side payments are not allowed, since they would be considered corrupt. Thus, the parties in the coalition can be rewarded only to the extent that measures they favor are passed. Each party favors some measures and opposes others, but each party is more strongly committed to some policies than others. In Table 13.10, the four measures expected to come before the next

---

[4]There is an important contrast of terminology between game theory and practical politics here. In the language of politics, a coalition between the two biggest parties, such as the Christian Democrat/Christian Social Union and the Social Democrats in Germany, would be called a grand coalition, whereas in cooperative game theory the term means a coalition of everybody in the game.

Table 13.9. Parties in the Parliament of Mitteleuropa.

Party	Vote	Position
Christian Conservative	40%	This party is strongly conservative on moral and family issues but moderate on economic issues, supporting measures favorable to small business and farmers.
Socialist Labor	40%	This party supports measures favorable to labor, and, in practice, generally supports central control of the economic system, but is neutral on moral and family value issues.
Radical	20%	This party favors free markets and limited government, but is extremely libertarian on family values and moral issues.

government are shown in the second column, and the relative strength of each party's support for each one of these measures is shown in the columns to the right. We may understand that "extremely favorable" indicates stronger support than "quite favorable," then "favorable" then "somewhat favorable," and similarly on the unfavorable side.

The rules of the game are that a coalition will pass any measure on which it has a clear position, that is, any measure on which at least two parties in the coalition are favorable or one is favorable and the other neutral. A set of measures passed is an *outcome* for this game. This may be a null set or a set of one or more measures passed. Thus, for example, the coalition of the Christian Conservatives and the Radicals (denoted {C, R}) will pass measures one and three: liberalized trade and an increase in the retirement age. We may say that {C, R} is effective for policies 1 and 3. The effectivity function for the various coalitions is shown in Table 13.11.

To determine what coalitions may be formed, we need to know the preferences of the parties over the outcomes of the game. To keep matters simple we need to consider only outcomes for which some coalition is effective, but there is another difficulty. Since we do not have numerical payoffs, it is hard to say what the preferences will be between *outcomes*, where some outcomes include more than one measure passed. For this example, we will assume that a party

Table 13.10.   Issues and Party Positions in Mitteleuropa.

Issue	Parties and Attitudes		
	C	S	R
(1) Liberalize International Trade	Somewhat favorable	Unfavorable	Extremely favorable
(2) Legalize same-sex marriage	Extremely unfavorable	Neutral	Quite favorable
(3) Increase in retirement age[5]	Somewhat favorable	Extremely unfavorable	Favorable
(4) Subsidies[6] to farmers and businesses that compete with imports	Quite favorable	Quite favorable	Extremely unfavorable

will prefer an outcome with one measure that it is quite favorable to an outcome with two measures that it is only somewhat favorable to. With this judgment and similarly interpreting Table 13.10 we summarize the preferences of the parties in Table 13.12.

We see that the only coalition that is stable is {C, S}. We see that, beginning from any other line, the Christian Conservatives and the Socialist Labor Party can both obtain an outcome they prefer by shifting to a coalition with one another and excluding the Radicals from power. In commonsense terms, the Socialists and the Conservatives can work together because each is willing to give the other a veto on the key issue the other wants to kill. (In real parliamentary governments, coalitions of traditionalist conservatives and

---

[5] In Mitteleuropa, as in several European countries, most pensions are public and raising the retirement age will cut the costs and thus avoid a tax increase.

[6] In case the word is unfamiliar, a "subsidy" is a direct payment from the government to a person or business, not in payment for any service the person or business delivers to the government, but simply to support that person or business. Subsidies to agriculture and to businesses that export or that compete with imports are quite common worldwide.

Table 13.11. Effectivity of Coalitions in the Parliament of Mitteleuropa.

	Coalitions	Issues Passed
(1)	{C, S, R}	1,2,3,4
(2)	{C, R}	1,3
(3)	{S, R}	2
(4)	{C, S}	4
(5)	Singleton coalitions	None

Table 13.12. Issues and Party Preferences in Mitteleuropa.

	Parties and Preferences		
Measures Passed	C	S	R
1,2,3,4	3	3	2
1,3	2	4	3
2	4	3	1
4	1	1	4

socialists have been formed in several countries in the 21st century, but have often resulted in a decrease in the voting margin for the members of the coalition).

In parliamentary systems, the formation of coalitions among different factions is formal and explicit, but also changeable. Coalitions of different factions are unavoidable in government — even kings and dictators are more supported by some factions than by others — but in other systems it may be less easy to observe, and expressions of support for some factions may even be forbidden. In still others factions join together in parties that are more or less permanent coalitions. This tends to be the case especially in English-speaking countries. This seems to be an area for NTU games,[7] but most real examples will not be even as simple as this complicated example.

6. CONCLUDING SUMMARY

The theory of cooperative games in coalition function form requires some assumptions that may seem oversimple: That the outcome for a coalition is a total numerical payoff that may be divided among the members in any way and that does not depend on decisions of anyone outside the coalition. This approach has two advantages. First, it leads to precise and sweeping conclusions, especially if we make the additional simplifying assumptions that lead to the Shapley value solution. Second, they bring side payments to the center of the theory, and side payments are important. But even when payoffs are numerical and transferable, we may need to allow for holdout behavior, a problem that sometimes seems to arise in real coalition formation. When the outcomes of the interdependent decisions of a coalition are not numerical payoffs or are not transferable, the decision-makers may nevertheless disagree in their preferences among the outcomes. In such a case we may use an effectivity approach, first determining the power of each coalition to narrow down the possible list of outcomes — perhaps to just one — and using that information we may consult the preferences of coalition members over outcomes and apply dominance reasoning to discover the core of the game in effectivity form.

---

[7]In 1962, political scientist William Riker applied the theory of games in coalition function form to the formation of political coalitions in his book, *The Theory of Political Coalitions* (New Haven: Yale University Press). His main point was that smaller winning coalitions will be preferred to larger ones, since the gains from winning are then divided among fewer winners. The consensus of his critics seem to be that such a tendency does exist, but that he has exaggerated it. In the example here, we see that all parties reject the coalition of all three, largely for that reason. The assumption that political payoffs are numerical and tradeable lead to the extremism of Riker's conclusion.

Q13. EXERCISES AND DISCUSSION QUESTIONS

Q13.1. A Prehistoric Game

Gung, Mog and Pok are prehistoric tribesmen planning a hunting expedition. They have different skills: Gung is very strong, Mog quick and agile, and Pok has great stamina. They have to decide which of them will collaborate and which (if any) are to be left to hunt independently, and how the catch is to be divided among those who collaborate. The payoffs to coalitions are shown in Table 13.13, with the payoffs measured by the expected value of the number of antelope they can catch. Assume this is a TU game with portions of antelope serving for side payments.

a. Construct a coalition function for this game, if possible. If so:
b. Determine whether the game is superadditive.
c. Describe the core of this game. Would a division of the catch in the proportions 4, 1, 1 (in the order GMP) be a stable side-payment schedule? Why or why not? Would a division of the catch in the proportions 2, 2, 2 (in the order GMP) be a stable side-payment schedule? Why or why not?
d. Determine the Shapley Value for this game.
e. Is the Shapley Value an imputation in the core of the game?

Table 13.13. Prehistoric Coalitions.

Coalition	Payoff
{GMP}	(6)
{GM}{P}	(4)(1)
{GP}{M}	(3)(1)
{PM}{G}	(3)(2)
{G}{M}{P}	(2)(1)(1)

Q13.2. An Exploitation Game

A and B are companies that sell their products to C, and A and B are both monopolies. Thus, their strategies are to raise their price or to hold the price at a lower, but still profitable charge. For C the strategies are to spend their whole budget or to hold back and buy the minimum — hold or spend. The payoffs in normal form are given in Table 13.14.

a. What is the Nash equilibrium for this game?
b. Construct an α-effectivity table for this game.
c. Is the strategy triple "hold, hold, spend" in the α-core of this game? Is any other strategy triple and outcome in the core?
d. Could the coalition of the two sellers, {A, B}, improve on the Nash equilibrium for this game?

Q13.3. Another Exploitation Game

Suppose instead that the payoffs in normal form are as shown Table 13.15. Notice that that the only change is in the lower right cell of the table.

Table 13.14. An Exploitation Game.

		C			
		Hold		Spend	
		B		B	
		Raise	Hold	Raise	Hold
A	Raise	5,5,5	5,4,6	7,7,3	7,3,8
	Hold	4,5,6	3,3,7	3,7,8	6,6,14

Table 13.15.   Another Exploitation Game.

		C			
		Hold		spend	
		B		B	
		Raise	Hold	Raise	Hold
A	Raise	5,5,5	5,4,6	7,7,3	7,3,8
	Hold	4,5,6	3,3,7	3,7,8	4,4,14

a.   Is the strategy triple "raise, raise, hold" in the α-core of this game? Is any other strategy triple and outcome in the core?

b.   Could the coalition of the two sellers, {A, B}, improve on the Nash equilibrium for this game?

Q13.4. A Reversed Rendezvous Game

Reconsider the Rendezvous Game, Table 13.6, but suppose that the preferences are reversed, so that A and B prefer "no rendezvous" and C prefers "rendezvous."

a.   Is {A, B} effective for "no rendezvous?"

b.   Is {C} effective for "rendezvous"

PART V

Advanced Topics

CHAPTER 14

N-Person Games

For applications of game theory to real problems, we will
often need to allow for more
than three players, and sometimes very many. Many of the
"games" that are most important in the real world involve

> **To best understand this chap-
> ter,** you need to have studied
> and understood the material
> from Chapters 1–8.

considerably more than two or three players — for example, economic competition, highway congestion, overexploitation of the
environment, and monetary exchange. So, we need to explore
games with many players. This can get very complicated. For example, if we have ten players, there are 10! =3,628,800 relationships
between them.[1] We will need to make some simplifying assumptions
to make analysis of games with many players feasible.

1. THE QUEUING GAME

Here is an example, a game with six players. As usual, we will begin
with a story. Perhaps you have had an experience like the one in this
story. Six people are waiting at an airline boarding gate, but the
clerks have not yet arrived at the gate to check them in. Perhaps
these six unfortunates have arrived on a connecting flight with a
long layover. Anyway, they are sitting and awaiting their chance to

---

[1] 10! is read "ten factorial" and is computed as follows: $10 \times 9 \times 8 \times 7 \times 6 \times 5 \times 4 \times 3 \times 2 \times 1$. The number of possible relationships in any game is $N!$, where N is the number of players.

check in, and one of them stands up and steps to the counter to be the first in the queue. As a result, the others feel that they, too, must stand in the queue, and a number of people end up standing when they could have been sitting.

Here is a numerical example to illustrate a payoff structure that might lead to this result. Let us suppose that there are six people, and that the gross payoff to each passenger depends on when they are served, with payoffs as shown in the second column of Table 14.1. Order of service is listed in the first column.

The gross payoffs assume, however, that one does not stand in line. There is a two-point effort penalty for standing in line, so that for those who stand in line, the net payoff to being served is two less that what is shown in the second column. These net payoffs are given in the third column of the table.

Those who do not stand in line are chosen for service at random, after those who stand in line have been served. (Assume that these six passengers are risk neutral.) If no-one stands in line, then each person has an equal chance of being served first, second,..., sixth, and an expected value payoff of

$$(1/6) * 20 + (1/6) * 17 + \cdots + (1/6) * 5 = 12.5.$$

In such a case the aggregate payoff is 75.

Table 14.1. Queuing Game Payoffs.

Order served	Gross Payoff	Net Payoff
First	20	18
Second	17	15
Third	14	12
Fourth	11	9
Fifth	8	6
Sixth	5	3

But this will not be the case. This game has a large family of Nash equilibria, depending on who stands and who sits, but we can show that there is no Nash equilibrium in which everybody sits, and in fact that a Nash equilibrium can only occur if four people are standing in line and two sitting.

We will show that by eliminating all other possibilities. First, "everyone sit" is not a Nash equilibrium, since an

Definition: *Expected Value* — (Repeated from Chapter 7). Suppose an uncertain event can have several outcomes with numerical values that may be different. The expected value (also known as mathematical expectation) of the event is the weighted average of the numerical values, with the probabilities of the outcomes as the weights.

individual can improve their payoff by standing in line, provided he or she is first in line. The net payoff to the person first in line is 18 > 12.5, so someone will get up and stand in line.

This leaves the expected value of the payoff at 11 for those who remain. (You should verify this by computing the expected value for your notes. The probability for each payoff will now be 1/5.) But we can also eliminate the five-sit, one stand possibilities. Since the second person in line gets a net payoff of 15, and since 15 > 11, someone will be better off to get up and stand in the second place in line.

This leaves the expected value payoff at 9.5 for those who remain. We can also show that this possibility, two sitting and four standing, cannot be a Nash equilibrium. Since the third person in line gets a net payoff of 12, and since 12 > 9.5, someone will be better off to get up and stand in the third place in line.

This leaves the expected value payoff at 8 for the three who remain. But this will not be a Nash equilibrium either. Since the fourth person in line gets a net payoff of 9, someone will be better off to get up and stand in the fourth place in line.

This leaves the expected value payoff at 6.5 for those who remain. Since the fifth person in line gets a net payoff of 6, no-one else will join the queue. If five or six people were standing in line, however, it would pay those at the end of the line to sit down. (*Exercise*: Verify this by computing the expected values. Hint: if there are 6 people in line the last person in line can get a payoff of 5 with probability 1.) Every assortment of strategies with four people in line (regardless of order) and two sitting is a Nash equilibrium, and no other assortment of strategies is. The total payoff is 67, less than the 75 that would have been the total payoff if, somehow, the queue could have been prevented.

Two people are better off — the first two in line — with the first gaining an assured payoff of 5.5 above the uncertain expected value payoff she would have had in the absence of queuing and the second gaining 2.5. But the rest are worse off. The third person in line gets 12, losing 0.5; the fourth 9, losing 3.5, and the rest get average payoffs of 6.5, losing 6 each. Since the total gains from queuing are 8 and the losses 16, we can say that, in one fairly clear sense, queuing is inefficient.

HEADS UP!

Here are some concepts we will develop as this chapter goes along:

N-Person Game: A game with N players is an N-person game. N may be any number, 1, 2, 3, or more, but usually refers to larger numbers.

Representative Agent: In game theory we may sometimes make the simplifying assumption that every agent chooses from the same list of strategies and gets the same payoffs in given circumstances. This is called a "representative agent" model or theory.

State variable: A state variable means a single number, or one of a small list of numbers, that together express the "state" of the game, so that a player who knows just the value of the state variable or variables has all the information she needs to choose a best response strategy.

Proportional game: A game in which the state variable is the proportion of the population choose one strategy rather than another is a proportional game.

2. SIMPLIFYING ASSUMPTIONS FOR *N*-PERSON GAMES

The previous section presents a "game" that extends the two and three person games in some important ways. The Queuing Game example also illustrates two common simplifying assumptions in *N*-Person games.

In the Queuing Game, all of the participants are assumed to be identical, to be *representative agents*. This illustrates one kind of simplifying assumption: The *representative agent model*. In this sort of model, we assume that all players are identical, have the same strategy options and get symmetrical payoffs. This does not mean that they end up in the same situation! As we saw in the Queuing Game, only one ended up first in line, and others were later in line and some still sitting. That's a key point of the representative agent model: *Even though the agents are identical, they may do different things in equilibrium.* The differences are a result of the Nash equilibrium in the game, *not* a result of any differences in the agents.

This "representative agent" approach shouldn't be pushed too far. It is quite common in economic theory, and economists are sometimes criticized for overdoing it. But it is useful in many practical examples, and the next few sections will apply it.

There is another powerful simplifying assumption in the Queuing Game. Notice that no passenger has to know any-

> **Definition:** *Representative Agent* — In game theory we may sometimes make the simplifying assumption that every agent chooses from the same list of strategies and gets the same payoffs in given circumstances. This is called a *representative agent* model or theory.

thing about the strategies the other passengers have chosen, who is first in line and who is second in line, and so on. All a passenger needs to know is the length of the line right now. If the line is short enough, the best response strategy is to get on line; otherwise, the best response strategy is to continue to sit. Thus, we could say that the length of the line is a *state variable*. A state variable is a single variable, or one of a small number of variables, that sum up the state of the game from the point of view of the representative agent. The state variable, or variables, are all that the agent needs to know in order to choose the best response strategy.

> **Definition:** *State Variable* — In this book, a state variable means a single number, or one of a small list of numbers, that together express the "state" of the game, so that a player who knows just the value of the state variable or variables has all the information they need to choose a best response strategy.

This terminology — *a state variable* — is usually used more narrowly in game theory. It is taken from the study of games that evolve continuously over time. Because these games are also based on the mathematical study known as *differential equations*, they are called *differential games*. One example of such a game is a pursuit game, in which one agent, the pursuer, wants to catch the other agent, the quarry, in the shortest possible time. In many such games, all that the pursuer needs to know is the distance between the pursuer and the quarry. The pursuer's best response is the response that makes that distance as small as possible. Similarly, all that the quarry needs to know is that same number, the distance from the pursuer. The quarry's best response is the response that makes that distance as large as possible. Thus, the distance serves as a state variable in a pursuit game of that kind. Of course, differential equations is a branch of calculus, and the mathematics required to analyze pursuit games is beyond the limits of this book. Students who are interested in differential games will need to take a more advanced course in game theory, as well as intermediate level mathematics. But the state variable idea itself requires little or no mathematics.

In the rest of this chapter, we will use the phrase "state variable" to refer to in any variable defined as above — a variable that sums up the state of the game so that players only need to know that state variable, and nothing else, in order to choose their best response strategy. I believe this terminology will be helpful, but do keep in mind that many other game theorists use it more narrowly.

The two assumptions, representative agents and a state variable, can complement one another as they do in the Queuing Game, and help us to think through complicated games with very large numbers of participants. They are powerful tools, and like all powerful tools, should be used carefully — a point we will return to at the end of the chapter.

3. GAMES WITH MANY PARTICIPANTS: PROPORTIONAL GAMES

The Queuing Game gives us one example of how game theory can be applied to larger numbers than two or three, and it provides some insights on some real human interactions. But there is another simple approach to multi-person two-strategy games that is closer to textbook economics, using the representative agent and state variable assumptions, and is important in its own right.

As an example, let us consider the choice of transportation modes — car or bus — by a large number of identical individual commuters. The basic idea here is that car commuting increases congestion and slows down traffic. The more commuters drive their cars to work, the longer it takes to get to work, and the lower the payoffs are for both car commuters and bus commuters. The commuters are representative agents — their payoffs vary in the same way as the number of cars on the road changes — and the state variable is the proportion of all commuters who drive cars rather than riding the bus. The larger the proportion who drive their cars, the slower the commute will be, regardless which transport strategy a particular commuter chooses.

Figure 14.1 illustrates this. In the figure, the horizontal axis measures the proportion of commuters who drive their cars. Accordingly, the horizontal axis varies from a lower limit of zero to a maximum of 1 (that is, 100%). The vertical axis shows the payoffs for this game. The upper (shaded) line shows the payoffs for car commuters. We see that it declines as the proportion of commuters in their cars increases. The lower, dark line shows the payoffs to bus

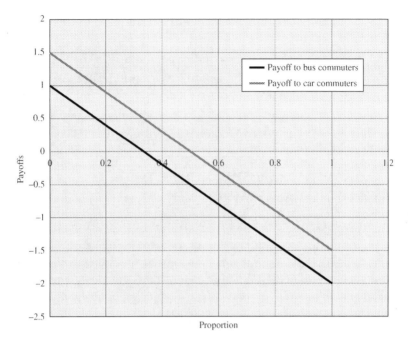

Figure 14.1. Payoffs in a Commuter Game.

commuters.[2] We see that, regardless of the proportion of commuters in cars, cars have a higher payoff than busses. In other words, commuting by car is a dominant strategy in this game. In a dominant strategy equilibrium, all drive their cars. The result is that they all have negative payoffs at −1.5, whereas, if all rode busses, all would have positive payoffs of 1. If all commuters choose their mode of transportation with self-interested rationality, all choose the strategy that makes them individually better off, but all are worse off as a result.

---

[2]For the sake of the example, payoffs were scaled so that the best payoff to bus riders is 1. Payoffs to bus commuters were computed as $1-3q$, and to car commuters as $1.5-3q$, where q is the proportion of car commuters. These numbers are arbitrary ones meant to express the ideas in the discussion. The idea comes from Thomas Schelling, *Micromotives and Macrobehavior* (New York: Norton, 1978) and Herve Moulin, *Game Theory for the Social Sciences* (New York: New York University Press, 1982), pp. 92–93.

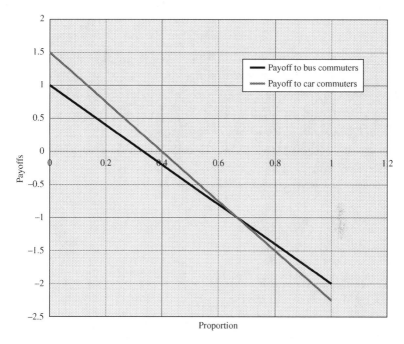

Figure 14.2. Payoffs in a More Complex Commuter Game.

This is a social dilemma, in that there is a dominant strategy equilibrium, but the choice of dominant strategies makes everyone worse off. But it probably is not a very "realistic" model of choice of transportation modes. Some people do ride busses. So let's make it a little more realistic, as in Figure 14.2.

The axes and lines in Figure 14.2 are defined as they were for Figure 14.1. In Figure 14.2, congestion slows the busses down somewhat, so that the payoff to bus commuting declines as congestion increases; but the payoff to car commuting drops even faster.[3] When the proportion of people in their cars reaches $q = 2/3$, the payoff to car commuting overtakes the payoff to bus-riding, and for larger proportions of car commuters (to the right of q), the payoff to car commuting is worse than to bus commuting.

---

[3] For this diagram, the payoff to bus commuters is unchanged, but the payoff to car commuters is calculated as $1.5-3.75q$.

Thus, the game no longer has a dominant strategy equilibrium. However, it has many Nash-equilibria. When 2/3 of commuters drive cars, that is a Nash-equilibrium. Here is the reasoning: Starting from 2/3, if one bus commuter shifts to the car, that moves into the region to the right of 2/3, where car commuters are worse off, so (in particular) the person who switched is worse off. On the other hand, starting from 2/3, if one car commuter switches to the bus, that moves into the region to the left of 2/3, where commuting by car is the best response, so, again, the switcher would be better off not to switch. The only proportion at which everyone is choosing a best response is $q = 2/3$.

This again illustrates an important point that, in a Nash-equilibrium, identical people may choose different strategies to maximize their payoffs. This Nash-equilibrium resembles some "supply-and-demand" type equilibria in economics, having been suggested by models of that type, but also differs in some important ways. In particular, it is inefficient, in this sense: If everyone were to ride the bus, moving back to the origin point in Figure 14.2 (as in Figure 14.1), everyone would be better off. The Nash equilibrium payoff is −1 (for both bus riders and car drivers) and the payoff to 100% bus riders is +1. As in a social dilemma, though, they will not do so when they act on the basis of individual self-interest without coordination.

This example is an instance of *the tragedy of the commons*. The highways are a common resource available to all car and bus commuters. However, car commuters make more intensive use of the common resource, causing the resource to be degraded (in this instance, congested). Yet the car commuters gain a private advantage by choosing more intensive use of the common resource, at least while the resource is relatively undegraded. The tragedy is that this intensive use leads to the degradation of the resource to the point that all are worse off.

In general, the tragedy of the commons is that common property resources tend to be overexploited and thus degraded, unless their intensive use is restrained by legal, traditional, or (perhaps) philanthropic institutions. The classical instance is common pastures, on which, according to the theory, each farmer will increase her herds until the pasture is overgrazed and all are impoverished. Most of the applications have been in environmental and resource issues. The recent collapse of fisheries in many parts of the world seems to be a clear instance of "the tragedy of the commons."

All in all, it appears that the Tragedy of the Commons is correctly understood as a multiperson social dilemma along the lines suggested in Figure 14.1, and, conversely, that the *N*-person social dilemma is a valuable tool in understanding the many tragedies of the commons that we face in the modern world.

4. HAWK VS. DOVE, REVISITED

In Chapter 5, we studied a "classical" two-by-two game called Hawk vs. Dove. The payoffs for that game are shown again, for convenience, in Table 14.2. This game is a biological application of game theory, but it may not seem quite right — after all, (1) hawks and doves are different species, and birds don't decide to be hawks or doves as strategies that they can change if they think the other strategy will pay better; and (2) hawks and doves are not interested in money payoffs, and we have no idea what subjective "costs and benefits" they might experience.

In fact, when game theory is applied in biology, it has to be interpreted a little differently. The perspective for biological applications of game theory is evolution and population biology. The payoffs are not in dollars or "utility," but in reproductive fitness. That is, the

Table 14.2. Hawk vs. Dove (Repeats Table 5.15, Chapter 5).

		Bird B	
		Hawk	Dove
Bird A	Hawk	–25,–25	14,–9
	Dove	–9,14	5,5

Table 14.3. Hawk vs. Dove as a Game Against Nature.

		Matching Bird		
		Hawk	Dove	Expected value payoff
Bird to be Matched	Hawk	−25	14	$-25p + 14(1 - p)$
	Dove	−9	5	$-9p + 5(1 - p)$
Probability		p	$1 - p$	

payoffs to a hawk or a dove are the bird's chances of surviving and leaving young. The greater the expected value of the number of young, the greater the payoffs. On the other hand, since population biology is concerned with whole populations of animals, there are always many more than two players. For the hawk vs. dove game, any play of the game is between just two birds, but the birds are matched at random from large populations. An individual bird — hawk or dove, as the case may be — is matched with another bird, who may be a hawk or a dove with probabilities that depend on the proportion of the population of hawks to the population of doves.

Suppose that hawks are more reproductively fit than doves. That means that, on the average, a hawk rears more baby birds to reproductive age than a dove does. That means that the population of hawks will grow faster than the population of doves, increasing the probability that a bird of either species will be matched with a hawk, and this might eventually tip the balance, making doves equally fit or more fit than hawks.

For a bird looking forward to her next match, we could think of the match as a "game against nature," since nature determines, at random with given probabilities, which type of bird will be the match. Table 14.3 shows the hawk vs. dove game from this perspective. We assume that the probability of being matched with a hawk is equal to the proportion of hawks in the whole population.

Thus, the payoff for each kind of bird depends on the probability of being matched with a hawk and thus on the proportion of hawks to doves in the population. This is shown by Figure 14.3. The probability

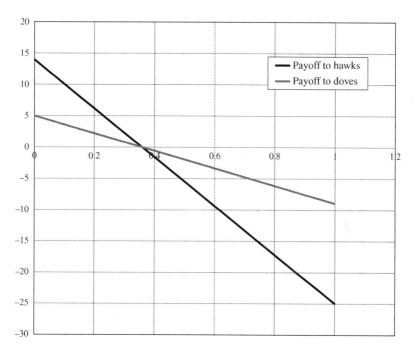

Figure 14.3. Payoffs to Hawks and Doves.

of being matched with a hawk (the proportion of hawks) is shown on the horizontal axis, and the payoffs to hawks and doves on the vertical axis. The payoff to hawks is shown by the solid dark line, while the payoff to doves is shown by the gray line. We see that, if there are fewer than 35% hawks in the population, populations of both species will grow, but the population of hawks will grow faster, so the probability of being matched with a hawk increases. This continues until the proportion of hawks reaches 35%.

If the proportion of hawks goes beyond 35%, doves are more reproductively fit than hawks. Both populations decline, but that of hawks declines faster. Thus, the proportion of hawks declines, again until it reaches 35%.

Assumption: *Types of Representative Agents* — In some game analyses we may assume that there are a small number of different types of representative agents.

In this game, the proportion of hawks is the state variable. "Hawk" is the best response strategy whenever the proportion is less than 35%, and "dove" whenever it is more than 35%. Thus, 35% hawks gives a Nash equilibrium — the only proportion at which every player is playing a best response strategy. Notice that this equilibrium corresponds to the mixed-strategy equilibrium of the original two-person game. This mixed-strategy equilibrium would be unstable if the agents could learn and adjust their strategies, but in the biological evolutionary interpretation, the mixed strategy equilibrium is stable.

This example gives a better flavor of the application of game theory in biology than the examples in earlier chapters, since the applications are to population biology and thus to more than two "players," even if only two are matched at any one time. (We will expand on that in Chapter 21.) It also gives us another example of the use of the simplifying assumptions, and particularly the state variable. In addition, we have a good example of how probabilities and expected values can enter into games of many players, by random matching of players to play simpler, perhaps two-by-two games. Finally, we see that the representative agent model can be modified to allow for more than one type of agent. Agents predisposed to play "hawk" are of a different kind (in population biology) than those predisposed to play "dove." In games played by human beings, different types of agents may have different tastes or different information or different command of resources, or even different kinds of rationality — though we will not get to that last one until Chapter 20.

5. PANIC BUYING

Recall that in Chapter 5, we treated panic buying as a two-person game. As explained there, "in March and April of the year 2020, in the early stages of the global pandemic of the COVID virus, problems of 'panic buying' emerged. That is, expecting that stores might be closed and merchandise unavailable, people bought more than they currently needed to obtain reserves against their future need.

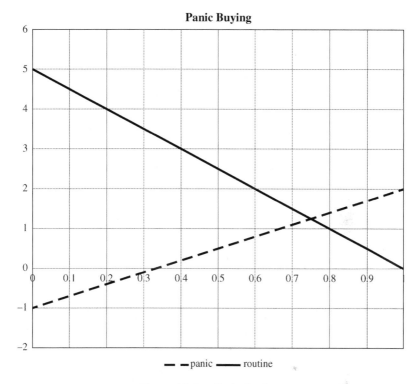

Figure 14.4. Panic Buying.

The result, though, was that the inventories of stores and wholesalers were exhausted, so that the merchandise was unavailable." But it was, of course, not a two-person game — millions of decision-makers were involved. Let us then think of it instead as an *N*-person game, a proportional game.

The strategies will be routine buying, only slightly in advance of need, and panic buying of much larger quantities in advance of need. Shoppers will be thought of as representative agents. The state variable is the proportion of all shoppers who engage in panic buying. If no or very few customers engage in panic buying, the payoff to routine buying is larger than panic buying — that is why routine buying is routine! As the proportion who panic buy increases, the payoff to panic buying increases, while that to routine buying decreases, as empty shelves become more common. The payoffs are shown in Figure 14.4. The payoff to routine buying is shown by the solid line, and the payoff to panic buying by the dashed line.

We see that there is a Nash equilibrium where nobody engages in panic buying. If just one customer deviates from that equilibrium by panic buying, that customer's payoffs go from a positive to a negative value. But there is also a Nash equilibrium where everyone engages in panic buying. If just one deviates to routine buying, their payoffs drop to zero. There are also many Nash equilibria where just q engage in panic buying, since both strategies pay the same in that case, but these are unstable equilibria in the following sense: If a few decision-makers deviate to panic buying then panic buying becomes the best response for all; while if a few deviate to routine buying then routine buying becomes the best response for all. We might describe the proportion q as a tipping point — on either side of it, the best response is at the extreme, where none of 100% of the population choose one or the other strategy as their best response. As was observed in Chapter 5, new information — such as news about the pandemic — could cause enough people to shift to panic buying that panic buying becomes the new equilibrium.

6. SUPPLY AND DEMAND

The simplifying assumptions we have looked at in this chapter did not originate in game theory, but in economics. The theory of supply and demand in microeconomics, for example, can be expressed in terms of representative agents and state variables. There are two types of representative agents, buyers and sellers, and the market price is the state variable for each of them. Sellers want to maximize their profit, so profit is their payoff. Buyers want to maximize the subjective satisfaction, or utility, that they obtain from all of the goods that they consume, within their limited income. This subjective "utility" is the buyer's payoff. Figure 14.5 shows the diagrammatic form of the theory of supply and demand, which will be familiar to students who have taken a course in microeconomic principles.

For this game theory text, we will need to make use of some ideas taken from the economic writings of the great French-speaking

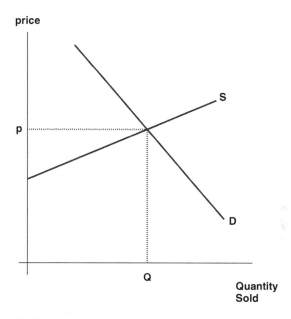

Figure 14.5. The Supply-and-Demand Diagram from Microeconomic Principles.

economist Leon Walras. In the Game of Supply and Demand, the rules of the game are determined by a process called tattonement. The tattonement begins when an auctioneer calls out a price at random. All of the buyers respond by indicating the quantities that they would want to buy at that price, and all of the sellers respond by indicating the quantities that they would want to sell at that price. The auctioneer adds up the quantities that people want to buy and the quantities that people want to sell. If the total quantity demanded is greater than the quantity supplied, the auctioneer tries again with a somewhat higher price. Conversely if quantity supplied is greater than quantity demanded. This trial and error adjustment of the price continues until the quantity that people offer to buy is the same as the quantity the others offer to sell. At that point, everyone buys and sells quantities they have announced at the price that has been finally arrived at. This is the "equilibrium" price, in supply and demand theory. The key point about the tattonement process is that no one buys or sells anything until the auctioneer arrives at a price that balances the offers to buy against the offers to sell. When the

Table 14.4.　Bidding Contingencies (Not a Payoff Table!).

		Strategy	
		Announce quantity on supply curve	Announce some other quantity
Contingency	Quantities balance	Maximum profit	Less than maximum profit
	Quantities unbalanced	Nothing	Nothing

auctioneer announces a price, how will a rational seller respond? To decide how he will respond, the seller will have to think in terms of two contingencies, as shown in Table 14.4. We see that the seller has a dominant strategy: To offer the amount that would maximize her profits at the announced price. Any attempt to "outsmart" the auctioneer by offering a different price can only fail. The same is true for buyers. Thus, the tattonement process leads to the supply-and-demand equilibrium via dominant bidding strategies.

The relationship between the price and the best-response quantity is the individual seller's "supply curve." The sum of all the individual sellers' supply curves is the market supply curve, line S in Figure 14.5. For an individual buyer, the quantity that maximizes his "utility" at a price and the best response quantity is his individual demand curve. The sum of all the individual demand curves is the market demand curve, line D in Figure 14.5. At the intersection of lines S and D, all suppliers are selling their best response quantities, and all buyers are buying their best response quantities. In those terms, the supply and demand equilibrium is a Nash equilibrium in an N-person game.

7. SUMMARY

Many important real-world games and strategic situations involve more than three agents, sometimes very large numbers of agents. If each agent tries to consider the strategies chosen by every other agent, as we have assumed in the previous chapters, the number of combinations of agents and strategies increases very much more rapidly than the number of the agents. This is a problem for the agents, and an even bigger problem for the game theorist.

In order to understand and analyze these very complicated games, many game theorists have found it necessary to make some simplifying assumptions. One simplifying assumption is that there is a single variable, or a small number of variables, which sum up the condition of the game in such a way that each agent can choose the best response strategy if they know the value of only that one variable or those few variables. Using a word common in the study of games that evolve over time, we can refer to these variables that sum up the condition of the game as "state variables," although we are using the term a little more broadly than it is usually been used in the past.

Another useful simplifying assumption is that all of the agents have the same payoff for the same strategy and the same value of the state variables. Then we will only have to analyze one decision for the whole population of agents. Even if this is not quite right, there may be only a few types of agents, with different information, preferences, or opportunities, and this is still much simpler than making a separate analysis for each of millions of different agents.

In this way, we can extend the analysis of social dilemmas to problems of congestion and the tragedy of the commons in which millions of people may interact, we can analyze interactions in whole populations of travelers, buyers, or hawks and doves, and we even find that the equilibrium of supply and demand from basic economics textbooks is an instance of Nash equilibrium with state variables and one or more type of representative agent.

Q14. EXERCISES AND DISCUSSION QUESTIONS

Q14.1. Patenting Game

Firms A, B, C, D, E, and F are all considering undertaking a research project that will lead to a patent for just one firm. The first firm to complete the research project will receive the patent. For each firm that undertakes the project the cost will be $210 million. Each firm that undertakes the research project will have an equal chance of finishing first and thus gaining a profit of $1 billion minus $210 million = $790 million. Those who undertake the research but do not finish first lose the $210 million stake. Those that do not undertake a research project get payoffs of zero.

For this problem the state variable is the number of firms that undertake a research project. Solve the problem along the lines of the Queuing Game, keeping in mind one important difference: In this game, those who "sit out" get a certain payoff of zero, while those who "jump in" get the uncertain payoff that has to be evaluated in expected value terms.

a. Compute the expected value payoff for those who undertake research projects for all values of the state variable from 1 to 6.
b. Determine how many firms undertake research projects at equilibrium.
c. The net social value of the research is the $1 billion value of the patent minus the total expenditure of the research. Compute the net social value of the research at equilibrium. Determine how the net social value of the investment changes as the state variable goes from 1 to 6.
d. Determine the number of firms doing research that corresponds to the maximum net social value. Defining an efficient situation as one in which net social value is maximized, comment on the efficiency of the equilibrium in this case.
e. Economists Robert Frank and Philip Cook argue that "winner-take-all" competition, in which only the one agent ranked first gets a payoff, is inefficient in that it leads people to commit "too much" resources to the competition. Comment on that idea in the light of this example.

Q14.2. Public Goods with N Agents

Consider the following N-person public goods contribution game. Each of the N agents can decide whether to contribute or not. Assume $N \geq 10$. The state variable is the number, M, of agents who contribute. For an agent who does not contribute the payoff is $1,000 + 100M$. For those who contribute it is $800 + 100M$, since the cost of a contribution is 200.

a. On a coordinate graph, draw the curve or line representing the payoff to a non-contributor as a function of the state variable M. (A spreadsheet XY graph utility will probably work for these purposes, but graph paper works, too.)

b. On the same graph, draw the curve or line representing the payoff to a contributor as a function of the state variable M. Determine whether the two curves intersect, and if so where; and draw conclusions with respect to the value of the state variable M at a Nash equilibrium.

c. Compare the result to a Prisoner's Dilemma.

Q14.3. Gone Fishin'

Swellingham is a fishing port near the Fishy Banks. There are N fishermen and potential fishermen in Swellingham, and N is very large. On any given day $M \leq N$ fishermen take their boats out to Fishy Banks to fish. The catch per boat per day is $100/M$ tons of fish. The price of fish on the world market is \$100 per ton. The cost to take a boat out on a particular day is \$200. Each agent has two strategies on a particular day: Fish or do not fish. The payoff to "do not fish" is always zero. Let M be the state variable for this problem.

a. Determine how the profit (per day) of a representative fisherman varies as M changes. This is the payoff to the strategy "fish."

b. On a coordinate graph, draw the curve or line representing the payoff to an agent who fishes as a function of the state variable M. On the same graph, draw the curve or line representing the payoff to a non-fisher.

c. Determine whether the two curves intersect, and if so where; and draw conclusions with respect to the value of the state variable M at a Nash equilibrium.

d. Compare the result to a Prisoner's Dilemma.

Q14.4. Medical Practice

Recall Problem 5 in Chapter 6. Doctors are considering whether to practice as ob/gyn's or to limit their practices to gyn. Let's extend that model to a market with N doctors qualified for ob and gyn practice. Assume $M \leq N$ practice both ob and gyn, while the rest practice only gyn.

The total revenue from gyn practice is Q, so the payoff to a doctor who practices only gyn is Q/N.

The revenue from ob practice is R, but those who practice ob pay a fixed overhead cost of S, which is the dollar equivalent of the subjective disutility of being on call 24/7, so the net payoff to a doctor with an ob-gyn practice is $Q/N + R/M - S$.

Let M be the state variable for this problem. Assume

$Q = 10,000,000$
$R = 1,000,000$
$S = 50,000$
$N = 50$

a. Determine how the revenue of a representative ob-gyn varies as M changes. This is the payoff to the strategy ob-gyn.
b. On a coordinate graph, draw the curve or line representing the payoff to an agent who practices ob-gyn as a function of the state variable M.
c. On the same graph, draw the curve or line representing the payoff to a doctor who practices gyn only.
d. Determine whether the two curves intersect, and if so where; and draw conclusions with respect to the value of the state variable M at a Nash equilibrium.

Compare the result to a Prisoner's Dilemma.

Q14.5. El Farol

Reread the El Farol Game, the Crowding Game from Chapter 6. Rewrite it as a proportional game with a very large number of participants.

a. What are the strategies in this game?
b. Are the players representative agents? Explain.

c. Is there a state variable? Explain.
d. Does this game have a dominant strategy equilibrium? If so, what is it? Why or why not?
e. Does the game have a Nash Equilibrium? If so, what is it? Why?
f. If so, is it efficient? Why or why not?

Q14.6. More Proportional Games

Drive Right, Heave-Ho, Drive On, and the Stag Hunt, like Hawk vs. Dove, are examples from Chapter 5. All can be rewritten as proportional games with very large numbers of participants.

For each of these games, indicate the state variable. Draw a diagram to represent the proportional game. Discuss the implications for application of the models.

Q14.7. Guacamole Valley

Farmers in the Guacamole Valley rely on water wells for irrigation. A farmer can dig to either of two aquifers, one deeper and the other shallower. Wells sunk to the deeper aquifer cost more, but there is more water down there. There are 100 farmers producing in the Guacamole Valley, and every single one of them must irrigate in order to produce. The payoff to a shallow well is

$$\text{Payoff}_s = 12 - 0.15Q$$

where Q is the number of shallow wells. The payoff to a deep well is

$$\text{Payoff}_d = 7 - 0.1R$$

where R is the number of deep wells.

Analyze this situation as an N-person game. What are the strategies? What is the state variable? What can you say about equilibrium?

Q14.8. Rockin' at Midsize

Midsize University is located in Midsize City in the Midwest, and Sheriff Horsepistol is responsible for law enforcement there. The

Table 14.5.   The Concert Crowding Game at Midsize University.

Contingency	Payoff to go	Payoff to don't go
If 0–3,999 other students go	2	0
If exactly 4,000 other students go	−2	0
If 4,001–4,999 other students go	−2	−1

Student Activity Board (SAB) at Midsize University has scored a wonderful coup. The most popular Rock'n'Roll demigod of the year is a Midsize alum (with a degree in civil engineering) and he has agreed to give a free concert for students at Midsize. But there is (of course!) a problem. All 5,000 students at Midsize want to attend, and the conference venue will seat only 4,000. If even 4,001 students attend, it will violate the fire laws. To make matters worse, Sheriff Horsepistol is already pretty angry at Midsize students, whom he regards as a rowdy lot. Perhaps he is right, considering that a civil engineering student went on to be a rock demigod. Anyway, Sheriff Horsepistol has made it clear that if there is any violation of the fire laws, he will stop the concert, arrest those present, and shut down all the off-campus hangouts where the current generation of Midsize students spend their leisure time.

The 5,000 Midsize students are participants in a large crowding game. Their strategies are to go to the concert or not to go. The payoffs are shown in Table 14.5. (The payoff to don't go in the last line is −1 because there are nothing else to do that night, because all the hangouts are closed.) Analyze this example as an N-person game, using the concepts of representative agents and state variables.

CHAPTER 15

Duopoly Strategies and Prices

One of the objectives of John von Neumann and Oskar Morgenstern in their great book, *The Theory of Games and Economic Behavior*, was to solve an unsolved problem of economic theory: Oligopoly pricing. "Oligopoly" means

> **To best understand this chapter,** you need to have studied and understood the material from Chapters 1–4. Some knowledge of the principles of economics will also be useful.

"few sellers," and this means a simple supply-and-demand approach would probably be too simple. With only a few sellers, the sellers may have some capacity to cut back on production and raise the price and the profit margin, as a monopoly would do. But to what extent? Would the oligopolists raise the price all the way to the monopoly level? From their point of view, that would be the cooperative solution to the game; but if they act non-cooperatively or competitively, the price may fall below the monopoly target, and might even fall to the competitive price level.

The problem was already of long standing. Since duopoly — a market with just two sellers — is the most extreme form of oligopoly, many studies focused on duopoly pricing. If that problem could be solved, then — probably — it would be fairly easy to move on to three, four, or N sellers. In this chapter, we will survey some of the traditional duopoly models, reinterpret them in game theory terms, and then explore some extensions.

1. COURNOT MODELS

The first contribution to our understanding of duopoly pricing came from a French mathematician, Augustin Cournot, in 1838. Cournot assumed that each firm would decide how much product to put on the market, and the price would depend on the total. In 20th century economics, the Cournot model is associated with the idea of an industry demand curve. (That was an idea that Augustin Cournot invented, though some English-speaking economists probably came up with the idea independently several years later.) An example is shown in Figure 15.1.

An "industry" is a group of firms selling the same or closely substitutable products. Thus, the total output of all firms in the industry is shown on the horizontal axis and the price prevailing in the industry is shown on the vertical axis. The downward sloping line is the "demand curve," and it can be interpreted — equally correctly — in two ways. First, given the total output of all firms in the industry, the corresponding point on the

HEADS UP!

Here are some concepts we will develop as this chapter goes along:

Duopoly: An industry in which just two firms compete for customers is called a "duopoly."

Cournot Equilibrium: When two firms each put a certain quantity of output on the market, and sell at the price determined by the market, the output is the strategy and the resulting Nash equilibrium is a "Cournot equilibrium."

Bertrand-Edgeworth Equilibrium: When two firms each set a price, and the firm with the cheaper price dominates the market, the price is the strategy and the resulting Nash equilibrium is a "Bertrand-Edgeworth equilibrium."

Reaction Function: The best response by one firm may be a mathematical function of the strategy (price or quantity sold) chosen by the other firm. This function is called a "reaction function."

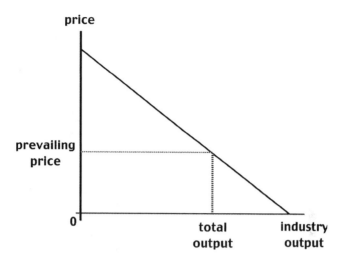

Figure 15.1. The Industry Demand Curve.

demand curve shows the price that will prevail in the industry. This is the interpretation used in Cournot models. Second, given the price in the industry, the distance from the vertical axis to the demand curve shows the maximum amount that can be sold in the industry. On either interpretation, the demand curve demonstrates the idea that a higher price will correspond to a smaller output sold.

Our duopoly industry will consist of two computer firms, MicroSplat and Pear Corp. The Cournot approach assumes that the two firms in the duopoly each decide how much to sell, put that quantity on the market, and sell it at whatever price results along the demand curve. (Notice, we are assuming that the two firms sell "homogenous products;" that is, that their products are perfect substitutes.) Thus, each firm has to make a guess — a "conjecture" — as to what the other firm will sell. Looking at Figure 15.2, suppose that Pear Corp.

Definition: *Conjecture* — A conjecture is a judgment on a question of fact, based on whatever evidence seems relevant, but which is inconclusive and might prove wrong.

conjectures that MicroSplat will put Q_1 units of output on the market. Then Pear can assume that they will have the segment of the industry demand curve to the right of Q_1 as, in effect, their own demand curve. Based on their conjecture about MicroSplat's plans, Pear understands that "their" demand curve has a zero point that corresponds

Definition: *Demand curve or function* — The relationship between the price of a good and the quantity that can be sold at each respective price is a demand relationship. It can be shown in a diagram as the demand curve, or mathematically as the demand function.

to Q_1 on the industry demand curve. In effect, Pear's demand is the residual, what is left over after MicroSplat maximizes its profits. Then Pear's problem is to adjust their output so that the industry price gives them the greatest possible profits.

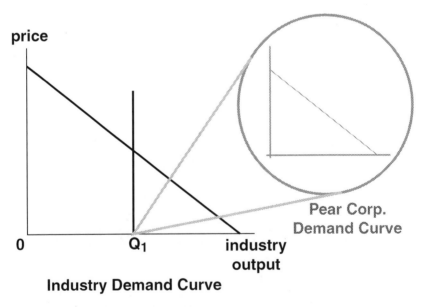

Figure 15.2. Firm 2's Conjecture and Estimated Demand Curve.

Cournot approached that problem using calculus, while basic economics textbooks often use a graphical approach. We will need to use a bit of that monopoly theory from principles of economics. We will apply the *law of one price,* which holds that every unit sold of a homogenous good must be sold for the same price. Also, we will need the concepts of *marginal cost* and *marginal revenue.*

> **Terminology:** *Marginal Revenue and Marginal Cost* — In monopoly theory, *marginal revenue* is the increase in revenue that results from the sale of one additional unit of output. *Marginal cost* is the increase in cost that results from the production of one more unit of output.

With a downward-sloping demand curve, a firm will have to cut its price a little bit in order to sell one additional unit of output. When we offset that price decrease (on all units sold) against the increase in sales, the increase in revenue will be less than the price at which the product is sold. The *marginal revenue* is the increase in revenue that results from the sale of one additional unit of output. The *marginal cost* is the increase in cost that results from the production of one more unit of output. Pear Corp. will act as a monopolist with its residual demand curve and adjust its output and sales in such a way that its marginal cost is equal to its marginal revenue.

Of course, MicroSplat will also have to try to conjecture about how much Pear will produce, and determine their own demand curve and profit-maximizing output in a similar way. Cournot's idea is that both companies are thinking in the same way, each making a conjecture about the output the other one will sell, and taking the remainder of the industry demand as their own, and choosing their profit-maximizing output accordingly.

Cournot's model predicts that the price and quantity sold will be intermediate between the monopoly price and quantity and the price and quantity that would be observed if there were perfect competition. This result can be generalized in several ways. We can

get rid of some of the simplifying assumptions, and we still
see the equilibrium price
somewhere between the competitive and monopoly prices.
The model can also be
extended to three, four or
more firms — and the more
firms we have in the marketplace, the closer the price
comes to the competitive
price. Best of all, this agrees
qualitatively with the evidence. When we observe
different industries and
compare them, we do seem to
find that, on the whole, oligopoly prices are lower than
monopoly prices, and the
more competitors there are,
the closer the price comes to
the marginal cost.

Cournot's model was the
first, and a very successful
model in many ways. All the
same, it had its critics, who
raised questions about its
logic and assumptions.

> **A Closer Look:** Augustin
> Cournot, 1801–1877
>
> Antoine Augustin Cournot was
> born in the French district of
> Franche-Compte in the town of
> Gray, France. He studied math-
> ematics at the Collège Royal in
> Besancon, the École Normale
> Supérieur in Paris, and the
> Sorbonne, winning recognition
> from the great mathematician
> Poisson. In 1838, while inspec-
> tor general of public education,
> he published *Recherches sur les
> principes mathematiques de la théo-
> rie des richesses* in which he
> discussed mathematical eco-
> nomics, and produced the first
> definition of demand functions.
> His theory of duopoly is recog-
> nized as an instance of the Nash
> equilibrium, so that some game
> theorists use the term Cournot-
> Nash equilibrium to refer to it.

2. BERTRAND MODELS AND NORMAL FORM GAMES

In a book review, Bertrand (1883) asked why the sellers would focus
on the **outputs** rather than compete in terms of prices. In game
theory terms, Bertrand is suggesting that the prices, and not the
outputs, would be the strategies among which the sellers would
choose. In Chapters 3 and 4, we had examples of duopoly pricing

games in which prices are the strategies as Bertrand argued. In those examples, we found that there is an unique Nash equilibrium at the lowest profitable price — that only the "competitive" price can be the Nash equilibrium. In a game of this kind, as long as there is any margin of price over marginal cost whatever, the best response is always to cut price below the other competitor.

In order to compare the Bertrand and Cournot approaches, we will construct an example in which prices and outputs are determined by the same demand curve, with prices as the strategies in one case and quantities as strategies in the other. Cournot approached that problem using calculus, while basic economics textbooks often use a graphical approach. We will

A Closer Look: Joseph Bertrand, 1822–1900

Joseph Bertrand was a prominent French mathematician and an opponent of the application of mathematics to the human sciences, including economics and psychology. In 1883, in a review of Leon Walras' *Théorie mathématique de la richesse sociale*, Bertrand also discussed Cournot's much earlier work in mathematical economics. In Bertrand's view it was quite mistaken, but Bertrand suggested instead that prices, not quantities, should be the strategic variables. This approach continues to be used in some work in mathematical economics to the present.

simplify the game as a game in normal form with just three strategies.

For our example, the industry demand (the total demand for the product of both firms together) will be as shown by Figure 15.3. This demand relation can be written in algebraic terms either as $Q = 9{,}500 - 5p$ or as $p = 1{,}900 - 0.2Q$, where Q is the total output of both firms and p is the price they both receive. Of course, profit depends on cost as well as demand. For each firm in this example, it is assumed that cost is $100 * Q$, where Q is the firm's own production and sales. Accordingly, the marginal cost is 100, regardless of the output sold. For a straight-line demand curve such as this, the

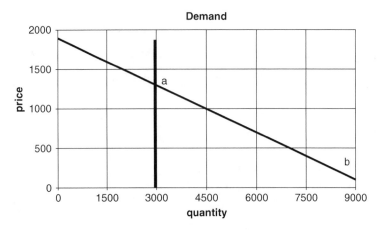

Figure 15.3. Demand for Oligopoly Example.

marginal revenue for a specific output is equal to the price at the zero output, but decreases with increasing output twice as fast as the price does. That is, the industry marginal revenue is MR = (1,900 − 0.4Q). Using the rule for maximum profits, we find that the monopoly profit-maximizing output would be 4,500 and the price would be 1,000.

We will limit MicroSplat and Pear Corp. each to just three strategies: Prices of 400, 700, or 1,000. As we saw, 1,000 is the monopoly price — total profits will be greatest if they both charge that price. In our example, this leads to payoffs as shown in Table 15.1. This will allow us to use the tabular form, that is, to represent this game in normal form as in the previous chapters. Payoffs are in millions. We see, as usual, that the only Nash equilibrium occurs when both firms charge the lowest price above marginal cost, in this case 400.

Let us see how this same example would appear in a Cournot analysis. For this analysis the quantity produced is the strategy. Once again, we will simplify by assuming only a limited number of possible strategies: Outputs of 2,000, 3,000, 4,000. Each firm will choose to produce one of those three quantities and the total amount for sale, on the industry demand curve, is the sum of the two quantities. For example, if MicroSplat chooses to produce and sell 2,000, and Pear

Table 15.1. Prices as Strategies and Payoffs to MicroSplat and Pear (In Millions of Dollars).

		Pear		
		400	700	1,000
MicroSplat	400	1.13,1.13	2.25,0	2.25,0
	700	0,2.25	1.8,1.8	3.6,0
	1,000	0,2.25	0,3.5	2,2

Table 15.2. Quantities as Strategies and Payoffs to MicroSplat and Pear.

		Pear		
		2,000	3,000	4,000
MicroSplat	2,000	2.2,2.2	1.8,2.7	1,2
	3,000	2.7,1.8	2.1,2.1	2,1.5
	4,000	1.4,2.8	1.5,2	1.2,1.2

chooses to produce and sell 3,000, the industry total is 5,000, and we see that the price in the industry will be $900. Reasoning along those lines, Table 15.2 shows the payoffs, in millions, that will result from every pair of output strategies the two firms may choose.

Again, payoffs are profits in millions, and the first number gives the payoff to MicroSplat. We see that the Nash equilibrium occurs when both firms sell 3,000 units. Indeed 3,000 units is a dominant strategy in this game. If we construct the best response functions for this example, we have Tables 15.3a and 15.3b.

The two tables are identical since the firms are symmetrical, though that might not always be so. We see that the *same* game (that is, the same cost and demand conditions) gives quite *different* results depending on whether the strategies are prices or quantities. But this is not a very satisfactory result. Shouldn't *both* prices *and* quantities play some role in a company's market strategies? There is a hint,

Table 15.3a. Best Response.

MicroSplat Strategy	Pear Best Response
2,000	3,000
3,000	3,000
4,000	3,000

Table 15.3b. Best Response.

Pear Strategy	MicroSplat Best Response
2,000	3,000
3,000	3,000
4,000	3,000

though, in a further criticism by an Irish economist named Edgeworth.

3. EDGEWORTH

Edgeworth (1897), agreed with Bertrand's criticisms of Cournot, but Edgeworth pointed out two complications. First, the sellers may have limited production capacity. In the examples in the previous section, at a price of 100 the total output sold is 9,000. If each company is limited to production of 4,000, then the price will not fall below 160, at which a total output of 8,000 can be sold. That's the "short run supply-and-demand" equilibrium price. Second, Edgeworth argued that a supply-and-demand price of 160 would not be stable either. Since each seller is selling all he

A Closer Look: Francis Ysidoro Edgeworth, 1845–1926

Edgeworth was born in Edgeworthtown, Ireland, of an Irish father and a Catalonian mother. He was educated by tutors and at Trinity College, Dublin, and Oxford, and after a long period of independent study of law, mathematics and statistics, became a very prominent economist and statistician. Edgeworth argued that exchange might not lead to any determinate outcome, and his 1897 critique of Cournot, although along the same lines as that of Bertrand, came to the conclusion that there might not be a determinate outcome.

can, the seller who charges a higher price will not lose customers to the lower price seller and may increase his profits by raising his price. Thus, no price is stable — either at or above the short run supply-and-demand level — and Edgeworth concluded that there will be no stable price, but rather price will be unpredictable and may fall anywhere over the whole range from the monopoly price down to the short run supply-and-demand price.

Edgeworth's reasoning is illustrated by Figure 15.4. Downward sloping line D is the industry demand curve that the two firms share. Each firm can produce up to a capacity limit of Q_0 units of output, at a cost of c per unit. The capacity limit for the industry as a whole is $2Q_0$. Thus, the industry "supply curve" is the right-angle formed by the cost line at c and the dark vertical line at $2Q_0$, and the supply-and-demand equilibrium price is p_0.

Now suppose that both firms are charging a price of p_1. Industry sales at that price are Q_1, so they split that output between them, each one selling $Q_1/2$ units of output. But if one of the two firms were to cut its price just slightly below p_1, it would be able to sell its entire capacity, Q_0, increasing its sales by $Q_0 - Q_1/2$. This lump

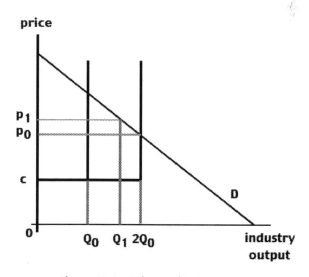

Figure 15.4. Edgeworth's Reasoning.

increase in sales at the cost of a very tiny cut in price will increase profits, so the best response to any price above p_0 will generally be a slightly lower price. It follows that no price above p_0 will be a Nash equilibrium. But p_0 may not be a Nash equilibrium either. At p_0, each firm is selling its capacity limit. If one firm increases its price above p_0, the other firm will be unable to take advantage of that by increasing its output — since it is already at its capacity limit — and therefore the firm that raises its price can act essentially as a monopolist with whatever is left of the market, and maximize its profits by restricting its output to less than Q_0. Edgeworth concluded that there may be no *predictable* price for a duopoly with a capacity constraint. Interpreting this in terms of game theory, we say that there is no Nash equilibrium in *pure strategies*, since every pure (price) strategy opens the firm to a counterstrategy that it, in turn, will want to counter.

In microeconomic theory, we define the short run as a period short enough so that the production capacity is fixed. In the short run, therefore, price is the only competitive strategy available to the companies. We might resolve the controversy between Cournot and Bertrand by saying that Cournot's model is applicable in the long run, when output capacity is variable, while Bertrand's analysis is applicable in the short run. But then, if Edgeworth is right, the short-run price has no stable equilibrium.

For now, though, we will ignore the capacity limits and return to the Cournot model, without some of the other simplifying assumptions made in the earlier sections of this chapter.

4. REACTION FUNCTIONS

In the examples in Section 2, we compared the Cournot model with the Bertrand model, limiting each firm to a few prices or quantities in order to apply the normal form game analysis. Of course, we should also allow for prices and quantities between the whole numbers, like \$750.98 or \$550.50. Let us see how that would work with a spreadsheet example. We return to the Cournot assumption that quantity sold is the strategy variable. For this example, the two companies in our duopoly game can choose any price and output within a profitable range. In general, there are infinitely many of them. For Cournot's example, each firm can produce zero or any positive quantity of output.

The two firms together determine the total amount on the market and the price is the corresponding point on the demand line. Thus, the demand curve for any one firm is the residual after the other firm has sold its offering. For example, suppose firm 1 sells 3,000 units of output. Then the zero point for Firm 2 is 3,000 total units of output, and the demand for Firm 2 corresponds to segment ab of the industry demand curve. Therefore, if MicroSplat sells 3,000 units of output, Pear's demand curve can be written in algebraic terms as $Q_P = 9,500 - Q_M - 5p$ or as $p = 1,900 - 0.2Q_M - 0.2Q_P$, where Q_M is MicroSplat's output, 3,000 units, and Q_P is Pear's output. We have $p = 1,300 - 0.2Q_P$.

Thus, each firm will choose the output that maximizes its profits given the output the other firm chooses, and depending on its own demand remaining after the other has sold the output it chose. We have assumed that cost is $100 * Q$, where Q is the firm's own production and sales, so that the marginal cost is a constant at 100. Pear's residual demand curve is can be written with the price as the dependent variable as $p = (1,900 - 0.02Q_M) - 0.2Q_P$. Thus, for Pear Corp the marginal revenue is MR $= (1,900 - 0.2Q_M) - 2 * 0.2Q_P$, which, simplified a little, is the same as MR $= (1,900 - 0.2Q_M) - 0.4Q_P$. So, MR = MC means $(1,900 - 0.2Q_M) - 0.4Q_P = 100$, and doing a little algebra, that is the same as $Q_p = 4,500 - 0.5Q_M$. This gives us the profit-maximizing output for Pear Corp. as a function of the output sold by MicroSplat. This function is called the *reaction function* for Pear Corp.

MicroSplat will also choose its output so as to maximize its profits, whatever output it may conjecture that Pear will sell, so we can find its reaction function in a similar way. Since the firms' cost and demand conditions are symmetrical, the reaction functions are also symmetrical. MicroSplat's reaction function is $Q_M = 4,500 - 0.5Q_P$.

Figure 15.5 shows the reaction functions for both firms in this example. The thin, solid line shows Pear's reaction to MicroSplat's output, which is shown on the vertical axis, and the thick, shaded line shows MicroSplat's reaction to Pear's output, which is shown on the horizontal axis. The point where the two reaction functions intersect is the Cournot-Nash equilibrium for this example, that is, the one point where each is choosing its best

Definition: *Reaction Function* — When strategies are chosen from a numerical scale, and each player's best response is expressed as a function of a strategy chosen by the other, this function is called a *reaction function.*

response to the other's strategy. At that point, each firm sells an output of 3,000 units, for a total industry output of 6,000. This corresponds to a price of 700. In this game the monopoly price would be 940 and the corresponding quantity sold would be 4,800. If the industry were "perfectly competitive," the price would be the same as the marginal cost, 100, which implies an industry output of 9,000. In the figure, the monopoly output is shown by the triangle, and the competitive output by the square, each divided equally between the

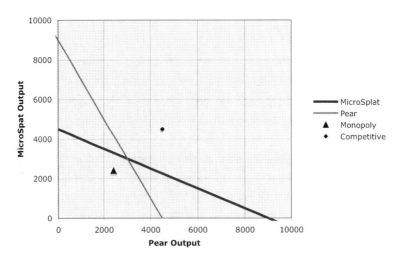

Figure 15.5. Reaction Functions.

two firms. These are not Nash equilibria, since they are not on the reaction functions.

This reaction-function technique can be used whenever two or more players in the game choose their strategies from some numerical scale.

Here we have applied the reaction function approach on the assumption that the quantity chosen is the strategy variable. When firms sell a homogenous product, the reaction function approach in inapplicable, since the best response to the other's price is always to undercut it, but by the least positive amount, which cannot be defined unless prices are limited to a finite list. Sometimes a model can be too simple. The challenge is to make the model just complicated enough. In much recent research, the duopoly model is extended to allow for one more complexity: Product differentiation. Modern economic theory usually assumes products are differentiated as a rule, but often relies on the idea that in the long run, free entry depresses profits to the lowest sustainable level. However, where entry is limited, the short-run condition that marginal cost equals marginal revenue continues to be applicable in the long run; that is, the oligopoly theory is the appropriate one. Accordingly, we continue to discuss a duopoly, but allow for product differentiation.

5. PRODUCT DIFFERENTIATION

In all of the pricing strategy models in this chapter so far, we assume that the rival firms sell identical products. In many industries, however, this assumption is not applicable. Different firms may sell products that are not perfect substitutes, so that some customers prefer one product rather than another. This is called "product

Definition: *Product Differentiation* — When different firms sell products or services that are not perfect substitutes, and make the distinction among the products a basis of promotion or an aspect of market strategy, we refer to this as *product differentiation.*

differentiation," and can be as important as pricing in the market strategies of many firms.

Suppose, instead, that we have two competing firms that sell differentiated products. The assumptions that lead to the Cournot model and to the Bertrand model are no longer applicable. If one firm limits its production, it will be able to sell primarily to the customers who prefer its own product rather than that of the other firm, and, presumably, charge a higher price accordingly. This is contrary to the Cournot model. On the other hand, if one firm charges a price above marginal cost, it will retain some of its customers even when the other firm charges a lower price. This contradicts the Bertrand model.

Product differentiation means that, in effect, each firm has a demand curve for its own output. Thus, the distinctions between the Cournot model and the Bertrand model disappear: By selecting a price, the firm also at the same time designates a quantity to be sold, and conversely. In effect, its strategy is a point on its demand curve. However, the demand curve for one firm will depend on the price asked by the other firm. If my rival raises his price, I can expect some of his customers to come to me and thus to have a favorable shift in my own demand curve. Thus, once again, I have to conjecture what my rival's price will be and try to choose my own price and quantity as my best response to it.

To keep things simple, we will continue to suppose that the two firms are identical in every way except for the characteristics of the product they sell. Figure 15.6 shows the individual demand curve for Firm 1. The solid curve is the demand curve if Firm 2 prices its product at $100, while the shaded curve is Firm 1's demand curve in case Firm 2 prices its product at $200.

Figure 15.7 shows how the quantity demanded[1] for Firm 1's product increases as Firm 2 raises its price. Suppose Firm 1 prices its

---

[1] Quantity demanded is computed according to the formula $Q = 10{,}000 - 50p + 20z$, where Q is the firm's quantity demanded, p is the firm's price, z is the rival firm's price. With linear demand functions such as these, the elasticity of demand for Firm 1's product depends on Firm 2's price. In the economics literature, it is

Demand Curves

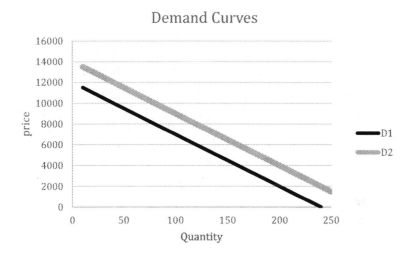

Figure 15.6.   Demand for Firm 1.

Firm 1 Quantity Demanded at $100

Figure 15.7.   Demand and the Rival's Price.

product at \$100. Firm 2's price is shown on the horizontal axis and Firm 1's quantity demanded is on the vertical axis.

Thus, each firm makes a conjecture as to his rival's price and quantity offered (and thus as to his own demand curve) and chooses his best response, which is the maximum profit along his expected demand curve.[2] As in the Cournot model, we can think of the best-response price as a function of the price charged by the other firm, that is, a reaction function. Figure 15.8 shows the reaction functions[3] for each firm. The price charged by Firm 2 is on the horizontal axis and that charged by Firm 1 is on the vertical axis, so that Firm 1's reaction function is shown by the solid curve while Firm 2's reaction function is shown by the shaded curve.

As before the intersection of the two reaction functions is the Nash equilibrium for this game of price competition. Using numerical computation, we can determine that each firm will ask a

---

common to make the assumption that $Q = \alpha p^{\beta} z^{\gamma}$ for appropriate constants α, β, γ. In this case, however, the price $p = (\frac{1}{1+\beta})MC$ becomes a dominant strategy, where MC is the firm's marginal cost. The linear formulation is a little simpler but also illustrates how reaction functions may be applied to this case.

[2] Of course, profits also depend on cost, and for this example each firm has a total cost of 100Q, so that the marginal cost is constant at 100.

[3] Derivation of the reaction functions requires a bit of calculus, and we will present it briefly in this footnote. For each Firm 1, the profit is $(p - 100)Q = (p - 100)$ $(10,000 - 50p + 20z)$. This is a quadratic expression in p, as follows:

$$\text{profit} = -50p^2 + 15,000p + 20pz - 1,000,000 - 2,000z$$

To determine the profit-maximizing price, we take the derivative with respect to p and set it equal to zero:

$$\frac{d\text{profit}}{dp} = -100p + 15000 + 20z = 0$$

Solving for p we have
$$P = 150 + 0.2z$$

This is Firm 1's reaction function. Firm 2's reaction function is

$$z = 150 + 0.2p$$

This method is discussed in more detail in the appendix to this chapter.

Reaction Functions

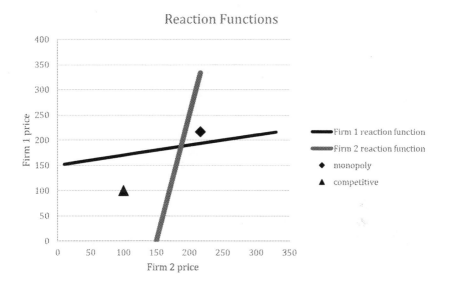

Figure 15.8. Price Reaction Functions.

price of $187.50, and this corresponds to the intersection. Each will offer 4,375 units for sale and receive a profit of $382,812.50.

What if, instead, the two firms were merged and run as a monopoly? Then *both* prices would be chosen so as to maximize the *total* profit of the two firms (that is, the two divisions of the merged firm) together. Here, again, the prices can be determined by numerical computation, and we find that profits are maximized if both divisions of the new monopoly firm charge prices of $216.67 for their distinct products. This is shown by the diamond in Figure 15.8. It is not a Nash equilibrium — it is off both reaction functions — but maximizes the profit of both of the firms together. At prices of $216.67, the two divisions of the merged monopoly firm would offer 3,500 units for sale and the profit from each division would be $408,333.33. Thus, we see that competition between these two firms would lead to some reduction of prices and profit margins, and some increase in their output, although even in the Nash equilibrium their prices remain above the marginal cost of $100 and their profits remain positive. If price competition were "perfect," so that each firm charged just its marginal cost, then the prices would be at 100 each, as shown by the triangle.

6. REACTION FUNCTIONS IN GENERAL

The reaction function approach can be applied in any case in which the strategy is a number to be chosen from some interval. Let us consider an example of the military defense policies of two neighboring countries. The Republic of Bogritania and the Pomegonian Peoples' State are the only two countries on the Isle of Isogonia. They are hostile, and each feels that for its defense, it must have an army of at least 10,000 soldiers plus half the number fielded by its rival. How large are the armies of the two countries in a Nash equilibrium?

The number of soldiers is the strategy for each of the two countries. (Although it is impossible to deploy half a soldier, they might choose to deploy a soldier for half a year.) Thus, the reaction functions are $q_A = 10000 + \frac{1}{2}q_b$, where q_A is the country's own army and q_B is the other country's army. We will approach the solution using algebra. Since each must be on its own reaction function, we may substitute

$$q_A = 10000 + \frac{1}{2}\left(10000 + \frac{1}{2}q_A\right)$$

$$= 15000 + \frac{1}{4}q_A$$

$$\frac{3}{4}q_A = 15000$$

$$q_A = 20000$$

Therefore, we have a Cournot-Nash equilibrium when each country deploys an army of 20,000; and if so, each country deploys its minimum of 10,000 plus 10,000 more representing half of the force deployed by the other country. The reaction functions for the two countries are shown as Figure 15.9.

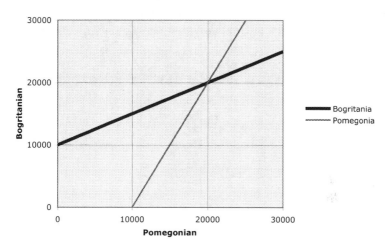

Figure 15.9. Reaction Functions for Army Sizes in Two Countries

Here is a further example to do with water for irrigation. For this example, two neighboring countries, Eastria and Westria, share a river that forms their mutual boundary. The Trickle River flows from north to south, and its waters are used for irrigation in the south. There are tributaries to the north in both countries, and both can withdraw water from the northern tributaries for irrigation there, but the more one of them diverts in the north the less will be available (in both countries) for irrigation in the south. If both countries pump out 1,000,000 kiloliters from the northern tributaries, about the same amounts will be available for irrigation of agriculture in the south. Whenever Eastria cuts its diversion below 1,000,000, Westria feels that its best response is to increase its diversion by half of the amount of Eastria's cut, while if Eastria increases its diversion, Westria's best response is to cut back their own diversion by half as much. Eastria's best responses to Westria's diversions are symmetrical.

The reaction functions are of the form $Q_W = 1,000,000 + \frac{1}{2}(1,000,000 - Q_E)$ and $Q_E = 1,000,000 + \frac{1}{2}(1,000,000 - Q_W)$, where Q_W is the diversion of water by Westria and Q_E is the diversion of water by Eastria. Once again, we can find the simultaneous solution of these two reaction functions by substitution and simplification:

$$Q_W = 1,000,000 + \frac{1}{2}\left(1,000,000 - Q_E\right) = 1,500,000 - \frac{1}{2}\left(1,000,000 - Q_W\right)$$

$$Q_W = \frac{1,500,000}{2} + \frac{1}{4}Q_W$$

$$\frac{3}{4}Q_W = \frac{1,500,000}{2}$$

$$Q_W = \frac{2}{3}1,500,000 = 1,000,000$$

In this case, the Nash equilibrium is that neither will deviate from the policy of extracting one million kiloliters, since only in

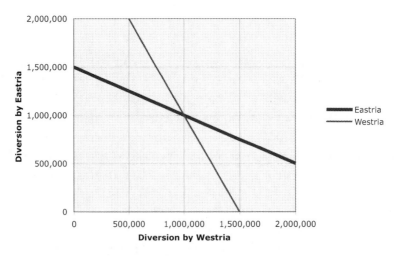

Figure 15.10.   Reaction Functions in a Water Game.

that case we have each country choosing its best response the other country's policy. The reaction functions are shown in Figure 15.10.

7. SUMMARY

It seems we have to acknowledge that the best pricing strategy will always depend a great deal on the specifics of the industry and its situation. There is no "architectonic," overarching model of rational pricing and output decisions. But there are principles that will have application in many cases, and so they are part of the tool-kit of any economist or businessperson concerned with pricing strategy. First, the concept of Nash equilibrium itself — that the rivals will each choose a strategy which is the best response to the other's strategy — will be applicable to all cases of price competition among rational, profit-seeking rival firms. Second, where the firms have control of the quantities they offer for sale but little control over the price, Cournot models and reaction functions are a good starting point. Where product differentiation is important, a more complex kind of reaction function will seem to be the best starting point. Where price competition is strong, the Bertrand model offers important insights, but limitations on productive capacity (or on inventories) may have profound and complicated impacts on pricing strategy.

APPENDIX: A MATHEMATICAL TREATMENT OF THE COURNOT MODEL

This appendix gives a discussion of the Cournot model in terms of calculus and marginal analysis. The discussion is more general as it makes fewer assumptions about the form of demand and cost functions. In addition, we show how the model is generalized to allow for differentiated products when the quantity offered is the strategy variable. To understand this appendix, the student needs some understanding of ordinary and partial differentiation and of the indefinite integral in calculus.

For the Cournot model in the main text of the chapter, the price prevailing in the industry is a function of the total output:

$$p = f(Q_A + Q_B) \qquad (15.A1)$$

where Q_A is the output of Firm A and Q_B is the output of Firm B. To allow for product differentiation, we instead assume

$$p_A = f_A (Q_A, Q_B) \qquad (15.A2a)$$

$$p_B = f_B (Q_B, Q_A) \qquad (15.A2b)$$

Here is the idea: Since the outputs of Firms A and B are not perfect substitutes, it seems that an increase in the output of Firm B may have less impact on the market price for Firm A than an equal increase in Firm A's output would have, and conversely. Equation (15.A1) is a special case of (15.A2), in which both functions are the same and Q_A and Q_B are combined by the sum operator.

Assume the costs are determined by

$$C_A = g_A (Q_A) \qquad (15.A3a)$$

$$C_B = g_B (Q_B) \qquad (15.A3b)$$

Each firm aims to maximize its profit:

$$\Pi_A = p_A Q_A - C_A \qquad (15.A4a)$$

$$\Pi_B = p_B Q_B - C_B \qquad (15.A4b)$$

When we apply calculus to the problem of finding a maximum, we rely on "necessary conditions" using the derivative. The intuition behind this is illustrated by Figure 15.A1. The plot shows how variable y changes as x changes. We want to find x_0, the value that corresponds to the largest value of y. We recall that the derivative of x with respect to y can be visualized as the slope of a tangent to the

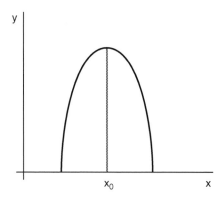

Figure 15.A1. Visualizing the Maximum.

curve. At the top of the curve, the tangent is flat, that is, the curve has a slope of zero. So, for a simple case like Figure 15.A1, the "necessary condition" for a maximum is that $\frac{dy}{dx} = 0$. There could be other values of x for which the slope is zero, but y is not at a maximum — for example, the slope would also be flat when y is at a minimum. So, the necessary conditions are not sufficient: We need some additional "sufficient" conditions. However, for the rest of this appendix we will not bother about that complication and will only explore the implications of the necessary conditions.

Maximization of (15.A4a) or (15.A4b) is a little more complicated. Without going into details, the necessary conditions are

$$p_A - Q\frac{\partial f_A}{\partial Q_A} - \frac{\partial g_A}{\partial Q_A} = 0 \tag{15.A5a}$$

In terms of economic theory, $\frac{\partial g_A}{\partial Q_A}$ is the firm's marginal cost, MC_A, and the marginal revenue is $MR_A = p_A - Q\frac{\partial f_A}{\partial Q_A}$, so (15.A5a) is equivalent to the familiar formula from economics, $MC = MR$.

Equation (15.A5a) can be rewritten as

$$\frac{\partial g_A}{\partial Q_A} = p_A\left(1 - \frac{Q_A\partial f_A}{p_A\partial Q_A}\right) \tag{15.A5b}$$

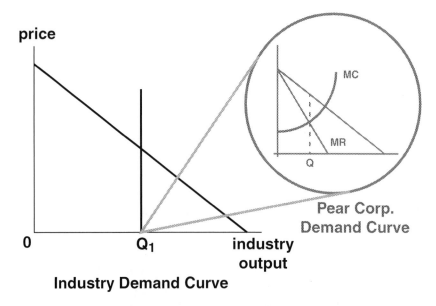

Figure 15.A2. Diagrammatics for the Calculus-Based Cournot Model.

In microeconomics, the *elasticity of demand* at a point on the demand curve is $\varepsilon = \frac{1}{\frac{Q_A \partial f_A}{p_A \partial Q_A}}$, so (15.A5b) yields

$$p_A = \frac{\partial g_A}{\partial Q_A}\left[\frac{\varepsilon}{1-\varepsilon}\right] = MC\left[\frac{\varepsilon}{1-\varepsilon}\right] \qquad (15.A5c)$$

Thus, the elasticity of demand determines the profit-maximizing markup of price over marginal cost. This idea is due to Abba Lerner and is called the Lerner Rule.

Figure 15.A2 shows the industry and individual firm demand curves as in Figure 15.2 in the text, with the individual firm's marginal cost and marginal revenue curves added in a slightly darker gray. The marginal revenue curve is the downward sloping line and the marginal cost curve is the upward sloping curve. Thus, as we would see in microeconomics, the firm will choose to produce Q, corresponding to the point where the marginal revenue curve cuts the marginal cost curve.

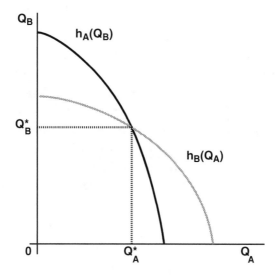

Figure 15.A3. Reaction Functions in a Nonlinear Example.

The profit-maximum condition is similar for Firm B. Solving for Q_A

$$Q_A = \frac{p_A - \dfrac{\partial g_A}{\partial Q_A}}{\dfrac{\partial f_A}{\partial Q_A}} = \frac{p_A - MC_A}{\dfrac{\partial f_A}{\partial Q_A}} \tag{15.A6}$$

since $\frac{\partial g_A}{\partial Q_A} = MC_A$. Differentiating (15.A6) we have

$$\frac{dQ_A}{dQ_B} = -\left[\frac{Q_A \dfrac{\partial^2 f_A}{\partial Q_B \partial Q_A} - \dfrac{\partial f_A}{\partial Q_B}}{(Q_A + 1)\dfrac{\partial^2 f_A}{\partial Q_A^2} - \dfrac{\partial f_A}{\partial Q_A}} \right] \tag{15.A7}$$

and, integrating this expression, we have the "reaction function" for Firm A,

$$Q_A = h_A(Q_B) \qquad\qquad (15.A8)$$

For practical purposes, as in the main text, $h_A(.)$, can often be found by substitution without going through the step of integration. Figure 15.A3 shows the reaction function for two firms. The Cournot equilibrium is at p_A^* and p_B^*, the intersection of the reaction functions. (We assume that the equilibrium is stable without investigating the conditions under which it will or will not be stable.)

As we have seen, the Cournot equilibrium in this case is a Nash equilibrium, and the Nash equilibrium with product differentiation can be thought of as a generalization of both the Bertrand and Cournot equilibria.

Q15. EXERCISES AND DISCUSSION QUESTIONS

Q15.1. Elevator Music

K*** and W§§§ are two soft-rock radio stations. They compete for an overlapping audience, but their styles are different, so that their products are, so to say, differentiated. Their profits from advertising are their payoffs, and the advertising revenue is proportional to the audience. Their strategies are to determine how much of their time to devote to commercials rather than music. If one of them cuts back on music to run more commercials, some of their audience will shift to the other station, and some will just turn off, and so the station's revenue can decline as a result. Taking this into account, each feels that the number of commercials per hour should be two-thirds of the other station's number per hour, plus 5. In a Nash equilibrium, what is the number of commercials per hour for each station? Explain in detail.

Q15.2. A Water Duopoly

Green Valley Water and Granite Slope Water are sellers of spring water in the Republic of Bogritania. Like mineral water sellers in many simple examples they have no costs so their payoffs are the revenues they obtain from selling water. The price of mineral water in Bogritania is $1,000 - q_v - q_s$, where q_v is the quantity sold by Green Valley and q_s is the quantity sold by Granite Slope. As a result:

Proposition A: Each firm's best response is to sell exactly half the amount left over by the other firm: For example, the best response q_v is $q_v = \frac{1}{2}(1000 - q_s)$.

a. Determine the Nash equilibrium for this duopoly.
b. Bonus question using calculus: Demonstrate why Proposition A must be true.

Q15.3. Another Water Duopoly

In the nearby Pomegonian Peoples' State, there are also two mineral water sellers, Brown Mountain and Limestone Hill. They too have no costs. Their water tastes sufficiently different so their products are differentiated. For each, the quantity demanded is $100 - 2p_A - p_B$, where p_A is the firm's own sales and p_B is the other firm's sales. We have

Proposition B: Each firm's best response price is determined by $p_A = 25 + \frac{p_B}{4}$.

a. Determine the Nash equilibrium prices for these two firms.
b. Bonus question: Demonstrate why Proposition A must be true.

Q15.4. Scholarship

West Philadelphia University and its rival Feehigh University compete to enroll many of the same students. Like most universities, they offer many "scholarships" that cover a fraction of the tuition. The objective is to enroll many students who would not enroll if they paid the full tuition, increasing "net tuition revenue." Thus, although different students pay different rates of net tuition, each university believes that it can enroll more students by offering more "scholarships," up to a point. Each university can choose a rate of discount — that is, total "scholarships" divided by the tuition that would be paid by enrolled students if there were no scholarships. Each university believes that its net tuition revenue is maximized when it gives scholarships at a discount rate that is 5% plus two thirds of the rate of discount chosen by the other.

a. Discuss this interaction as a non-cooperative game.
b. Use and define the game-theoretic terminology for this example, derive a solution, and explain.

CHAPTER 16

Rationalizable Strategies

While the Nash equilibrium is the fundamental concept of non-cooperative game theory, it is not the last word. This chapter and the next will explore some extensions of the reasoning behind Nash

> **To best understand this chapter,** you need to have a good understanding of the material from Chapters 1–5 and 7–8.

equilibrium. Like the Nash equilibrium, these concepts draw on the idea that a rational decision-maker will choose a best response to the strategy that the decision-maker believes that the counterpart will choose. In this chapter, we will allow for the possibility that the decision-maker may have mistaken beliefs about the counterpart's strategy, something that cannot occur in a Nash equilibrium. Nevertheless, we will rely on a common assumption of non-cooperative game theory: The common knowledge of rationality. That is, we assume not only that each decision-maker is rational, but that each knows that the other is rational, and each knows that the other knows that she is rational, and so forth.

1. THE RADIO FORMATS GAME AGAIN

While the Nash equilibrium is a key topic in non-cooperative game theory, it assumes that each player knows enough about the other players' strategies to choose a best response to them. This may be a problem in some games. In this chapter, we revisit that problem and consider how the players in the game might think it through.

The Radio Formats Game from Chapter 4, Section 3 provides a good example for our purposes. It is shown in normal form in Table 16.1. To choose a best response, a decision-maker has to rely on a conjecture as to which strategy the counterpart will choose. In games with dominant strategies, this is not very important. In a game with a dominant strategy, as we saw in Chapter 3, a decision-maker has a relatively easy decision to make. Let us call the decision-maker A and her counterpart or opponent B. Decision-maker A, being rational, wants to choose her best response to the strategy chosen by B. Since her best response is the same, regardless of the strategy chosen by B, A does not need to know anything about B's strategy choice in order to make their choice. They do not even need to know whether B is rational.

In a game like the Radio Formats Game, this is no longer true. Neither player has a dominant strategy. For W***, either Top 40 or Blend may be a best response, depending on the strategy chosen by K†††. Conversely, any of the three strategies could be a best response for K†††, depending on the strategy chosen by W***. Thus, in order to choose a strategy, the manager of W*** must make a conjecture as to which strategy the manager of K††† will choose. Moreover, the manager of W*** knows that the manager of K††† will base her decision on her best guess as to which strategy the manager of W*** will choose. How are we to break out of this circle?

In this case the problem can be simplified a little. Some strategies can be ruled out. The manager of K††† might reason "W***

Table 16.1. A Radio Formats Game (Repeats Table 4.3, Chapter 4).

		K†††		
		Top 40	Classic Rock	Blend
W***	Top 40	30,40	50,45	45,40
	Classic Rock	30,60	35,35	25,65
	Blend	40,50	60,35	40,45

might choose Top 40 or Blend, but never Classic Rock. After all, Classic Rock is not a best response to any strategy that I might choose, so the manager of W***, being rational, will never choose it." Indeed, the manager of K††† can go still further. They may reason "In fact, W*** will not choose Top 40, either, even though it is a best response to Blend; but Blend is my best response only if W*** chooses Classic Rock, and being rational, they know I know that and so I will never choose Blend, so it will never be a best response for them to choose Top 40." Thus, the manager of K††† can conclude, "Indeed W*** will choose Blend, and so I will choose my best response to Blend, which is Top 40." *And the manager of W*** knows this,* since they know that the manager of K††† is rational, so they can reason "The manager of K††† thinks I will choose Blend, so they will choose their best response, Top 40; therefore, my best response is Blend." Similarly, the manager of K††† can reason, "The manager of W*** expects me to choose Top 40, so they will choose their best response, Blend; accordingly, I will choose my best response to Blend, which is Top 40." And when they play their strategies, they discover that they are right.

Strategies that can be justified by reasoning along these lines — "They think I will choose strategy i, so they will choose strategy j, which is their best response to strategy i; therefore, I will choose strategy k, my best response to strategy j" — are called *rationalizable strategies.* This is a technical term in game theory. It does not mean a strategy that is rationalized by just any argument you might come up with. To qualify the strategy as a rationalizable strategy, the argument must make use of the idea that each decision-maker is rational and knows that the other one is rational also. Thus, the reasoning begins from an assumption that I will choose a particular strategy that it could be reasonable for me to choose, in that it would be a best response to one of the strategies I might face. I then reason that my counterpart will choose their best response to it, and I choose my best response to their best response. So, in the Radio Formats Game, by choosing "Blend, Top 40" the stations have both chosen rationalizable strategies.

2. A METHOD FOR FINDING RATIONALIZABLE STRATEGIES

Rationalizable strategies seem complex and can be hard to follow, but they are a little simpler than they may at first seem. There is a straightforward way to find all of the rationalizable strategies in a game in normal form such as the Radio Formats Game. To see that we will need a little more terminology.

Any strategy may be a best response to some strategy that the other player or players might choose, or it may not. If strategy *i* is *not* a best response to *any* strategy that the other player or players might choose, then it is an *irrelevant strategy*.[1]

Let us take another look at the second row of the Radio Formats Game, the strategy of Classic Rock for station W***. We see that this strategy is not a best response to any of the strategies that K††† might choose. Thus, Classic Rock is an example of an irrelevant strategy from the point of view of W***, in the Radio Formats Game as shown in Table 16.1.

Since we know that W***, as a rational decision-maker, will never choose Classic Rock, we might drop that strategy out of the game, producing a new and smaller *reduced game*. The reduced game is shown in Table 16.2. When we examine this game, we see that in the reduced game, Blend is an irrelevant strategy for K†††. Once again, we eliminate the strategy Blend for K†††, leaving the second-stage reduced game shown in Table 16.3. Now, examining this second-stage reduced game, we see that in this game, Top 40 is an irrelevant strategy for W***. Once again, we eliminate it, leading to the third-stage reduced game in Table 16.4. In the third round reduced game, once again, we find that Classic Rock is an irrelevant strategy for K†††. Yet once more we eliminate it, leading to the fourth-stage reduced game in Table 16.5.

---

[1] There is no widely accepted term in game theory for a strategy with this property. This meaning will be adopted for the term "irrelevant strategy" throughout this book.

Table 16.2. A Reduced Radio Formats Game.

		K✝✝✝		
		Top 40	Classic Rock	Blend
W***	Top 40	30,40	50,45	45,40
	Blend	40,50	60,35	40,45

Table 16.3. A Second Round Reduced Radio Formats Game.

		K✝✝✝	
		Top 40	Classic Rock
W***	Top 40	30,40	50,45
	Blend	40,50	60,35

Table 16.4. A Third Round Reduced Radio Formats Game.

First payoff to W***, second payoff to K✝✝✝.		K✝✝✝	
		Top 40	Classic Rock
W***	Blend	40,50	60,35

Table 16.5. A Fourth Round Reduced Radio Formats Game.

First payoff to W***, second payoff to K✝✝✝.		K✝✝✝
		Top 40
W***	Blend	40,50

Of course, this fourth-round reduced game has no irrelevant strategies, so we can go no further. What we have done in this example could be called the *iterated elimination of irrelevant strategies*, IEIS.[2] When we apply IEIS to a game in normal form, the strategies that remain, and cannot be eliminated, are the rationalizable strategies for the game. We see, then, that the Radio Formats Game has only two rationalizable strategies: Blend for W*** and Top 40 for K†††.

Of course, these strategies are the Nash equilibrium strategies for the Radio Formats Game. And this is true in general: Nash equilibrium strategies are always rationalizable. But, as we will see, the converse may not be so in more complex games.

3. DOMINATED STRATEGIES

In this chapter and the previous chapters, we have studied a number of examples in which one strategy (call it Strategy A) always gives a greater payoff than another strategy (which will be called Strategy B) for any strategy chosen by the other player or players. Then we say that strategy A *dominates* strategy B and that strategy B is a *dominated* strategy. Sometimes we stress that the payoffs to Strategy A are greater than those to strategy B by saying that strategy A *strongly* or *strictly* dominates strategy B, and that strategy B is *strongly* or *strictly dominated*.

Here is a key point to keep in mind: *If a strategy is dominated it is always irrelevant*. This must be true since the payoff to the dominant strategy A is always greater than the payoff to dominated Strategy B, so strategy A will always be a better response to any strategy than Strategy B is. Strategy B will never be the best response. However, as we have seen, we may also have irrelevant strategies that are not dominated by any other pure strategy.

---

[2]In the research literature of game theory, this method is more often called "iterative elimination of dominated strategies." If we allow for mixed strategies (Chapter 8), then there is no difference, since every irrelevant strategy will be dominated by some mixed strategy. But this novel terminology allows us to define rationalizable strategies without reference to mixed strategies, for a somewhat simpler discussion.

Since a dominated strategy is always irrelevant, dominated strategies will be eliminated in the process of IEIS. In 1953, John Nash and Lloyd Shapley took advantage of this idea and solved a simplified poker game by iterative elimination of all dominated strategies. For years afterward, this was a standard method in game theory. It was not until the 1980's that Bernheim and Pearce defined rationalizable strategies and derived the method of IEIS to discover them.

4. THE NUTTER GAME REVISITED

An important application of the concepts of Nash equilibrium and dominant strategies in economics is the theory of price decisions in oligopolies, and in the simplest case, duopolies. We owe this application to Professor Warren Nutter of the University of Virginia, and we read about "Nutter games" of pricing strategies in Chapters 3 and 4. In Chapter 3, allowing only two prices — monopoly and competitive — we saw that the competitive price would be a dominant strategy. However, reconsidering this in Chapter 4, with three strategies, we saw that there might not be a dominant strategy, but that the competitive price would be the unique Nash equilibrium. (These models are again revisited in Chapter 11, with quite different results). Here we take another look at the 3-strategy Nutter game.

Let us find the rationalizable strategies by applying IEIS to this game, shown in Table 16.6. We see that, for both companies, a high

Table 16.6. The Three-Strategy Nutter Pricing Game Once Again (Reproduces Table 4.8).

		Grossco		
		Low	Medium	High
Magnacorp	Low	20,20	80,10	90,5
	Medium	10,80	60,60	150, 15
	High	5,90	15,150	100,000

price is never a best response — the high price strategy is an irrelevant strategy. For each company, we can eliminate this strategy to
create a reduced game. Since the game is symmetrical, and the strategy is irrelevant for both players, it does not matter in which order
we eliminate it. Once we have eliminated the high price strategy for
each player, we have the reduced game shown in Table 16.7.

We are not finished quite yet. Examining Table 16.7, we see that
the medium pricing strategy is irrelevant for each of the two companies. In fact, the low price strategy is a dominant strategy in the
reduced game. Conversely, the medium price strategy is a dominated strategy, and we recall that a dominated strategy always is
irrelevant. Anyway, we can eliminate it for each of the two players,
giving rise to the reduced game shown as Table 16.8.

What we see in Table 16.8 is that the only rationalizable strategies in this game are the low pricing strategies, and once again, the
rationalizable strategies correspond to the Nash equilibrium for this
game. This seems an important insight about price competition:
When one company can gain against its rival by asking a slightly
lower price, only the lowest profitable price, the competitive price,

Table 16.7.   A Reduced Nutter Pricing Game.

		Grossco	
		Low	Medium
Magnacorp	Low	20,20	80,10
	Medium	10,80	60,60

Table 16.8.   A Further Reduced Nutter
Pricing Game.

		Grossco
		Low
Magnacorp	Low	20,20

is rationalizable; and as a result, it will always be chosen at a Nash equilibrium.

5. THE RETAIL LOCATION GAME REVISITED

Another, relatively complicated game from Chapter 4 was the retail location game. Here, we found that there could be two Nash equilibria. Let us apply the rationalizable strategies idea to this game. Table 16.9 shows the game in normal form with best responses underlined for convenience.

As we see from the underlines, this game has two Nash equilibria. Whenever one of the two companies locates in Upscale Mall and the other locates in Center City, we have a Nash equilibrium. Now let us investigate the rationalizable strategies for this game, using the method of iterated elimination of irrelevant strategies. Examining Table 16.9 with the underlines, we see that Snugburb is an irrelevant strategy for Nicestuff. Thus, we eliminate it, leaving the reduced game shown as Table 16.10.

In the first-stage reduced game shown in Table 16.10, we see that Uptown is now an irrelevant strategy for Wotchaneed's. Thus, we eliminate it, giving rise to the second-stage reduced game shown as Table 16.11.

Table 16.9. The Retail Location Game with Underlines (Approximately reproduces Table 4.11, Chapter 4).

		Wotchaneed's			
		Upscale Mall	Center City	Snugburb	Uptown
Nicestuff Stores	Upscale Mall	3,3	10,9	11,6	8,8
	Center City	8,11	5,5	12,5	6,8
	Snugburb	6,9	7,10	4,3	6,12
	Uptown	5,10	6,10	8,11	9,4

Table 16.10.   A Retail Location Game, First-Stage Reduction.

		Wotchaneed's			
		Upscale Mall	Center City	Snugburb	Uptown
Nicestuff Stores	Upscale Mall	3,3	10,9	11,6	8,8
	Center City	8,11	5,5	12,5	6,8
	Uptown	5,10	6,10	8,11	9,4

Table 16.11. A Retail Location Game, Second-Stage Reduction.

		Wotchaneed's		
		Upscale Mall	Center City	Snugburb
Nicestuff Stores	Upscale Mall	3,3	10,9	11,6
	Center City	8,11	5,5	12,5
	Uptown	5,10	6,10	8,11

Table 16.12.   A Retail Location Game, Third-Stage Reduction.

		Wotchaneed's		
		Upscale Mall	Center City	Snugburb
Nicestuff Stores	Upscale Mall	3,3	10,9	11,6
	Center City	8,11	5,5	12,5

In this second-stage reduced game, we see that Uptown is an irrelevant strategy for Nicestuff Stores. Eliminating it gives us the third-stage reduced game shown in Table 16.12.

In the third-stage reduced game, we see that Snugburb is an irrelevant strategy for Wotchaneed's. Accordingly, we eliminate it to obtain the fourth-stage reduced game shown in Table 16.13.

Examining the fourth-stage reduced game, we see that it cannot be reduced any further. There are no irrelevant strategies in this

Table 16.13.   A Retail Location Game, Fourth-Stage Reduction.

		Wotchaneed's	
		Upscale Mall	Center City
Nicestuff Stores	Upscale Mall	3,3	10,9
	Center City	8,11	5,5

game. Thus, both strategies — Upscale Mall and Center City — are rationalizable strategies for both players in this game. In fact, we could have been certain at the start that both of those strategies are rationalizable, since they correspond to the Nash equilibria, and strategies that correspond to Nash equilibria are always rationalizable. The elimination of irrelevant strategies has served to assure us that there are no other rationalizable strategies in this game.

For a game like this, with two or more Nash equilibria, we encounter two new problems. First, rationalization will not be sufficient to assure us that the players will find a Nash equilibrium. Any combination of two of the strategies "Upscale Mall" and "Center City" will be rationalizable. For example, the manager of Nicestuff Stores might reason, "Wotchaneed's thinks that I will choose Upscale Mall, and their best response to that is Center City, so I will choose my best response to that, which is Upscale Mall." At the same time the manager of Wotchaneed's reasons, "Nicestuff thinks I will choose Upscale Mall, so they will choose their best response to that, which is Center City; accordingly, I'll choose my best response to Center City, Upscale Mall." Thus, they both choose Upscale Mall, and dividing the market between them, can barely cover their fixed costs, so that each has a much worse payoff than would be the case at either of the Nash equilibria. Of course, once they have played the game, each will discover that his rationalization has been mistaken, but by then (according to the rationalizable strategies approach) it will be too late, as the strategies have already been chosen.

In the Nash equilibrium these mistakes could not occur, since by definition each one chooses the best response to the strategy the other actually chooses. But in a case like this, a unique decision that cannot be reversed if a mistake is made, will the decision makers actually have information enough to realize a Nash equilibrium? Perhaps not! But in this — as in most of the theory of games in normal form — we assume that the two players cannot communicate before they make their decision. Such communication might be illegal — it might constitute a conspiracy in restraint of competition — but if they can communicate, communication could make a difference. This will be further considered in the next chapter. For now, we will explore a few more examples of irreversible decisions without communication.

6. RATIONALIZABLE STRATEGIES WITHOUT NASH EQUILIBRIA

As we have seen, every Nash equilibrium strategy is rationalizable. However, while Nash equilibrium in pure strategies may not exist, we will nevertheless have rationalizable strategies. Here is an example to illustrate this.

Laura and Mark own adjacent properties in Surfy City, a seaside resort. Laura's property is an unoccupied lot, while Mark has a building in need of renovation. Each is considering four different uses to which the property might be put. Those are their strategies, and the payoffs are as shown in Table 16.14.

Examining this game, we find that it has no Nash equilibria in pure strategies. Nevertheless, let us apply the method of iterative elimination of irrelevant strategies and see whether there are any rationalizable strategies. For Laura, we see that "mini-golf" and the "go-cart track" are irrelevant strategies. Accordingly, we drop those two strategies and obtain the reduced game in Table 16.15.

Now we see that, in the reduced game, bar and shop are dominated strategies for Mark. They are thus irrelevant, and so we can eliminate them, leaving the game shown in Table 16.16.

Table 16.14. A Surfy City Real Estate Game.

		Mark			
		Bar	Restaurant	Office	Shop
Laura	Parking	7,6	9,11	10,9	9,8
	Mini-golf	5,7	6,6	5,6	5,6
	Go-cart track	8,6	7,4	7,5	8,7
	Cinema	9,7	11,7	8,8	6,7

Table 16.15. A Reduced Surfy City Game.

		Mark			
		Bar	Restaurant	Office	Shop
Laura	Parking	7,6	9,11	10,9	9,8
	Cinema	9,7	11,7	8,8	6,7

Table 16.16. A Further Reduced Surfy City Game.

		Mark	
		Restaurant	Office
Laura	Parking	9,11	10,9
	Cinema	11,7	8,8

All of the strategies remaining in this game are best responses to some strategy that the other person might choose, so the elimination of irrelevant strategies has gone as far as it can. As we have seen, all of these four strategy combinations are rationalizable.

For example,

1. Laura reasons, "He thinks I think he will choose office, and that I will therefore open a parking lot. Accordingly, he will build

a restaurant. Thus, I will maximize my payoff by building a cinema.

2. Alternately, Laura reasons, "He thinks I think he will choose a restaurant, so that I will build a cinema. Thinking that, he will choose office. Accordingly, I will open a parking lot."

3. Mark reasons, "She thinks I think she will build a cinema, so I will operate an office complex. But that will lead her to open a parking lot, and I will respond by opening a restaurant."

4. Alternatively, Mark reasons, "She thinks I think she will open a parking lot, so that I will operate a restaurant. On that basis she will choose a cinema. Thus, I will open an office complex."

Depending on which of the first two rationales Laura adopts and which of the two rationales Mark adopts, any combination of the cinema, parking lot, restaurant, and office complex might be built.

This is not a very satisfactory result! But if it really is true that people have to make their decisions without any information other than the payoffs and strategies in the game, and if they cannot correct their mistakes, a complicated game like the Surfy City Real Estate Game presents us with some very difficult and deep problems.

Chapter 8 introduces the concept of mixed strategy equilibria, that is, equilibria with randomized strategies. We learn there that every game must have at least one mixed strategy solution, and Surfy City is no exception. That would be difficult to calculate from Table 16.14. (Linear programming would be useful for that). However, we need not. We recall that the game in Table 16.16. is strategically equivalent to the one in 16.14: Any Nash solution to the game in 16.14. must also be a solution to 16.16. Accordingly, using 16.16, we calculate that the game has a mixed strategy equilibrium in which Laura chooses a parking lot with probability $\frac{2}{3}$ and builds a cinema house with probability $\frac{1}{3}$, while Mark chooses between a restaurant and an office building with probability $\frac{1}{2}$ each.

Notice that *any* mixed strategy that plays rationalizable strategies with some probabilities is rationalizable, since, for example, Mark might reason that "Laura expects me to play $\left(\frac{1}{2}, \frac{1}{2} \right)$, so she will choose the mixed strategy payoffs $(\frac{1}{3}, \frac{2}{3})$. That makes the expected value payoffs to me of the restaurant and office strategies the same, namely $8\frac{1}{3}$, so any probability mixture of them will also be the same, and is a (weak) best response for me." Thus, a game like this one has an infinite number of rationalizable strategies. But only the mixed strategy equilibrium is a Nash equilibrium.

7. RATIONALIZABLE STRATEGIES AND NASH EQUILIBRIUM AGAIN

As we saw above, Nash equilibrium strategies are always rationalizable. But, as we saw in the Surfy City example in the previous section, the converse is not always true — pure strategies may be rationalizable even when there is no pure strategy Nash equilibrium. This is another complexity we need to be careful about. In some exceptional cases, there may be both an unique Nash equilibrium and non-equilibrium strategies that are rationalizable. Here is another example to illustrate this.

Kacy and Lee are classmates from high school who have both enrolled as Freshmen at Concordville College, a small liberal arts college that does not compete in "big-time" sports but encourages its students to compete on a club basis in one or another of the minor sports the college offers. Kacy and Lee each plan to join one club team, and while they would enjoy participating in the same sport, their preferences among sports are somewhat different, so it might not work out. Besides, the competition might interfere with their long-standing friendship. Their preferences are represented by the arbitrary payoff numbers in Table 16.17, where a larger number indicates a more preferred alternative for each of the two student-athletes.

Underlining, or a similar method, will show that there is only one Nash equilibrium in this game, where both Kacy and Lee choose Lacrosse. We may apply IEIS to determine which strategies are rationalizable. For Kacy, Squash is dominated by rowing, and so is irrelevant. For Lee, Rowing is dominated by Squash, so it is

Table 16.17.   Sports at Concordville.

		Lee			
		Archery	Lacrosse	Rowing	Squash
Kacy	Archery	9,4	7,4	5,5	6,9
	Lacrosse	6,7	9,9	3,3	5,5
	Rowing	8,7	8,6	10,4	7,6
	Squash	6,9	7,5	5,5	3,8

Table 16.18.   Sports at Concordville, Reduced.

		Lee		
		Archery	Lacrosse	Squash
Kacy	Archery	9,4	7,4	6,9
	Lacrosse	6,7	9,9	5,5
	Rowing	8,7	8,6	7,6

irrelevant for Lee. Eliminating these irrelevant strategies, we obtain the reduced game shown as Table 16.18. But we see that each of these strategies is a best response to a strategy that the other Freshman might choose — none is irrelevant! So, this game cannot be reduced any further, and all of these six strategies, Archery, Lacrosse, and Rowing for Kacy and Archery, Lacrosse, and Squash for Lee, are rationalizable.

For example, Kacy might reason "Lee thinks I will choose Archery, since I always beat him at that in High School, so he will choose his best response, which is Squash. Therefore, I will choose my best response to Squash, which is Rowing." At the same time Lee might reason, "Kacy thinks I will choose Rowing, and challenge him at that, so he will choose his best response to Rowing, which is Archery. Accordingly, I will choose my best response to Archery, which is Squash." Of course, both will find that their rationalizations were wrong. The point is that rationalization can lead to mistakes *even when there is an unique Nash equilibrium.* We should also observe that IEIS will not always find the Nash equilibrium, even when it is unique. As a method for finding the Nash equilibrium, IEIS is heuristic; but when reduction of the game by IEIS is possible, it will simplify the game in useful ways, as the last three examples illustrate.

8. CONCLUDING SUMMARY

The rationalizable strategies approach is based on the idea that, in some games, the participants know *only* that both players are rational, and have common knowledge of rationality, and what the strategies and payoffs are, but have no opportunity to correct their mistakes. That is particularly likely if the game is played just once and there is no communication. But, at the same time, the rationalizable strategy approach complements other approaches such as mixed and pure strategy Nash equilibrium. To find the rationalizable strategies in a game in normal form we may eliminate irrelevant strategies, and do so iteratively as long as there are irrelevant strategies in each reduced game. Since the reduced game obtained by iterative elimination of irrelevant strategies is strategically equivalent to the original game, we can simplify a complex game by eliminating its non-rationalizable strategies before applying the other solution concepts to the reduced game.

Q16. EXERCISES AND DISCUSSION QUESTIONS

Q16.1. Location, Location, Location

Once again, consider the location game in Chapter 4, Exercise 4.2. For this example, we have two department stores: Gacey's and Mimbel's. Each store will choose one of four location strategies: Uptown, Center City, East Side or West Side. The payoffs are shown in Table 16.19.

Table 16.19. Payoffs in a New Location Game (Reproduces Table 4.14).

		Gacy's			
		Uptown	Center City	East Side	West Side
Mimbel's	Uptown	30,40	50,95	55,95	55,120
	Center City	115,40	100,100	130,85	120,95
	East Side	125,45	95,65	60,40	115,120
	West Side	105,50	75,75	95,95	35,55

a. Apply IEIS to determine the rationalizable strategies in this game.
b. Compare the result with the Nash equilibrium, if any.

Q16.2. A Restaurant Game

There are two restaurants in Fahrview Mall, "Casa Sonora" and "Tanaka and Lee." While the chefs are trained in different culinary traditions, each can cook in a variety of styles and those are the strategies shown in the normal form game below. The markets for the different styles are partly different but may overlap, as, for example, Cantonese and Mexican are likely to share the Family dinner market, while tapas and sushi could both appeal to the "something-with-the-beer" crowd. Payoffs Shown in Table 16.20 are the best guess profitabilities of the two restaurants, rated on a scale from 1 to 5.

a. Analyze this game, using the concept of Nash equilibrium.
b. Apply IEIS to determine which strategies are rationalizable.
c. Comment on the advantages and disadvantages of the Nash equilibrium concept, and of rationalizable strategies, and illustrate your comments by reference to this game.

Table 16.20. Payoffs for Restaurants.

		Casa Sonora		
		Mexican	Tapas	BBQ
	Mongolian	4,1	3,4	2,2
Tanaka and Lee	Cantonese	1,1	2,4	2,3
	Sushi	5,2	2,1	3,3

Table 16.21. An Athletic Conference Choice Game.

		Southpaw			
		A	B	C	D
	A	8,7	6,6	4,7	2,8
Topnotch	B	5,4	10,9	4,5	3,6
	C	6,9	5,4	9,10	7,3
	D	5,8	4,7	5,6	6,4

Q16.3. Athletic Conference

Topnotch University and Southpaw State are athletic rivals. Each is considering joining a new conference, which means scheduling most of their games with teams in that conference. They would like to continue to schedule one another, and that will be easier if they are in the same conference, but each has other rivalries they would like to maintain. Their strategies are the conferences they might choose: The Alpine Conference (A), the Big North (B), Central Colleges (C), or the Dust Bowl Conference (D). Payoffs are shown in Table 16.21, on the usual scale of 1 to 10.

a. What Nash equilibria exist, if any?
b. Apply IEIS to determine which strategies are rationalizable.

Table 16.22.   Media Payoffs.

		C								
		News			TV			Web		
		B			B			B		
		News	TV	Web	News	TV	Web	News	TV	Web
A	News	5,5,5	6,10,6	6,12,6	6,6,10	7,7,7	7,12,10	6,6,12	7,10,12	7,8,8
	TV	10,6,6	7,7,7	10,12,7	7,7,7	6,6,6	7,12,7	10,7,12	7,7,12	10,8,8
	Web	12,6,6	12,10,7	8,8,7	12,7,10	12,7,7	8,8,10	8,7,8	8,10,8	7,7,7

Q16.4. Advertising Media

Firms A, B and C are each considering which of three advertising media to choose. The choices (strategies) are newspapers, TV, and the Web. For the purposes of these firms, some media are better than others, but any medium is best when only one of the three firms chooses it, and less effective as more of the firms use it. The payoffs are shown in Table 16.22.

a. Determine all rationalizable strategies in this game.
b. What are the Nash equilibria in pure strategies, if any? How are they related to the rationalizable strategies?
c. Give an example of the reasoning that would support one of the rationalizable strategy solutions.

CHAPTER 17

Trembling Hands and Correlated Strategies

As we observed in Chapter 16, Nash equilibrium is not the whole of non-cooperative game theory. This chapter considers two kinds of extensions of Nash equilibrium, designed to solve some of the

To best understand this chapter, you need to have a good understanding of the material from Chapters 1–9 and 16.

problems we have seen in applications of Nash equilibrium. First, as we recall, there may be more than one Nash equilibrium. In such a case we would like to "narrow the field," and (ideally) to select one of the Nash equilibria as the one that rational decision-makers would arrive at. We may be able to do this by making a small change in the assumptions about rationality — perhaps making the assumptions a little more "realistic." This is called a *refinement* of the Nash equilibrium. Some refinements have been discussed in other chapters, and will be reviewed along with one particularly important refinement. Second, again supposing there are two or more Nash equilibria in pure strategies, the players may randomize their strategies — not independently, but in such a way that their strategies are correlated. These *correlated strategy equilibria* will be discussed in this chapter. We will begin with a puzzle encountered in the exercises in the first two chapters: Weak dominance.

1. WEAK DOMINANCE

Recall the sibling rivalry game from Chapter 2, Exercise 2.1 and Chapter 5, Exercise 5.1. Two sisters, Iris and Julia, are students at Nearby College, where all the classes are graded on the curve. Since they are the two best students in their class, each of them will top the curve unless they enroll in the same class. Iris and Julia each have to choose one more class this term, and each of them can choose between math and literature. They're both very good at math, but Iris is better at literature. Each wants to maximize her grade point average. Their game in normal form is shown in Table 17.1. This game illustrates a complication we have mostly avoided in the other chapters: If Iris chooses math, then math and literature are *equally* good responses for Julia, with each providing a payoff of 3.8. However, if Iris chooses literature, then math is the better response for Julia, for a payoff of 4.0. Similarly, if Julia chooses literature, then literature and

Table 17.1. Grade Point Averages for Iris and Julia (Reproduces Table 2.9, Chapter 2).

		Iris	
		Math	Lit
Julia	Math	3.8,3.8	4.0,4.0
	Lit	3.8,4.0	3.7,4.0

math are equally good responses for Iris, while if Julia chooses math, then literature is a clear better choice for Iris, for 4.0 rather than 3.8. Thus, for Julia math, and for Iris literature, are instances of *weakly dominant strategies*.

This is a complication we need to be careful about. Remember the definition of a dominated strategy from the Chapter 3: "Whenever one strategy yields a higher payoff than a second strategy, regardless which strategies the other players choose, the second strategy is dominated by the first, and is said to be a dominated strategy."

Definition: *Strictly Dominated Strategy* — Whenever one strategy yields a payoff strictly greater than that of a second strategy, regardless which strategies the other players choose, the second strategy is strictly or strongly dominated by the first, and is said to be a strictly or strongly dominated strategy.

We can distinguish two closely related concepts, depending on how we interpret "higher payoff" in that definition. If the payoff to the first strategy is always strictly greater than the payoff to the second strategy, we say that the first strategy strictly or strongly dominates the second strategy. All of our other examples in Chapters 3 and 4 are examples of strictly dominated strategies. The other possibility is a weakly dominated strategy. The first strategy weakly dominates the second strategy if the payoff to the first strategy is at least as great as the payoff to the second strategy, and sometimes, but not always, strictly greater.

We have, then, weakly dominant strategies for both sisters: Math for Julia and literature for Iris. And indeed, these strategies form a Nash equilibrium, as a strictly dominant strategy equilibrium would. But the reverse, where Julia chooses literature and Iris chooses math, is also a Nash equilibrium. The strategies (math, lit) give both girls perfect 4.0 averages, while the other Nash equilibrium, (lit, math), leaves Julia with only a 3.8. Nevertheless, neither sister has any incentive to shift away from (lit, math), making it a Nash equilibrium. This is a strange equilibrium not so much because Julia is worse off as because neither girl has a positive reason for choosing these strategies. If Iris unilaterally switches to lit, she is just as well off, with a 4.0; and similarly, if Julia switches unilaterally to math, she is no worse off either, with a 3.8. Nevertheless, it is a Nash equilibrium, since each student is choosing her best response (there is no better response) to the strategy chosen by the other. A Nash equilibrium of this sort is called a *weak Nash equilibrium.*

> **Definition:** *Weakly Dominated Strategy* — Whenever one strategy yields a payoff no less than a second strategy, regardless which strategies the other players choose; and the payoff to the first strategy is strictly greater than the payoff to the second strategy for some strategies the other players might choose, the second strategy is weakly dominated by the first, and is said to be a weakly dominated strategy.

Here is one more example of weakly dominant strategies and weak Nash equilibrium. For this example, two side-by-side restaurants have to decide what varieties of food to feature. Both have southern cooks. Each of the two can feature Cajun cooking, barbecue, or both. The game is shown as Table 17.2.

For this game, we can see that the weakly dominant strategy for both restaurants is to offer both cuisines. Moreover, the game has

Table 17.2. Payoffs for Restaurants.

		Susanna's Diner		
		Cajun	Barbecue	Both
Sweet Pea Inn	Cajun	7,7	8,8	6,9
	Barbecue	8,8	7,7	6,9
	Both	9,6	9,6	6,6

five Nash equilibria, where one or both restaurants offer both cuisines; and all of these five Nash equilibria are weak.

One further caution is in order. When we apply iterated elimination of irrelevant strategies, IEIS, we have seen that strictly dominated strategies are always irrelevant, and so can be eliminated, because a strictly dominated strategy can never be a best response. However, as we have seen, a weakly dominated strategy may be a best response, and so should not be eliminated — unless we have some additional reason to do so. One possible reason will be taken up in the next section.

2. REFINEMENT: A TREMBLING HAND

As we have seen in Section 1, the Sibling Rivalry game between Iris and Julia has two Nash equilibria, but one of them makes less sense than the other. It is the weak Nash equilibrium of the two. In the game of restaurants and cuisines, there are five Nash equilibria, but all are weak. This is a fairly common problem in more complex applications of game theory. Where there are two or more Nash equilibria, some of them may seem unreasonable. This is possible because the definition of Nash equilibrium captures only one aspect of rationality, the "best response" criterion. We might be able to add some other rational considerations that would eliminate the unreasonable Nash equilibria. These additional tests of rationality are called *refinements* of Nash equilibrium, and refinements of Nash equilibrium fill a large and important category of advanced game theory.

In this case, the additional test of rationality is the fail-safe test — why choose a strategy that can do worse, but can never do better, than another? In other words, why choose a weakly dominated strategy? In the grade point game, the choice of a lit course could make Julia worse off (with a 3.7) if Iris made the mistake of enrolling in the lit course at the same time. But (we have assumed) Julia doesn't have to worry about that, because Julia knows Iris is rational and will not make a mistake that reduces her (Iris') GPA. Perhaps it would be more realistic for Julia to assume that it is *very probable* that Iris will choose the equilibrium strategy, but not quite certain. Suppose, for example, that the probability that Iris will choose her best response is 95%, but there is a 5% probability that she will choose the wrong strategy. In that case, Julia's expected payoff for choosing "lit" is $(0.95)(3.8) + (0.05)(3.7) = 3.798$. On the other hand, by choosing "math," Julia can do no worse than 3.8, so she will choose math.

So, assuming that Julia allows for a very small probability that her sister will make a mistake, the weak, unreasonable Nash equilibrium is eliminated. This assumption is called "the trembling hand assumption." The idea is that the agent may "tremble" when choosing the best response, accidentally choosing a wrong response. It is a good example of a refinement of Nash equilibrium, since it assumes a slightly different kind of rationality and, as a result in this example, narrows the possibilities to a single equilibrium that seems to be more reasonable than the other.

Let's see what happens if Iris assumes that Julia could have a trembling hand — Iris guesses that Julia will choose an equilibrium strategy with 90% probability and the wrong strategy with 10% probability. Then the expected value for Iris if she chooses math is $(0.9)(4) + (0.1)(3.8) = 3.98$, but Iris can guarantee herself a 4.0 by choosing lit — so she chooses literature, and the two students each get their 4.0.

In general, we suppose that each player in the game assumes that the other players will very probably choose their best responses, but assigns some small positive probability to the possibility that they might make a mistake. Then let the probability of an error decrease step by step toward zero. If there is one equilibrium or more that is never chosen when the probability is small enough (but positive), then eliminate it (or them) from consideration. The equilibrium that remains is a trembling hand stable Nash equilibrium. If we make the probabilities of mistakes small enough, the trembling hand equilibrium will be one of the Nash equilibria in the game. The trembling hand Nash equilibrium "refines" the Nash equilibrium concept by assuming that the agents choose failsafe strategies that are also best responses.

For another example, recall the Market Entry Game from Chapter 2, as represented *in normal form*. This is shown in Table 17.3. Once again, we see that this game has two Nash equilibria: First, where Bluebird chooses "enter" and Goldfinch chooses "*If Bluebird enters* then accommodate; *If Bluebird does not enter*, then do business as usual," and second, where Bluebird chooses "don't" and Goldfinch chooses "*If Bluebird enters* then initiate price war; *If Bluebird does not enter*, then do business as usual." This second Nash equilibrium is questionable, though, since Goldfinch has nothing to lose by adopting its other strategy of accommodating Bluebird in case they do enter. It is a weak Nash equilibrium. Moreover, if Goldfinch thinks there is just a small probability that Bluebird will enter, despite the threat of losing 5, then Goldfinch's best response is definitely to

Table 17.3. The Market Entry Game in Normal Form (reproduces Table 2.1, Chapter 2).

		Goldfinch	
		If Bluebird enters then accommodate; *if Bluebird does not enter*, then do business as usual.	*If Bluebird enters* then initiate price war; *if Bluebird does not enter*, then do business as usual.
Bluebird	Enter	3,5	–5,2
	Don't	0,10	0,10

choose the strategy "*If Bluebird enters* then accommodate; *if Bluebird does not enter*, then do business as usual." Suppose the probability that Bluebird will enter is 0.05. Then Goldfinch's expected value payoff for the accommodation strategy it is $0.95 * 10 + 0.05 * 5$, while for the price war strategy is $0.95 * 10 + 0.05 * 2$. Accordingly, when we apply the trembling hand refinement, we conclude that the first of the two Nash equilibria is the one that will occur.

Applying some ideas from Chapter 9, we recall that the first equilibrium is also the subgame perfect equilibrium for this game. For any game in extensive form that has a subgame perfect equilibrium, the subgame perfect equilibrium will be a trembling hand perfect equilibrium when the game is expressed in normal form. At the same time the trembling hand equilibrium is applicable to games that do not have subgames or subgame perfect equilibria. And it serves to eliminate some weak equilibria that do not seem reasonable, as in the Sibling Rivalry example. These are strong arguments for the importance of the trembling hand assumption. For these reasons it has come to be the first thought of most game theorists when there are problematic Nash equilibria. A few years ago, at a prominent university, a computer science professor announced the grading for the class: "The grading curve was set by giving the highest score on the final an A, and then adjusting all lower scores accordingly. The students determined that if they collectively boycotted, then the highest score would be a zero, and so everyone would get an *A*. Amazingly, the students pulled it off."[1] The resulting game had two Nash equilibria: Where everyone boycotts and one where no-one boycotts. But the outcome where everyone boycotts is a weak Nash equilibrium and most game theorists "would say that the first equilibrium, where no one takes the exam, is unlikely to result because it is not 'trembling hand perfect,' an idea that helped … Reinhard Selten win the Nobel Memorial Prize in Economics." But this may not be the last word, all the same, and Section 4 will give an example with a somewhat less reassuring outcome. First,

---

[1] Rampell, Catherine, Gaming the System, *The New York Times*, available at: https://economix.blogs.nytimes.com/2013/02/14/gaming-the-system/, as of December 26, 2021, originally posted on 2/14/2013.

however, in the next section, we review and compare some other refinements of Nash equilibrium.

3. OTHER REFINEMENTS

The Trembling Hand assumption is not the only refinement of Nash equilibrium. Indeed, we have seen several other possible refinements in the preceding chapters: Payoff dominance, risk dominance, and strong Nash equilibrium are other possible refinements. Let us briefly recall these ideas:

Payoff dominance: (Chapter 5) If Nash equilibrium A results in higher payoffs to all players than equilibrium B, or to higher payoffs to some and no lower payoffs to others (that is, A is a Pareto-improvement relative to the B) then A is payoff dominant relative to B. On its face, it seems that rational agents would recognize this and choose the payoff dominant equilibrium.

Risk dominance: (Chapter 5) If Nash equilibrium A exposes each decision-maker to the risk of the maximum loss and equilibrium B does not, then B is risk-dominant relative to A. If decision makers are untrusting, then the risk dominant equilibrium might be chosen, and there is some experimental evidence that this may happen.

Strong Nash equilibrium. If two or more decision-makers can form a coalition and shift from equilibrium A to equilibrium B, making at least one of the members of the coalition better off and none worse off, then A is *not* strong. However, if no such potentiality exists, then A is strong.

This is a little embarrassing — which of these concepts should we use, in any particular case? That (of course) is why the agreement between trembling hand perfection and subgame perfection is important: They are based on quite different ideas, but give the same results where both are applicable. If all of our refinements agreed in that way then the problem of multiple Nash equilibria would largely be solved. But we learned in Chapter 5 that payoff dominance and risk dominance can disagree, and the next section gives an example in which both dominance and strong equilibrium concepts disagree with the perfection refinements.

4. TWO RESTAURANTS, AGAIN

Let us apply trembling hand reasoning to the game between the two restaurants choosing their cuisines. We look again at Table 17.2 for the game in normal form. The manager of Sweet Pea Inn knows that "both" is Susanna's best response to any strategy that Sweet Pea Inn might choose. Suppose, nevertheless, that Susanna's plays "Cajun" with probability 0.05, "barbecue" with probability 0.05, and "both" with probability 0.9. Then the expected values of the three strategies for Sweet Pea Inn are

Cajun	$0.05(7) + 0.05(8) + 0.9(6)$	$=$	6.15
Barbecue	$0.05(8) + 0.05(7) + 0.9(6)$	$=$	6.15
Both	$0.05(9) + 0.05(9) + 0.9(6)$	$=$	6.3

Thus, when Susanna has a "trembling hand," Sweet Pea Inn's best response is the strategy "both." Doing a few experiments (or a little algebra) we find that this is true for any probabilities of mistakes smaller than 0.05. Since the game is symmetrical, the same is true for Susanna's Diner if Sweet Pea Inn has a "trembling hand." Thus, the game has only one "trembling hand" stable Nash equilibrium, and that is the strategy "both."

Unlike the other examples, trembling hand stability does not eliminate the unreasonable equilibrium, 6,6, but rather eliminates all of the other equilibria that make at least one of the two restaurants better off. Indeed, the trembling hand stable equilibrium in this example results in the worst possible payoffs for both players. That is, every other outcome, including four Nash equilibria, is payoff and risk dominant relative to the perfect equilibrium, and the other four equilibria are strong, while the perfect equilibrium is not.

Where does this leave us? So far as theory is concerned, the existence of several competing refinements is an unsolved problem.

However, strong equilibria are an important step toward the theory of cooperative games. Pragmatically, risk and payoff dominance have the advantage of simplicity. In the sibling rivalry game, (math, lit) is both payoff and risk dominant, and that is probably enough reason to predict (math, lit) as the option they would choose; but furthermore, that equilibrium is both trembling hand perfect (as we have seen) and strong. There are some practical problems that seem to correspond to risk dominant equilibria, where the risk dominant equilibrium prevents initiatives that could make all better off. On the other hand, it is hard to find practical applications of the trembling hand equilibrium, complex as it is, and while subgame perfect equilibrium does have some practical applications, it is, if anything, even more complex. These two refinements may set a standard of rationality that few real humans put into practice. Nevertheless, if we can better understand how these high standards of rationality influence the decisions of the rational agent, we may be able to increase the rationality of our own decisions.

5. A CONFESSION GAME

Rog, Barry and Dave are members of the notoriously mischievous Blanchard School boygang, and they have been at it again. As a Hallowe'en stunt, they TP'ed the principal's house. Everybody knows they are the likely villains, so they will all be punished severely — if nobody confesses. On the other hand, if one or more of the boys confesses to the mischief, he will get a reduced punishment, as a reward for honesty, and the others will get off. Counting a severe punishment as –2, a reduced punishment as –1, and getting away with it as 0, the payoffs are shown in Table 17.4.

This game has similarities both to coordination and anticoordination games. There are three Nash equilibria — the three cells in which just one boy confesses and the others go unpunished. As we have seen before, there could be a problem with a game like this. Rationalization will not be useful. If the boys have no clue as to which Nash equilibrium will occur, they may guess wrongly. For example, if Rog and Barry assume that Dave will confess, but Dave

Table 17.4. Payoffs in a Confession Game.

		Dave			
		Confess		Don't	
		Barry		Barry	
		Confess	Don't	Confess	Don't
Rog	Confess	−1,−1,−1	−1,0,−1	−1,−1,0	−1,0,0
	Don't	0,−1,−1	0,0,−1	0,−1,0	−2,−2,−2

assumes that Rog will confess, they all choose "don't" and all get severe punishments, the −2 payoff. Unless they have some clue, some "Schelling focal point," there is a real danger that they might fail to get any efficient equilibrium at all. Refinements do not help, either. In particular, when the probability of a tremble is small enough, none of the equilibria can be eliminated by a trembling hand. Each boy may be tempted by the fail-safe (risk-dominant) strategy of volunteering to confess, though, if all three of them do it, this is inefficient and not a best-response equilibrium. In addition, there is a feeling that it is unfair for one boy to be punished for what all three did.

But these naughty boys have a solution. They will "draw straws" to determine which boy is to confess. Three straws of different length are held by one of the boys so that only one end is showing and the others cannot tell which straw is shortest. Then the other two boys each draw one straw, and the boy who held the straws is left with the other one. The boy who holds the shortest straw must confess. This way, each boy has an equal chance of being the one to be punished.[2]

This solution to the problem is interesting in several ways.

---

[2]The boy who holds the straws has a strong incentive really to conceal which is shorter. If he is careless and holds the straws so that the others can see which is shortest, he will certainly be left with the short straw and be punished.

- It is very much like a mixed strategy, in that the decision is made at random, assigning probabilities to strategies.

  o The probability of $(-1, 0, 0)$, $(0, -1, 0)$ and $(0, 0, -1)$ are each 0.333
  o The probabilities of all other outcomes are zero.

- It is different from mixed strategy Nash equilibria in that the probabilities are assigned to joint strategies, whereas in a Nash equilibrium, players assign probabilities independently to their own strategies.

  o For a mixed-strategy equilibrium, each boy chooses a probability of 0.293 for "confess."
  o As a result, there is a probability of more than one-third that none will confess, resulting in severe punishments all around.

- Once the straw is drawn, it provides a Schelling focal equilibrium — all have the same expectations as to who will confess. Suppose, for example, that Rog draws the short straw. Rog cannot benefit by cheating on the agreement. The other boys are expecting Rog to confess, so they will not confess, and that means that if Rog cheats by not confessing, he will get a -2 payoff instead of a -1. So, Rog's best response is just to put some newspapers down the seat of his pants and march off to take his punishment. Similarly, the other two boys have nothing to gain by confessing, since that would get them -1 instead of 0.
- Before the straw is drawn, each boy has an expected value payoff of $-1/3$ — better than he can get by volunteering to confess, the fail-safe strategy.
- Since each boy has the same probability of having to confess and the same expected value payoff, this solution is "fair" in a way that the pure strategy Nash equilibria are not.
- Even if the probabilities would not have been equal, the result would still have provided a Schelling focal equilibrium for the original game. Suppose, for example, that Barry held the straws

and was able to cheat, so that Rog had a higher probability of getting the short straw. Even if Rog suspected what was going on, once he had the short straw confession would still be his best response.

• Unlike the Nash equilibrium itself and the refinements, it relies on some communication among the players before the play of the game. If communication is not feasible, as in the Prisoner's Dilemma, this approach would not be applicable. This is a half-step toward cooperative solutions, but only a half-step since the three boys ultimately chose their best responses.

Drawing straws in this game is an example of a *correlated equilibrium*. A correlated equilibrium is a new kind of solution to a non-cooperative game. In general, a correlated equilibrium is an arrangement that assigns probabilities to the joint strategies in the game that correspond to Nash equilibria of the game.[3] Of course, the probabilities of choosing the joint strategies have to add up to one. If there is only one equilibrium (for example, as in the Prisoner's Dilemma) then that equilibrium is also the only correlated equilibrium, since the probability assigned to it must be one. If one equilibrium is payoff dominant over the others, as in the Heave-Ho game, once again, the payoff dominant equilibrium will be the only correlated equilibrium, since there is no point in assigning any probability to a dominated Nash equilibrium. However, correlated equilibria can make a big difference in a coordination or anticoordination game like the Confession Game. Since (as we have seen) each player is choosing their best response in a correlated equilibrium, a correlated equilibrium is a non-cooperative

---

[3]The idea of correlated equilibrium was originated in Luce, R. Duncan and Howard Raiffa, *Games and Decisions* (New York: Wiley and Sons, 1957). This text follows their discussion. Aumann, Robert J., Subjectivity and correlation in randomized strategies, *Journal of Mathematical Economics*, **1**, pp. 67–96, extended the correlated equilibrium solution, showing that in some cases, the correlated equilibrium is better than any pure strategy Nash equilibrium. This more powerful solution requires that the players be given private messages in a somewhat complicated way, and will be discussed in a later section.

Table 17.5. Drive On (Reproduces Table 5.4, Chapter 5).

		Mercedes	
		Wait	Go
Buick	Wait	0,0	1,5
	Go	5,1	−100,−100

equilibrium; and in a game with two or more Nash equilibria in pure strategies, there may be infinitely many correlated equilibria, corresponding to different assignments of the probabilities (although it may be that only one way of assigning the probabilities will be recognized as fair by the players).

This is important because it increases the role of *coalitions* in non-cooperative games. We saw in Chapter 6 that a coalition may form in a non-cooperative game of 3 or more persons, provided that the joint strategy of the coalition consists of best-response strategies for the individuals. That is, the coalition forms to enforce one out of a number of possible Nash equilibria. What Barry, Dave and Rog have done in this example is to form the grand coalition in their game, in order to avoid the danger of inconsistent guesses in their coordination-anticoordination game. The coalition has chosen a correlated, probabilistic joint strategy rather than a joint pure strategy, because it seems to them to be more fair — and otherwise it would be harder to come to an agreement.

For yet one more example, remember the Drive On Game from Chapter 5, Section 4. the payoff table for that game is reproduced as Table 17.5. We recall that this is an anticoordination game. Suppose, as we suggested in Chapter 5, that a traffic light should be installed at the intersection of Pigtown Pike and Hiccup Lane. Then we will have a correlated strategy solution as each person stops or goes accordingly as he has a red or a green light.

6. SOME APPLICATIONS

Here is another example, a real-world business case! New Zealand telecommunications companies Teamtalk Ltd and MCS Digital Ltd.

Table 17.6.   Arm-Wrestling to Settle a Dispute.

		MCS Digital	
		Pursue lawsuit	Concede
Teamtalk	Pursue lawsuit	–2,–2	2,–1
	Concede	–1,2	0,0

were embroiled in a lawsuit. Instead of taking it to court, they arm-wrestled to decide who would win the issue.[4] "'Sure, losing hurts but not nearly as much as paying lawyers bills,' defeated Teamtalk Chief Executive David Ware told Reuters." The idea seems to be that lawyer's bills would have made even the winner worse off if both continued to pursue the lawsuit. If both pursue the lawsuit, then each has a chance of winning — but rather than bring chance explicitly into the model we just look at the expected value payoff, net of lawyer's fees, in that case. This leads to a payoff table like the Table 17.6.

The payoffs of –2, –2 if both pursue the lawsuit reflect the expected value of winning and getting a settlement less lawyer's fees. This game is a Hawk vs. Dove game with two Nash equilibria, each at a cell where one contestant concedes. Arm-wrestling is a correlated strategy equilibrium. The probabilities of winning may not be exactly 50–50 — in fact the two men probably had different estimates of those probabilities — but they need not be 50–50 exactly. All that we require is that each executive had a better expected value with the arm-wrestle than with the lawsuit. Once the arm-wrestle is done, it establishes a Schelling focal point, and the game is settled.

Let's take another look at the Retail Location Game, Table 4.11 in Chapter 4. It is reproduced in Table 17.7. We recall that this game has two Nash equilibria, where one store chooses Center City and the other chooses Upscale Mall. Could there be a correlated strategy

---

[4]BBC News, available at http://news.bbc.co.uk/2/hi/asia-pacific/2836069.stm as of 3/13/2013.

Table 17.7. A Retail Location Game (Reproduces Table 4.11, Chapter 4).

		Wotchaneed's			
		Upscale Mall	Center City	Snugburb	Uptown
Nicestuff Stores	Upscale Mall	3,3	10,9	11,6	8,8
	Center City	8,11	5,5	12,5	6,8
	Snugburb	6,9	7,10	4,3	6,12
	Uptown	5,10	6,10	8,11	9,4

Table 17.8. A Retail Location Game, Fourth-Stage Reduction (Reproduces Table 16.13, Chapter 16).

		Wotchaneed's	
		Upscale Mall	Center City
Nicestuff Stores	Upscale Mall	3,3	10,9
	Center City	8,11	5,5

solution? Of course, there is. As a first step it might be helpful to consider only the rationalizable strategies, so we simplify the game by using the interative elimination of irrelevant strategies. Thus, we get the reduced game at Table 16.13, reproduced in Table 17.8. The two department stores can then choose a correlated strategy based on this reduced game. If they can arrange to choose between (Center City, Upscale Mall) and (Upscale Mall, Center City) with probabilities $\frac{1}{4}$, $\frac{3}{4}$, then they will have the same expected value payoff, $9\frac{1}{2}$.

Correlated equilibria are not a new discovery. People seem to have made decisions by such methods as drawing straws or drawing a "black ball" from an urn for thousands of years before game theory was conceived. Stop-lights at a traffic intersection began to be used a few years before von Neumann began to write on gambling games. Nevertheless, understanding how these practices solve problems that arise in non-cooperative game theory extends our theory in useful ways. It underlines the importance of communication and information in strategic choices and refines our understanding of the sort of information that may be helpful. It may also point the way to new applications in practice, as the next section suggests.

7. A MORE ADVANCED CORRELATED STRATEGY EQUILIBRIUM[5]

Here is an example that illustrates the key role of information in correlated strategy solutions for some more complex games. It is another location strategy game. Lotsa 'Lectronics and Yall-Com Computers are both retail chains with distinct but overlapping product lines. Both are considering opening new locations in the suburbs of Gotham City. Both are considering locations in Groundswell Mall, but they are also considering stand-alone locations in the Route 40 commercial corridor. If they both locate in the mall, they can both do fairly well, attracting consumers who are happy to shop both stores and others who are shopping for other items at the Mall, but if one company knows the other will locate at the Mall, the stand-alone location will be the more profitable one as it will establish a local monopoly for its entire product line. The payoffs are shown in Table 17.9.

We see that this game is an anticoordination game, and as such has two pure strategy Nash equilibria. At each of them, one company locates at the Mall, and the other at the stand-alone site, and does relatively well. There is also a mixed-strategy equilibrium at which each company chooses the mall with probability $\frac{5}{7}$. The expected-value payoff in this case is 9.14, which is better than the payoff from being the only store at the Mall. However, the two companies can clearly do better than that with correlated strategies. If they flip a coin to decide between the two pure-strategy Nash equilibria, they can avoid the bad outcome where both locate at stand-alone sites, and get expected value payouts of 9.5 each.

---

[5]This section adapts Aumann's 1974 contribution. See footnote 3.

Table 17.9. A Mall Location Game.

		Yall-Com	
		Stand-alone	Mall
Lotsa	Stand-alone	2,2	12,7
	Mall	7,12	10,10

(*Exercise*: Verify all of this). But in that case, there is also no probability of the outcome where both companies locate in the Mall, which is the cooperative solution of this two-party game.

But there is a third party in the game — Groundswell Mall, which would like to have both stores locate at the mall. The only way that the Groundswell company can influence their decision is by sending one or more of the store chains a "special invitation." The "special invitation" doesn't provide any special advantages for the company that gets it, nor does it cost the Groundswell Company anything. In game theory we would say that the "special invitation" is *cheap talk,* and for that reason we would not expect that it will influence the decision of the two stores. But it can't do any harm, so why not send a "special invitation" to both chain store companies?

But that is not what Groundswell Mall does. Having read the works of Robert Aumann, they realize that they can make the "special invitation" a signal for this anticoordination game in a way that improves the outcome for everybody. Groundswell Mall will randomize their invitation strategy, sending "special invitations" to both companies with probability $\frac{1}{2}$, sending a "special invitation" only to Lotsa 'Lectronics with probability $\frac{1}{4}$ and sending a "special invitation" only to Yall-Com Computers with probability $\frac{1}{4}$. The probability that they send no special invitations at all is zero. Groundswell announces the probabilities it will use, and requests each company to keep it confidential if they do get a "special invitation."

The chain store companies will now have a bit more information that they might use to choose their locations. Suppose that each company considers making its decision according to Rule R: "If I receive a 'special invitation,' choose the Mall, and otherwise choose the stand-alone location."

Now suppose that Lotsa knows that Yall-Com is making its choice according to Rule R. If Lotsa does not get a special invitation, they know that only one thing could have happened: Yall-Com got the only "special invitation" and will locate in the Mall. Lotsa's best response is the stand-alone location — following Rule R. If Lotsa does get a "special invitation," then they know that one of two things has happened: Either they got the only "special invitation" or both companies got "special invitations." They know one thing that did not happen: Yall-Com did not get the only "special invitation." Knowing that, they must revise their estimates of the probabilities. The revised probabilities respectively are $\frac{1/4}{1-1/4} = 1/3$ that Lotsa got the only "special invitation," and $\frac{1/2}{1-1/4} = 2/3$ that both companies got "special invitations." Thus, if they follow Rule R and locate at the Mall, their expected value payoff is $(1/3)(7)+(2/3)(10) = 9$. On the other hand, if they deviate from Rule R, and choose the remote location, their expected value payoff is $(1/3)(2)+(2/3)(12) = 8\frac{2}{3}$. Once again, Lotsa's best response is to play according to Rule R. We see that Lotsa's best response if Yall-Com plays Rule R is also to play Rule R. The converse is also true — if Lotsa plays Rule R, then Yall-Com's best response is to play according to Rule R.

In effect, by sending "special invitations" on a randomized basis, Groundswell Mall has introduced a new strategy and Nash equilibrium to the game between Lotsa 'Lectronics and Yall-Com Computers. It is also a payoff-dominant equilibrium, since the expected value payoffs for the two companies are both $1/4(7)+1/2(10)+1/4(12) = 9\frac{3}{4}$, better than they can do with a flipped coin correlated strategy solution. Groundswell Mall also does better, since they will get at least one store and have a 50% chance for both. Otherwise, they would only get one, since the stores would flip a coin to decide.

This is an important step forward. What we have found is that if a signal is supplied by a third party, then in some games it is possible to obtain a correlated strategy solution that is better than *any* Nash equilibrium and better than the average of any set of Nash equilibria. To do this, however, the third party may have to fine-tune the signal so that each of the players gets a different signal, and that they do not know what signal the other player gets. This is a tricky problem, and there are few hypothetical applications. It may be that there are (as yet) no real-world applications. However, there is some reason to think that if the same players play the same game over and over, they could learn to play according to a correlated strategy equilibrium with each using cues and clues to anticipate the other player's choices.[6]

8. MORE ON INFORMATION AND EQUILIBRIUM

Let's consider a game of congestion with just three players. (Realistically congestion would be a phenomenon of much greater numbers, but that would belong in Chapter 14. Nevertheless, we can learn something from this simplified example.) Motorists A, B, and C can get from Ytown to Zburg by taking Route 1 or Route 2, and those are their strategies. Route 1 tends to be faster, but if all three take it, it will be slow because of congestion.[7] The game is shown in normal form in Table 17.10. Payoffs are time spent making the trip, so they are negative numbers, time lost. We see that there are three Nash equilibria, where two of the motorists choose Rt. 1 and the remaining one chooses Rt. 2. But which is to choose Rt. 2? Without some signal to provide a focal point, the three motorists may not be able to arrive at an efficient Nash equilibrium.

The Traffic Authority would like to help. To do this, they erect an electronic sign just before the branch that leads to Rt. 1 and Rt. 2. The sign can display one of two messages: "Rt. 1 is faster" or "Rt. 2 is faster." (Perhaps they will use data on the number of cars that entered the routes previously or rely on statistical analysis of data on past use of the routes.) Once again, suppose the motorists

---

[6]See Roger A. McCain, *Game Theory and Public Policy* (Elgar, 2009), Chapter 5 for more detail.

[7]This is suggested by an example in Pigou, A. C., *Economics of Welfare* (London: Macmillan, 1920).

Table 17.10. Payoffs in a Congestion Game.

		C			
		Rt. 1		Rt. 2	
		B		B	
		Rt. 1	Rt. 2	Rt. 1	Rt. 2
A	Rt. 1	−20,−20,−20	−10,−10,−10	−10,−10,−10	−5,−15,−15
	Rt. 2	−10,−10,−10	−15,−15,−5	−15,−5,−15	−30,−30,−30

decide according to rule R: "Choose the route that the sign says is faster." Then, if the message recommends Rt. 1, they all take that route, with resulting payoffs of −20 each. If then one deviates to Rt. 2, against Rule R, the resulting payoff is −10, which is better. If the message recommends Rt. 2, then following Rule R they all take that route, with payoffs of −30 each. If one deviates against Rule R to Rt. 1, the payoff is −5, and so following Rule R can never be a best response.

Instead suppose then that the motorists decide according to Rule R': "Choose the route that the sign says is slower." Once again, if all choose that rule, they will all choose the same route, with bad consequences. In short, the information provided by the sign does not help, because all decision-makers get the same information. In order to assure that a coordination equilibrium will occur, the authority needs to send different messages to different decision-makers, as a traffic light does. Suppose, for example, that there were a way for the Traffic Authority to send different messages to different motorists. The authority might send the message "Rt. 1 is better" to two of the three motorists, at random, but not to the third. If then the motorists all choose according to Rule R*: "If I get a message, then choose Rt. 1," then the efficient Nash equilibrium will occur.

What this example underlines is that information *can* support an efficient Nash equilibrium in an anticoordination game, but it may be necessary to fine-tune the content and especially the distribution of the information. It may be necessary for different decision-makers to get different information, but information that is correlated so as to support strategy decisions that are appropriately correlated.

9. SUMMARY

The theory of games in normal form provides a rich tool-kit for the analysis of strategic interactions. The Nash equilibrium, in particular, has wide applications. However, it is not the only tool for the analysis of non-cooperative games.

As discussed in Chapter 3, a dominant strategy equilibrium is a powerful result. As is always the case, powerful tools need to be used with caution. When we allow for weakly dominated strategies, the case is less clear. On the other hand, we may be able to refine the concept of Nash equilibria by allowing for a very small probability that the players may choose wrong strategies. If each player allows for that possibility, assigning a very small probability that the other player will have a "trembling hand" and choose the wrong strategy, then they will avoid weakly dominated strategies and some unreasonable Nash equilibria are eliminated. This is a good example of "refinement" of Nash equilibrium, an important area of research in advanced game theory.

Coalitions can play a part in non-cooperative games, but only if all members of the coalition are choosing their best responses. The coalition may choose a joint mixed strategy, however. This is a correlated equilibrium, and it requires that the members of the coalition find some method of jointly choosing strategies at random with specific probabilities. Some traditional methods are choosing the short straw or allocating scarce resources by lottery. In some cases, it may make sense for the coalition to designate one person to choose for them. Correlated equilibria may provide symmetrical solutions to coordination games in which the payoffs would otherwise be unequal, as in the Confession Game. When a trusted third party signals both players in different, appropriate ways a correlated strategy solution may improve on any Nash equilibrium or weighted average of Nash equilibria.

Of course, best responses and the Nash equilibrium are at the basis of all these conceptions; but for many purposes, Nash equilibrium is the starting point, not the destination.

Q17. EXERCISES AND DISCUSSION QUESTIONS

Q17.1. El Farol

Refer to Chapter 6, Section 4. The payoffs are shown in Table 17.11.
    Propose a correlated strategy equilibrium solution to this game.

Q17.2. Medical Practice

Refer to Problem 6.5 in Chapter 6. Doctors are considering whether to practice as ob/gyn's or to limit their practices to gyn. Propose a correlated equilibrium solution to this game.

Q17.3. Government Reorganization Plan

Generous Metals Corp. is unable to pay its debts to bondholders and pensioners and thus on the verge of bankruptcy. Because Generous is "too big to fail," the government has proposed a "reorganization plan." Under the plan, both bondholders and pensioners would

Table 17.11. Payoffs in a Crowding Game.

		Carole			
		Go		Home	
		Barb		Barb	
		Go	Home	Go	Home
Amy	Go	-1,-1,-1	2,1,2	2,2,1	0,1,1
	Home	1,2,2	1,0,1	1,0,1	1,1,1

Table 17.12. Government Reorganization Plan.

		Bondholders	
		Accept	Reject
Pensioners	Accept	46,44	30,47
	Reject	50,25	22,20

sacrifice some of the funds they are owed. Each can accept or reject the reorganization plan. If either rejects, the bankruptcy will be arbitrated by a law court. If only one rejects and contests the reorganization, they can gain some advantage; but if both reject the plan, the legal costs of the bankruptcy proceedings will leave them both worse off. The payoffs are shown in Table 17.12, in terms of the proportion of their debts they will be able to collect ("cents on the dollar"). A spreadsheet is likely to help with this relatively complex problem.

a. Discuss this example as an application of correlated equilibrium theory.
b. Is there anything the government can do to improve the outcome, other than forcing the two sides to accept the reorganization plan whether they want to or not? What?
c. Compare and contrast this example with the example of Teamtalk vs. MCS Digital in the chapter.

Q17.4. Subcontractors

Morris and Neal are computer scientists with different, complementary specialties. Bigscale Business Software (BBS) would like to recruit them both, but each is thinking of setting up as an independent businessman. If one works for BBS, the other, as an independent

Table 17.13. Payoffs to Computer Scientists.

		Neal	
		BBS	Independent
Morris	BBS	140,130	110,150
	Independent	160,105	55,50

businessman, will benefit as a subcontractor. The payoffs, in annual incomes, are shown in Table 17.13.

Discuss this example as an application of correlated equilibrium theory. (For this unsymmetrical example, a spreadsheet is suggested.)

CHAPTER 18

Voting Games

In an election, a group of people decide some issue by counting votes. Elections lend themselves to the scientific metaphor of game theory: Like games, elections have known

> **To best understand this chapter,** you need to have studied and understood the material from Chapters 1–6.

rules, and there is usually a definite winner and loser or losers. The strategy in an election game includes a decision how to cast one's vote, although it may be more complex than that. In this chapter, we sketch some basic ideas in the game-theoretic analysis of voting. We begin, as usual, with an example. In the example, the executive committee of a sorority have to decide how much to spend on a party.

In discussing elections, the payoffs will not be numerical since the alternatives may be complex and not measurable on any single numerical scale. Instead, we borrow an idea from economics and assume that the voters have preferences over the alternatives they are voting on. The preferences will often disagree, but voters will choose their strategies so as to obtain alternatives that are more preferred.

1. PARTY! PARTY! PARTY!

Signa Phi Naught sorority is planning a big party, and the executive committee has to decide how much to spend from the sorority treasury. The members of the executive committee are Anna, Barbara, and Carole — A, B, and C, as usual. They can spend $1,000, 1,500, 2,000, 2,500, or 3,000. The three committee members' preferences

426 *Game Theory (Fourth Edition)*

HEADS UP!

Here are some concepts we will develop as this chapter goes along:

Majority: More than half of the votes cast.

Plurality: More votes than any other candidate or alternative.

Single-peaked Preferences: If there is some dimension on which each voter's preference ranking first increases to a "best" and then decreases, the voter's preferences are said to be "single-peaked."

Median Voter: If one voter's peak preference lies between one-half of the voters who prefer more and one-half who prefer less, that person is the median voter and her vote can be decisive.

Strategic Voting is voting for something other than one's first preference, in the hope of improving the outcome of the voting from one's own point of view.

Naïve Voting is voting for one's first choice regardless of the consequences.

The Condorcet Rule, proposed by the Marquise de Condorcet, is that a candidate who would win a two-way naïve vote against every other candidate should be the candidate selected.

Preference Voting is a scheme in which the voters vote for more than one candidate or alternative, ranking them in order of the voter's preference.

---

are shown in Table 18.1. For example, Anna's top preference is to spend $1,500, while $1,000 is her least preferred alternative — fifth out of five. The preferences are shown in a diagram in Figure 18.1. Since first preference is highest, it is shown at the top of the diagram.

Anna moves that funds for the party be appropriated in the amount of $1,500. Barbara and Carole each propose an amendment to the motion. Barbara's amendment is that the amount be increased to $2,000. Barbara and Carole both vote for it, since they both prefer $2,000 to $1,500. ($2,000 is Barbara's first preference and Carole's

Table 18.1. Preferences for the Three Members of the Executive Committee.

	$1,000	$1,500	$2,000	$2,500	$3,000
A	Fifth	First	Second	Third	Fourth
B	Third	Second	First	Fourth	Fifth
C	Fifth	Fourth	Third	First	Second

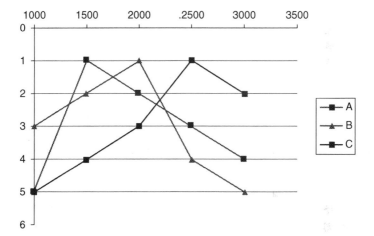

Figure 18.1. Preferences of Members of the Executive Committee for Spending Levels.

third, but $1,500 is Barbara's second preference and Carole's fourth). Carole's amendment is to increase the amount still further, to $2,500. Anna and Barbara both vote against it, since they both prefer $2,000 to $2,500 ($2,000 is Anna's second preference. $2,500 is Anna's third preference and Barbara's fourth.) Thus, the final decision is for a $2,000 expenditure on the party.

Barbara has gotten her way in this vote. Why? There are two interrelated reasons. First, if we start from the left in Figure 18.1, the preference level for each person rises (not necessarily at a steady rate, but without reversals) until the top preference is reached, and then the preference level declines, again without any reversals. When

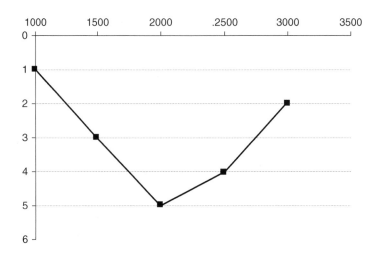

Figure 18.2.   Yolanda's Preferences.

preferences have this property, they are said to be *single-peaked*. For a contrast, look at Yolanda's preferences in Figure 18.2. Yolanda is not on the executive committee, so she doesn't have a vote, but she prefers a cheap party. At the same time, she feels that if more than $1,000 is spent, then the sorority might as well go all the way and spend the maximum of 3,000, which is her second preference. Thus, Yolanda's peak preference is ambiguous: $1,000 is a peak, but the second preference of $3,000 is a sort of a peak, too — at least relative to the fourth and fifth preferences, $2,500 and $2,000. It is a "local peak." With two local peaks, Yolanda's preferences are not single-peaked. In general, preferences with more than one local peak are not single-peaked, and to say that preferences are single-peaked is to say that there is only one *local* peak.

Because the members of the steering committee all have single-peaked preferences, we can pick out each person's peak preference without any ambiguity, and this also means that any movement toward a person's peak preference will make that person better off, in preference terms.[1]

Table 18.2. Peak Preferences.

Person	Peak preference
Anna	$1,500
Barbara	$2,000
Carole	$2,500

Here is the second reason why Barbara gets her way: When we list the peak preferences, they are as shown in Table 18.2. We see that Barbara's peak preference is at the middle. Barbara is the *median person* in this vote. She and Anna will agree in support of any movement toward her first preference, $2,000, and together they have a majority. She prefers spending less than Carole, so she and Carole will agree to support any movement toward $2,000, from below, and,

Terminology: *Majority* — Remember, a *majority* means more than half of the votes. If there are more than two alternatives, it may be that none of them gets a majority. In that case, the one that gets the most votes is said to have a *plurality*. These terms are sometimes confused in casual discussion, but in this textbook, we use the term "majority" in its strict meaning — more than half.

again, together they have a majority. Taking both sides together, any movement toward Barbara's peak preference will have majority support. That's the advantage of being the median person.

In general, the median person is the person who prefers a quantity higher than half-minus-one of the voters, and also lower than half-minus-one of the voters. Thus, any movement toward the median person's first preference from below will be supported by all the voters whose first preference is greater than hers; so, with her vote as well, it will be supported by a majority. Similarly, any movements toward the median person's peak preference from above will be supported by all of those whose first preference is less than hers, so with her vote (once again!) it will have majority support. In general, so long as all voters have single-peaked preferences, a move toward the median person's peak preference will be preferred by a majority of the voters. Thus, the median person's preferences are decisive, under majority rule.

This example illustrates some important aspects of the game-theoretic analysis of voting. First, votes can be thought of as strategies in a game, and majority rule (or some alternative voting scheme) defines the rules of the game. On the other hand, the sorority party example includes some important simplifying assumptions. The simplifying assumptions are that preferences are single-peaked and that the issues are laid out along a single dimension from less to more. We have also assumed that each member of the executive committee simply votes for the alternative she prefers. But that isn't always the rational choice of strategies, as the next example shows.

2. THE THEME FOR THE PARTY

The executive committee of Signa Phi Naught were unable to decide among three possible themes for decoration and costumes at the party. The three themes they were considering are Roaring Twenties (R), Swiss Alps (S), or Tropical Island (T). The executive committee have decided to put the issue before the whole membership of the sorority for a vote. There are three types of voters among the sorority membership, with different preferences among the three alternatives. We shall call the types X, Y, and Z. The preferences of the three types, and the proportion of the votes cast by each, are shown in Table 18.3.

Now, suppose each type votes for their first preference. The Tropical Island alternative has the plurality, but for 60% of the voters, this is the worst of the three alternatives. This is a fundamental problem of plurality voting — it can, and often does, conflict with majority rule, in that the winning alternative is opposed by a majority of the voters. If the executive committee of Signa Phi Naught want to have a majority decision, they will have to choose election

Table 18.3. Preferences for Themes for the Party.

Themes

Voter type	R	S	T	Voter percentages
X	First	Second	Third	31
Y	Second	First	Third	29
Z	Third	Second	First	40

rules that are a little more complex than plurality voting. One possibility is that they could have a runoff election — that is, hold two rounds of voting, with the alternatives in the second round of voting being the two that got the most votes in the first round. If there is a majority for one of the alternatives in the first stage, the second stage is cancelled. In this case — still assuming that the voters just vote for their first preferences — the first round will eliminate S, and R will win in a runoff election between R and T.

Definition: *Runoff Election Rule* — An election may be conducted in two stages, with the candidates in the second stage limited to the first two finishers in the first stage. The second stage is called a "runoff." This procedure is used in some American states and in French presidential elections, among others.

3. STRATEGIC VOTING

But, if the voters are rational game-players, they may not vote for their first preferences. Suppose — for example — that the election is held in a single stage and the alternative that obtains the most votes is adopted (plurality rule). Suppose also that types Y and Z vote for their first preferences. Then, by voting for their *second* preference, type X can improve their outcome from their third to their second preference. This is an example of strategic voting. Whenever a voter votes for something other than his or her first preference, in order to get a better overall outcome, we say that the voter is voting *strategically*. When the voters simply vote for their first preference in each round, we would say that they vote *naïvely*.

In fact, type Y could also benefit by voting strategically, in some circumstances. Let us look at the game of plurality voting as a game in normal form. The strategies will be to vote for the first or the second preference. We will ignore the possibility of voting for the last preference because it can never lead to an improved outcome in this game. Payoffs are listed in the order X, Y, Z.

Remember that, since the payoffs are preferences, smaller numbers are better in this game. Examining Table 18.4, we observe that there are three Nash equilibria. They are shaded in gray in Table 18.4. What we have in the left-hand panel is a coordination game of a type we already know is difficult. All of the lightly shaded cells are equilibria, but which will

> **Definition:** *Strategic and Naïve Voting* — *Strategic voting* is voting for something other than one's first preference, in the hope of improving the outcome of the voting from one's own point of view. *Naïve* voting is voting for one's first preference regardless of the outcome.

occur? There is some possibility that the type X and Y voters will guess wrong and choose first, first, or indeed second, second, and end up with their third preferences anyway! Nor is that the end of

Table 18.4. Plurality Voting in the Party Theme Game.

		Type Z			
		First		Second	
		Type Y		Type Y	
		First	Second	First	Second
Type X	First	3rd,3rd,1st	1st,2nd,3rd	2nd,1st,2nd	1st,2nd,3rd
	Second	2nd,1st,2nd	3rd,3rd,1st	2nd,1st,2nd	2nd,1st,2nd

the story. Another possibility is that types X and Y might form a coalition, caucus and take a vote among themselves, and coordinate their strategies on the alternative that gets a majority in their caucus. That will be alternative R, Roaring Twenties. Since it is a Nash equilibrium and the caucus vote makes it a Schelling focal point, the coalition will not need any enforcement to make this the outcome.

It would be no better if the election had been conducted with a runoff to assure a majority decision. Notice that the runoff election is not immune to strategic voting, either. By voting for their second preference at the first stage, type Z can assure that they get their second preference instead of the third. On the first stage, Swiss Alps would win with 69%, so no runoff would be needed.

4. VOTING PROBLEMS AND CRITERIA

The Party Theme Game illustrates the wide range of problems and possibilities in real elections. On the one hand, plurality voting is not very satisfactory. It can leave the majority very dissatisfied, and the outcome can be unpredictable. In the Party Theme Game, any outcome is possible, depending on who votes strategically. Coalitions and caucus-votes are also common in real elections. On the other hand, to get a majority decision will require some more complicated voting scheme, such as a runoff election. But runoff elections, too, may be influenced by strategic voting. Worse still, it is not obvious what outcome is the "correct" one. When there is no majority in naïve voting, which outcome represents "majority rule?"

This is not a new question! In the mid-1700's,[2] the Marquis de Condorcet proposed a partial answer to the question. He proposed that electoral schemes ought to be set up in such a way that an alternative that would win a two-way naïve vote against every other

---

[2]This period is sometimes known as "the enlightenment," because of the ferment of new ideas and the belief in progress that were widespread at the time. Enlightenment ideas were very influential in the formation of the United States. The Marquis de Condorcet was one of the most important French thinkers of the enlightenment.

alternative should be the one chosen. In the Party Theme Game, for example, S defeats R 69-31 and S defeats T 60-40. Thus, S is the Condorcet alternative. But not all voting games have a Condorcet alternative.

In the mid-20th century, Nobel Laureate economist Kenneth Arrow proposed a more complete list of criteria for a good election scheme — and proved that no election scheme can possibly realize all of them.[3] In order to understand the point, we need to look at the list in detail; and indeed, the list is of interest in itself. Arrow's criteria include:

A Closer Look: The Marquis de Condorcet 1743–1794

Marie-Jean-Antoine-Nicholas de Caritat, Marquis de Condorcet, was a major figure of the enlightenment, a mathematician who made important contributions to integral calculus and probability as well as originating the mathematical study of elections. Although a supporter of the French Revolution, he opposed the Jacobins, was imprisoned by the Jabcobin government in France, and died while in custody, from unknown causes.

1. *Efficiency:* There should be no alternative that every voter prefers to the winning alternative.
2. *Completeness:* The voting scheme leads to complete and consistent ranking of all alternatives.
3. *Neutral:* The ranking of any two alternatives depends only on the preferences of the voters between those two alternatives (as with Condorcet's criterion).
4. *Non-dictatorial:* There is **no** one person whose preferences decide the election regardless of the preferences of others.[4]

---

[3] In fact, Arrow's proposition is even broader. It applies to all mechanisms for social choice — including, for example, market mechanisms.

[4] Recall that, in Section 1, Barbara got her way in the election to determine how much to spend, because she was the median person — but if the preferences of Anna and Carole had been different, the outcome could have been different!

Arrow's "general impossibility theorem," which states that there is no way of voting or more generally of deciding issues that satisfies all these criteria, has had a great influence on systematic thinking about elections, including game theoretic thinking. The third assumption, also known as "the independence of irrelevant alternatives," has been the most controversial. It essentially rules out strategic voting. As we have seen, strategic voting will be common, but there is no reason to think that makes things any more predictable — quite the contrary.

Much of the controversy on the Arrow Theorem has focused on the possibility that a small sacrifice on one of Arrow's criteria — for example, a less than complete or less than neutral scheme[5] — might give us satisfactory overall results.

We should observe that Arrow's result does not say that all four criteria can never be satisfied in any election. It says that there are some cases, in which people have certain preferences, for which one or more of the four conditions will fail. Some schemes might satisfy all those criteria for some, or even for very many, distributions of preferences. For example, we have seen that a runoff election with strategic voting results in a win by the Condorcet candidate in the Party Theme Game — but as we will see in the next section, the results can also be the opposite. We might settle for an election scheme that would satisfy Arrow's criteria for a very large proportion of the preferences that we are likely to find in practice. Research in this direction has made some progress.[6]

---

[5]Robert's Rules of Order, a guide to parliamentary procedure used by many American voluntary organizations, prescribes that, if there is no majority, the ballot be repeated time and again until there is one. Presumably this relies on strategic voting in later rounds to produce a majority. But Robert's Rules do not define a neutral scheme, since the agenda order of presentation of issues can definitely influence the outcome.

[6]Dasgupta, Partha and Eric Maskin, The Fairest Vote of all, *Scientific American*, **290**(3) (March, 2004), pp. 92–97; Dasgupta, Partha and Eric Maskin, On the robustness of majority rule, *Journal of the European Economic Association*, **6**(5) (September, 2008), pp. 949–973.

5. ALTERNATIVE VOTING SCHEMES

Many alternative electoral schemes have been proposed. Here are a few.

1. *Plurality rule.* Although it is problematic, many elections are conducted according to plurality rule, including British Parliamentary elections, where it is known as the "first past the post" system.
2. *Runoff elections.* If there is no majority on the first round, a second election is held between the winner of the plurality and the second-place finisher. As we have seen, a runoff can produce a majority for one candidate, but runoff elections are highly strategic and can be unpredictable.
3. *Preference voting.* One possible problem with conventional voting is that it provides information only on the first preference. Preference voting schemes are designed to obtain more information

A Closer Look: Kenneth Arrow 1921–2017

A native New Yorker, Kenneth Arrow did his undergraduate work at the City College of New York, receiving a degree of Bachelor of Science in Social Science but a major in Mathematics. His graduate work was at Columbia. After obtaining an M.A. in mathematics, he switched to economics, but his dissertation research was delayed by the war and other concerns. Completed in 1951, his dissertation, *Social Choice and Individual Values,* had immediate and lasting impact. His pathbreaking work on competitive equilibrium under uncertainty has had equal impact on economics and on innovation in financial markets. He shared the Nobel Memorial Prize in Economics in 1972 with John Hicks for his "pioneering contributions to general economic equilibrium theory and welfare theory."

from voters, by allowing the voters to rank the alternatives in order of their preferences.

a. Borda Rule. Proposed by another French Enlightenment figure, the chevalier de Borda, the Borda Rule is one such. The Borda Rule has the voter assign points to each alternative in inverse proportion to their preference ranking. Thus, in a three-way election, the first preference gets 3 points, the second preference 2, and so on. The alternative with the greatest sum of points wins. In the Party Theme Game, alternative T gets 5 points, R 6 and S 7; so, S, the Condorcet alternative, wins. But this will not always be so!

b. Preference voting with a single transferable vote, also known as an instant runoff. Each voter ranks all the alternatives from most to least preferred. If the first preferences produce a majority, then it is adopted. If not, then the alternative getting the fewest first preference votes is eliminated and its supporters' votes transferred to their second preference listing. If (with more than two alternatives) this does not

---

A Closer Look: De Borda and the Borda Count

In 1770, the French mathematician Chevalier Jean-Charles de Borda (1733–1799) proposed a new method of counting votes: The voter would rank the candidates or alternatives according to their order of preference. If there were N alternatives the first preference would get N points, the second $N - 1$, and so on. The alternative with the most points would win.

Born in Dax, France, de Borda was a military engineer and inventor who perfected instruments of navigation, waterwheels, and pumps and, in between, found time to participate in the American war of independence. He was among the creators of the metric system of measurements.

A Closer Look: *Election Laws in American States*

Election laws vary widely in American states, and some have changed in recent years. In California and Washington State, in the so-called "Jungle Primary," candidates of any party or none may compete at the first stage, and at the second stage, the election is between the top two finishers at the first stage, even if one has won a majority in the first stage. Thus, it is possible that the winner of a majority in the first stage could lose at the second stage, though this does not seem to have happened as yet. Maine has adopted ranked-choice voting with a single transferable vote. Alaska has adopted a two-stage election process in which candidates from any party or none can be candidates at the first stage, and the top four vote-getters advance to the second round, in which the vote is a ranked-choice vote with votes counted by the single transferable vote system. See Richard A. Pides, "More Places Should Do What Alaska Did to Its Elections," *New York Times*, February 16, 2022.

produce a majority, then the procedure is repeated until a majority is obtained. In the Party Theme Game, the "instant runoff" produces the same result as a two-stage runoff, but that will not always be true.

c.   Condorcet tournament. In this approach, all two-way matches are evaluated, with a vote counted for one or the other dependent on which is ranked higher by each voter. If there is a Condorcet candidate, that alternative is declared the winner. Since there is not always a Condorcet Candidate, this will not decide all elections. When there is no Condorcet candidate, there is something of a tie among all those who win at least some of their pairings. In that case another method must be used to break the tie — perhaps a Borda ranking or "instant runoff."

4.   Approval voting. In this case there is no ranking of one alternative against another, but each voter simply votes "yes" or "no" for

each alternative. The alternative that gets the largest number of "yes" votes wins. Approval voting is highly strategic, since there is no obvious (naïve) answer to the question of how many alternatives to approve. In the Party Theme Game, for example, suppose type Z approves its first and second choice, while type Y approves only their first choice. This is a Nash equilibrium, regardless of what X does. (See Table 18.5.) Theme S is chosen. Thus, in this game, strategic approval voting yields the Condorcet alternative. But that will not always be true.

While there is much to be learned about these and other alternative voting schemes, it seems clear that there is no one scheme that is clearly best, or even satisfactory, in all circumstances.

A Closer Look: Some Key Research on Voting Schemes

Strategic voting sometimes seems underhanded, and can be expected to lead to violations of Arrow's Neutrality criterion, which is a commonsense idea of "good outcomes" for an election. Would it be possible to design or discover a scheme in which people would not want to vote strategically? In separate research, closely related to Arrow's, Alan Gibbard, a University of Michigan philosopher, and Mark Satterthwaite, a Northwestern University management specialist, proved that there can be no such system.

Apparently, we must give up some part of Arrow's criteria in some elections. But this problem emerges in only some elections, not all. Is there an election scheme that will produce good results according to the Arrow criteria in most elections, or in more elections than most other schemes? Eric Maskin, who had done his doctoral study under Arrow, and Partha Dasgupta have explored this and found that a Condorcet Tournament "works well" in that sense it satisfies the commonsense criteria whenever *either* plurality rule *or* the Borda count does, thus doing better than either.

Table 18.5.   Approval Voting in the Party Theme Game.

		Type Z			
		Approve first		Approve first and second	
		Type Y		Type Y	
		Approve first	Approve first and second	Approve first	Approve first and second
Type X	Approve first	3rd,3rd,1st	1st,2nd,3rd	2nd,1st,2nd	2nd,1st,2nd
	Approve first and second	2nd,1st,2nd	tie*	2nd,1st,2nd	2nd,1st,2nd

Note: *The tie is between R and S, first and second for X and Y and second and third for Z. But either X or Y can assure the first preference by shifting unilaterally to "first," and Z can assure second preference by shifting unilaterally to "first, second." Therefore, the tie is not a best response for anybody and cannot be a Nash equilibrium.

But all of our examples so far have been "made up." Do these difficulties arise in the "real world?" We will look at some cases, and find that they do indeed.

6. CASE: THE FINNISH PRESIDENTIAL ELECTION[7]

The Republic of Finland is located in northern Europe, on one of the Scandinavian peninsulas, sharing its main common border with the Russian Republic and sharing borders also, in the far north, with Sweden and Norway. Urho Kekkonen was President of Finland for 25 years beginning in 1956. In Finland, in 1956, presidential elections were[8] conducted by a 300-member Electoral College. The Finnish Electoral College was unlike the Electoral College in the USA (among other ways) in that there would be a runoff if there were no majority in the first round of voting.

---

[7]This example is from George Tsebelis, *Nested Games: Rational Choice in Comparative Politics* (University of California Press, 1990), pp. 2–4.

[8]Finnish presidential elections are now conducted by popular vote, with a second "runoff" round in case there is no majority on the first round.

Table 18.6. Parties and Preferences in the Finnish Presidential Election.

Party	Kekkonen	Paasikivi	Fagerholm	Votes
Agrarian	First	Second	Third	88
Communist	First	Third	Second	56
Conservative (major faction)	Third	First	Second	77
Conservative (minor faction)	Second	First	Third	7
Socialist	Third	Second	First	72

In 1956, four parties participated in the Finnish presidential election: the Agrarian (Farmers') Party, the Communists, Conservatives, and Socialists. The Communists, as the smallest of the four, did not propose a candidate, but the other three parties did. Kekkonen was the candidate of the Agrarians, J.K. Paasikivi, the incumbent, was the candidate of the Conservatives, and K.-A. Fagerholm the candidate of the Socialists. Table 18.6 shows the preferences of the four parties, as reconstructed from the evidence, and the votes each party could cast in the Electoral College. We see that the Communists, without a candidate of their own, favored Kekkonen,[9] and that the Conservatives were somewhat divided on their second preference, with a minority preferring the Agrarian Kekkonen over the Socialist Fagerholm.

Now let us see what happens if each party votes naïvely in each round. In the first round, Kekkonen gets 56 + 88 = 144; Paasikivi 84; and Fagerholm 72. The runoff would then be between Kekkonen and Paasikivi, and Kekkonen would get the same 144 votes, while Paasikivi would get 156. Thus, Paasikivi wins — the worst outcome from the point of view of the Communist party.

---

[9]At that time, the Russian Republic was part of the Soviet Union, and relations with Finland's huge neighbor were the major problem for the Finnish government. Finland had been invaded by the Soviet Union 15 years before, and Finland had successfully defended its independence although it had lost some territory. It seems likely that the Communists saw Kekkonen as being more likely to maintain friendly relations with the Soviet Union than Fagerholm would have been.

Game Theory (Fourth Edition)

But the Communists did not vote naïvely. Instead, on the first round, 42 of the 56 Communist electors voted for Fagerholm. This could not have been a matter of different individual preferences. Communist parties at that time worked on the basis of "democratic centralism," meaning that individual members obeyed the party decision once it was made. In any case, the result was to swing the election to Kekkonen. In the first round Fagerholm got 114 votes, more than Paasikivi's 84, eliminating Paasikivi. Thus, the runoff election was between Kekkonen and Fagerholm. In the runoff, the Communists voted unanimously for Kekkonen and Kekkonen won, 151 to 149.

We see that a small party was able to manipulate the runoff electoral scheme by strategic voting. What about the other criteria and schemes?

1. Condorcet. Paasikivi defeats Kekkonen 156–144 and Fagerholm 165–135, so Paasikivi was the Condorcet candidate.
2. Borda Rule. Paasikivi wins with 628 points to 595 for Kekkonnen and 577 for Fagerholm.
3. Single transferable vote. In the two-round runoff, the Communists were able to manipulate the system by changing the preferences they expressed between the first and second rounds. With an "instant runoff," that is, single transferable vote, they cannot do that. In order to get Fagerholm into the runoff, at least 13 Communists have to list him as their first preference rather than Kekkonen, but those preferences mean that they are counted as voting for Fagerholm after Paasikivi is eliminated, and Fagerholm wins 162 (72 Socialist, 77 Conservative, and 13 Communist votes) to 138. Since the Communists preferred Fagerholm somewhat to Paasikivi, they probably would have done this. Notice that Fagerholm would lose in a two-person race with either of the two other candidates.[10]

---

[10]The socialists could have prevented this by strategically voting for Paasikivi over Fagerholm, as indeed they could have in the actual election but did not. It seems that they did not vote strategically because they were responsible to their party

4. Approval voting. There are plural Nash equilibria, so the outcome is unpredictable.

7. CASE: THE AMERICAN PRESIDENTIAL ELECTION OF 1992

As the Finnish example illustrates, strategic voting is especially likely where there are more than two alternatives and complicated voting rules. The United States, too, has an Electoral College that decides Presidential elections, and it is part of a quite complex electoral routine that takes place every fourth year. The authors of the American Constitution had hoped that this Electoral College would keep political parties out of presidential elections, but in fact political parties have dominated the Electoral College since the election of John Adams in 1796. Since the 1850's, only nominees of the Republican or Democratic Parties have been elected, but there have usually been "third parties" that got some popular vote so that simple majorities in the popular vote have been uncommon. However, in the 20th century, the Electoral College result always agreed with the plurality in the popular vote.[11] When there were three or more significant candidates, however, the results have been less predictable. The election of William Clinton in 1992 provides an important example. In a three-way race with the incumbent George H. W. Bush and billionaire Ross Perot, as the candidate of the reform party, Clinton received the plurality with 43% of the popular vote. Bush obtained 37% and Perot 20%. In Table 18.7, we see an example based loosely on this 1992 election. It is only loosely based because we are making the simplifying assumption that there were only three

---

members, who would not have tolerated it. That is, their electoral game was imbedded in a larger game of party politics in which a strategic vote was not a best response. Clearly, it is very difficult for a party to vote against its own nominee.

[11] There were exceptions in the elections of 1876 and 1888, and in 1876 some states sent more than one delegation to the Electoral College, so that the result was really decided by the Congress. In 1888, the popular vote was extremely close. In 2000, a decision of the Supreme Court decided the election and in 2016, the candidate with the plurality in the popular vote was defeated in the electoral college.

Table 18.7. Preferences in a Simplification of the 1992 US Presidential Election.

	Ranking for			
	Clinton	Bush	Perot	Percentage of electorate
Democrats	1st	2nd	3rd	43
Republicans	3rd	1st	2nd	37
Reformers	2nd	3rd	1st	20

types of voters, and that each voter of a particular type had the same preferences as every other of the same type, especially for second and third preferences. We don't know that — in this case we have little evidence on second and third preferences — and reality was probably more complicated. But the example is consistent with what we do know about the 1992 election and, as we will see, it is complicated enough!

The fifth column gives the assumed proportion of the electorate who is of each of the three types. These are the proportions in the actual popular vote, rounded off to percentages. We have assumed that the Reformers favored Clinton over Bush — some important leaders of the movement made that fairly clear — and that the Republicans preferred Perot to Clinton, as being less liberal, while the Democrats would have chosen Bush over Perot, perhaps seeing Perot as unpredictable. These last two are guesses for the sake of the example.

In this election, as in the Finnish one, no one of the candidates gets a majority. If everyone votes naively, then Clinton wins the plurality (as in fact he did). A question is this: why did the Republicans not vote strategically? Had they voted for Perot rather than Bush, Perot could have been elected, and the Republicans would have had their second preference rather than their third. But perhaps they had the longer-term strength of their party in mind. That is, it may be that for the Republicans, the 1992 election game was imbedded in a larger game, in such a way that a win for Perot would have made it much more unlikely that the Republicans would come back.

This election differs from the Finnish one, in that there was no provision for a runoff election. In fact, of course, the popular vote does not decide the issue, but the Electoral College does, and Clinton got a 370-168 majority in the Electoral College. This can happen because a candidate who gets the plurality in a state (or in a congressional district, in the case of two states) gets all the electoral votes for that state (congressional district). Thus, a candidate who gets 50.01% of the vote in every state (district) would get 100% of the electoral vote.

What about the other schemes and criteria?

1. Condorcet. Clinton defeats Bush, 63-37; Bush defeats Perot 80-20, and Perot defeats Clinton 57-43. This cycle means that there is no Condorcet candidate in this race. Cycles like this play an important part in recent analytical thinking about elections, especially the Arrow Impossibility Theorem. The cycle in this example reflects the assumption that all Republicans would vote for Perot over Clinton. This may not have been true in the real world. If as many as 25% of Republicans would have gone for Clinton in a Clinton-Perot race, then Clinton becomes the Condorcet candidate, but the extreme assumptions were chosen to illustrate the important possibility of a cycle, and the fact that there may not be a Condorcet candidate in some elections.

2. Borda Rule. With naïve voting, Bush wins 36%-34%-30%, in percentages, with Perot last; but if 50% of Democrats strategically (or sincerely!) shift their second-place listing to Perot, then Clinton wins 34%-33%-33%.

3. Two-round runoff. With naïve voting, Clinton wins 63-37 in the runoff, but Republicans can shift it to their second preference, Perot, by strategically voting for him on the first round.

4. Single Transferable Vote. Same as two-round runoff.

5. Approval voting. The Nash equilibrium is that the Republicans approve both Bush and Perot, Democrats and Reformers approve only their own candidates, and Perot wins with 57% of the population approving against 43% for Clinton and 37% for Bush.

A Closer Look: The Electoral College in the United States

The American electoral college allocates a number of votes to each state based on its Congressional representation. If the electoral college cannot produce a majority for any candidate, then the election is decided by the House of Representatives, with each state casting one vote. This has only happened in 1824.

The drafters of the American Constitution seem to have adopted the Electoral College with some reluctance, thinking that there would be no candidate well known in all sections of the far-flung American Republic. Instead, they envisioned compromises among the different sections and saw the Electoral College as a way of working them out.

Proposals to eliminate the College have come and gone over the years. Following the election of 2000, it was often argued that it would be impractical to eliminate the College, since (1) it would require a constitutional amendment; and (2) this would be opposed by the smaller states, which also are more numerous, because the College favors the smaller states.

It is true that the smaller states get more votes in the College relative to their population, but it is **not** clear that the College always favors smaller states. In the election of 1960, the candidates were John Kennedy and Richard Nixon, and while Kennedy won both the popular and electoral votes, his margin was much wider in the Electoral College (almost 50%) than in the popular vote (one-tenth of 1%). He did this by winning big states by small margins and losing small states by big margins. Under the winner-take-all system of allocating electoral votes, Kennedy got 100% of the votes of the big states he won, no matter the margin.

In that case, clearly, the Electoral College worked against the smaller states, and at that time, much of the interest in revising the electoral system came from them (and from Republicans). Two small states, Nebraska and Maine, did eliminate the winner-take-all system (which is a matter of state law), instead awarding one electoral vote to the winner of a majority in each Congressional district and two to the winner of the state-wide majority. But it is not clear that this worked to their advantage either. In the election of 2008, Nebraska awarded four votes to Senator McCain, but one to Senator Obama, arguably diluting Nebraska's influence.

In a spirit of truth-in-authorship, I should say that I have favored elimination of the Electoral College and election of the President by popular vote since 1960, and have been persuaded by Professor Eric Maskin's reasoning that Americans should adopt preference voting with a Condorcet Tournament as the first stage in the counting. But, then, my friends all think I'm pretty weird.

8. SUMMARY

Elections can be predictable if (1) there are only two alternatives under consideration, or the alternatives can all be ordered as more and less; and (2) all voters have single-peaked preferences over the alternatives. In that case, naïve voting will make the median voter's preferences decisive, and there is little scope for strategic voting. With three or more alternatives that cannot be ordered as more or less, the possibilities emerge that no one alternative can get a majority, and that strategic voting can influence the results. In some cases, strategic voting with multiple Nash equilibria may give a majority to any of the alternatives, depending on which equilibrium is realized. There are a variety of schemes for finding a majority in such cases, but no one of them clearly works better than the others in every case. There could even be controversy about what the "right" outcome of the election would be. If there is one candidate who will get a majority against all the rest in two-way elections, a "Condorcet candidate," then it seems we might want that candidate to win the election — but there may not be a Condorcet candidate, and even if there is, election procedures simpler than holding all two-way elections will sometimes miss the Condorcet candidate. The more extensive (but still reasonable) list of criteria proposed by Kenneth Arrow can also be shown to be impossible. There is no mechanism that always satisfies all of the Arrow conditions. All in all, election procedures present trade-offs between different disadvantages, and there is no universal answer to "what is the best election procedure."

Q18. EXERCISES AND DISCUSSION QUESTIONS

Q18.1. Impasse in the Faculty Senate

(This is a true story, with the names changed to protect the more or less innocent). At a Faculty Senate Meeting, Prof. Gadfly brought a complex motion that was mostly supported by a large majority, but that had two controversial paragraphs. Some Senators proposed amendments that would delete one or the other, and some proposed an amendment that would eliminate both. The chair, Prof. Marian from the Library School, put the three amendments to a vote together. The vote was 5-5-5. Prof. Marian then cast the tie-breaking vote for the amendment that would have eliminated both paragraphs, but Prof. Mugwump, the parliamentarian, pointed out that a majority vote would be necessary to amend the motion, and 6 votes is not a majority. Analyze Prof. Marian's problem using concepts from this chapter. What would you advise Prof. Marian to do?

Q18.2. Language

The European Language Club has to elect a new president. There are three candidates: Jean-Jacques, Francesca, and Angela. The club has three major factions consisting respectively of students majoring in French, Italian, and German. The election will be held by majority rule. In case no candidate has a majority in the first stage, there will be a second-stage runoff election. The voting proportions of the three factions and their first and second preferences are shown in Table 18.8.

Table 18.8. Language Club Preferences.

Faction	Percentage	First	Second
French	0.4	Jean-Jacques	Angela
Italian	0.25	Francesca	Jean-Jacques
German	0.35	Angela	Francesca

Table 18.9. Weekend Trip Factions.

	Options			
Factions	Lancaster Cty	Cape May	AC	Proportion
Bald Eagles	2	1	3	25%
Silver Foxes	3	2	1	35%
Gray Marauders	1	2	3	40%

a. Which candidate will win if all factions vote naively?
b. What if the Germans vote strategically?

Q18.3. Weekend Trip

Old And Restless (OAR) is a retired people's club that arranges trips and other recreation for its members and they are planning their next weekend trip. There are three factions: the Bald Eagles, who prefer nature-oriented destinations, the Silver Foxes, who prefer destinations with a lively nightlife, and the Gray Marauders, who like shopping destinations. The alternatives are (1) Lancaster County; (2) Cape May; and (3) Atlantic City. Their preferences are shown in the Table 18.9.

a. Is there a Condorcet alternative?
b. Suppose the decision is made by single transferable vote or "instant runoff" rules.

Table 18.10. Donation Factions.

	Charity			
Faction	P	S	H	Number in faction
A	First	Third	Second	18
B	Second	First	Third	12
C	Second	Third	First	10

    i.   If all vote naively which will win?

    ii. What if there is strategic voting?

c. Suppose the decision is made by the Borda Rule

    i.   If all vote naively which will win?

    ii. What if there is strategic voting?

Q18.4. Beneficiary

The Beneficent and Vigilant Order of Herbivores (BVOH), a vegetarian social club in a suburb of Philadelphia, will stage its annual fund-raising dinner, featuring tasty vegetarian Indian cuisine. They have to decide which charity to support with the proceeds, and because of their bylaws they cannot compromise, so that the proceeds will be donated to one charity or the other. The alternatives are (1) the Society for the Protection of Animals (P); (2) Strep Throat Research Fund (S); and (3) Support for the Homeless Endowment (H). The 40 members of BVOH are divided into three factions, and their preferences are shown in the Table 18.10.

a. Is there a Condorcet candidate? Which? Explain.

b. If the vote is decided by plurality rule, which alternative will win? Will some factions vote strategically? Which?

CHAPTER 19

Social Mechanism Design

Games such as poker, golf and "soccer" football are defined by their rules, and in many cases, there are international commissions or associations to define and sometimes revise the rules. Many of the

To best understand this chapter, you need to have studied and understood the material from Chapters 1–9 and 12.

metaphorical "games" we have considered in this book, such as economic competition and environmental pollution, have "rules" more or less defined by law or public policy. Those who make the laws and policy rules — legislators, judges, citizens — should be concerned that the rules have good results. But the results will depend on how people respond and interact with one another once the rules are in place. Some kinds of rules may lead to unanticipated bad results if people act non-cooperatively, choosing a best response to the actions of others without efficiently or equitably coordinating the action of a whole group. Since there is evidence that people often (though perhaps not always) behave non-cooperatively, it seems reasonable to try to design rules that can achieve their objectives even when people do indeed act non-cooperatively. Rules with that property are said, in the jargon of economic game theory, to be "incentive compatible," meaning that no rational self-interested person has an incentive to act against the objectives of the rules. Since the rules can be thought of as defining a social mechanism, this process is called "social mechanism design."[1]

In 2007, the Nobel Memorial Prize in Economics was awarded to three economists (two of them game theorists) for foundational work in social mechanism design. The three were Leonid Hurwicz (1919–2008), Eric Maskin, and Roger Myerson. "Mechanism design, Professor Maskin explained,[2] can be thought of as the 'reverse engineering part of economics.' The starting point, he said, is an outcome that is being sought, like a cleaner environment, a more equitable distribution of income or more technical innovation. Then, he added, one works to design a system that aligns private incentives with public goals." In 2012, the Nobel went to Lloyd Shapley and Alvin Roth for further work on social mechanism design, and in 2020, Robert Wilson and Paul Milgrom received the honor for their work in designing auction mechanisms in particular. Since the 1990s, we have gained experience in putting these ideas to work in real institutions, and there have been good results when the institutional design is followed closely. We will begin the chapter with the work for which Shapley and Roth were honored, which provides a good example of the ideas both for the role of game theory and the successful use of new proposals on social mechanism design.

1. MATCHING

Coalitions are central to cooperative games, and many of our everyday activities can be thought of as formation of small coalitions:

---

[1] Of course, this program has deep and distinguished roots in the tradition of social philosophy. Without returning to ideal societies in classical Greek philosophy, we see in social mechanism design strong echoes of Jeremy Bentham's "utilitarian" philosophy of law and of the rule-utilitarianism of the great philosopher of liberty, John Stuart Mill. However, the application of game theory contributes both a more precise conception of rational self-interest and the contrast of cooperative and non-cooperative solutions.

[2] *New York Times*, October 16, 2007, p. B1.

Dating and marriage, job seeking and applications to schools and colleges, registration for courses for the term, formation of business partnerships and sports teams are well-known examples. Many of these have the following characteristics: (1) effective coalitions will be small, in some cases just two; (2) agents are of two or more types, and an effective coalition requires just one of each; (3) there are idiosyncratic differences among agents within each type; that is, no two people are just alike; and (4) benefits of the coalition depend on how well the individuals complement one another, and may vary depending on the individuals in the small coalition. Marriage is clearly an example along these lines, and this is the classical matching problem: We may find that two people who are not matched with one another, but matched to two other people, but who both would

> **HEADS UP!**
>
> Here are some concepts we will develop as this chapter goes along:
>
> **Mechanism Design:** When we treat the desired outcome of the game as a given and try to discover what rules will give that outcome as a Nash equilibrium, we are engaging in mechanism design.
>
> **Incentive Compatible:** If the desired outcome results from the rules of the game with non-cooperative play, the outcome is incentive compatible.
>
> **Type:** Agents may be of different types, and so may respond differently to the same rules.
>
> **Revelation:** Successful mechanism design may require that people be given incentives to reveal information they have that the designer and authority lack, including the type the agent is, when the agent knows this but the authority does not.

prefer to be matched to one another. This is an imperfection! If we could create a list of matches without that kind of imperfection, it would be a solution in the *core* of the matching game. (See Chapters 12–13).

In a 1962 collaboration[3] with David Gale (1921–2008), Lloyd Shapley proposed an algorithm with which agents making non-cooperative decisions (choosing according to their own preferences) would end with matches in the core of the matching game. In 1984, Alvin Roth discovered[4] that the American Medical Association had been using an equivalent system to match interns to positions, with good success. Here is an example to illustrate the algorithm.

There are four young men who aspire to be apprentice carvers and four master carvers, each looking for one apprentice. They all have different preferences as to whom they would wish to associate with. In a multi-stage application process, one side (proposers) apply to individuals on the other side (responders) in the order of the proposers' preference. The most preferred responder gets the first application from each proposer. Each responder gives a *tentative* acceptance to the one they most prefer. If some are unmatched then the process continues for another round. At each stage, any unmatched candidates submit applications to the masters, in order of the candidates' preferences. Each Master *tentatively* accepts the one he most prefers. But, on a subsequent stage, the Master can

A Closer Look: Lloyd Shapley

Shapley (1923–2016) was a mathematician and the son of astronomer Harlow Shapley. As a graduate student at Princeton when the work of von Neumann and Morgenstern was new, he did early research in cooperative game theory, and is also remembered for deriving the Shapley value solution for cooperative games with side payments.

---

[3]Gale, David and L. S. Shapley, College admissions and the stability of marriage, *American Mathematical Monthly*, **69** (1962), pp. 9–14

[4]Roth, Alvin E., The evolution of the labor market for medical interns and residents: A case study in game theory, *The Journal of Political Economy*, **92**(6) (December, 1984), pp. 991–1016.

accept a more preferable candidate and bump the one he has tentatively accepted.

The master carvers are John, Will, Abe and Mark. The aspiring apprentices are Davey, Ronnie, Jemmy, and Tommy. The preferences of the masters as to which apprentice they would have are shown in Table 19.1 and the apprentices' preference for their masters are shown in Table 19.2. The steps in the algorithm that matches apprentice candidates to masters is shown in Outline 19.1. As you read over the outline, double-check that each step follows the preferences of the applicants and masters.

Table 19.1. Masters' Preferences.

Master carvers' preferences

Preference rating	John's preferences	Will's preferences	Abe's preferences	Mark's preferences
1	Ronnie	Jemmy	Jemmy	Jemmy
2	Jemmy	Tommy	Ronnie	Davey
3	Tommy	Ronnie	Davey	Tommy
4	Davey	Davey	Tommy	Ronnie

Table 19.2. Apprentices' Preferences.

Apprentice candidates' preferences

Preference rating	Davey's preferences	Ronnie's preferences	Jemmy's preferences	Tommy's preferences
1	John	Will	John	Abe
2	Abe	John	Abe	John
3	Will	Mark	Mark	Will
4	Mark	Abe	Will	Mark

Outline 19.1. The Matching Algorithm in this Example.

Stage 1. The candidates submit applications
       a. John gets applications from Davey and Jemmy and rejects Davey.
       b. Will gets Ronnie
       c. Abe gets Tommy
       d. Mark gets nobody
Stage 2. Davey applies to his second choice — Abe.
       e. Abe drops Tommy and tentatively accepts Davey
Stage 3. Tommy applies to his second choice, John, and is rejected.
Stage 4. Tommy applies to his third choice, Will
       f. Will drops Ronnie for Tommy
Stage 5. Ronnie applies to his second choice — John
       g. John drops Jemmy for Ronnie, his first choice
Stage 6. Jemmy applies to his second choice, Abe,
       h. Abe drops Davey for Jemmy
Stage 7. Davey applies to his third choice, Will, and is rejected.
Stage 8. Davey applies to Mark and is accepted with a sigh of relief.

The final matches are shown in Table 19.3. By elimination we can see that there is no pair both of whom would prefer to be matched but who are not matched. John is matched with Ronnie, his first choice. Ronnie has his second choice but Will, his first choice, is matched with Tommy, whom Will prefers to Ronnie.

Table 19.3. The Final Matches.

Matches	
John	Ronnie
Will	Tommy
Abe	Jemmy
Mark	Davey

Will is matched with Tommy, his second choice, but Jemmy, his first choice, is matched to Abe, whom Jemmy prefers to Will. Will is Tommy's third choice, but his higher preferences, Abe and John, are each matched with apprentices they prefer to Tommy. Abe is matched with Jemmy, his first choice. Abe is Jemmy's second choice, but his first choice, John, is matched with Ronnie, whom John prefers to Jemmy. Mark and Davey are matched. Davey is Mark's second preference, but his first preference, Jemmy, is matched to Abe, whom he prefers. Mark is Davey's lowest preference, but the other three all are matched with candidates they prefer to Davey.

A contrast will help to make the significance of this result more clear. Suppose instead that the master's acceptances were not tentative but final: An approved application becomes a contract and is binding on both. Let us call this a commitment algorithm. At the first stage, then, John faces a dilemma. It is easy for him to reject Davey, his least preferred apprentice, but should John accept Jemmy, his second preference, or reject him in the hope that Ronnie will be available on a second round? Will faces a somewhat similar dilemma, as Ronnie is his third choice, but he could still do worse. Suppose, reasonably, that John "plays it safe" by accepting Jemmy but Will rejects Ronnie. Then, having been rejected, Ronnie, Tommy and Davey submit their backup applications, Davey to Abe, Ronnie to Mark (as John, his second preference, is no longer accepting applications) and Tommy to Will (for the same reason). Will will accept Tommy, his first choice. Mark would reject Ronnie — he cannot do worse — but Abe now faces a dilemma whether to reject Davey, his second-worst choice. Suppose he gambles by rejecting Davey. Then, on the third round, Davy applies to Mark — the only available master who has not rejected him — and Ronnie applies to Abe. Both are finally accepted. The matches then are as shown in Table 19.4. But we see that John and Ronnie would prefer to be matched, rather than in the matches they have in Table 19.4. Thus, the matches that emerge from this commitment algorithm may not be in the core of the matching game.

The reason for this unstable outcome is that at the start, John had to guess what future applications he might get if he turned

Table 19.4.   An Inefficient Match.

Matches

John	Jemmy
Will	Tommy
Abe	Ronnie
Mark	Davey

Jemmy down, and he made the reasonable, fail-safe guess by accepting his second choice. An advantage of the deferred-acceptance algorithm is that no-one has to guess, since acceptance is tentative and not binding until every position is filled. Moreover, for this discussion, we have assumed that the candidates applied in the order of their preferences. But they, too, have to rely on guesswork. Suppose that Jemmy is concerned that John, his first choice, might reject him and so he first applies to his second choice, Abe. That could lead on to yet a different set of matches. This unpredictability is another disadvantage of the commitment algorithm that the deferred acceptance algorithm avoids.

And this sort of guesswork will be familiar to some of the students reading this book. Students applying to universities or to graduate programs are often advised to include at least one "insurance school" that is pretty sure to admit the student. And why apply to Worldsgreatest University if it is likely to reject you?

As noted above,[5] the placement of medical interns in the USA has been organized by a process equivalent to the deferred-acceptance algorithm, and this has been refined in the 1990s by a consulting game theorist. In Britain, medical intern placements took place on a regional basis. Some regions adopted a deferred-acceptance algorithm while others did not. The ones that did not

---

[5]This paragraph draws extensively on Roth, Alvin E., What have we learned from market design? *The Economic Journal*, **118**(527) (2008), pp. 285–310.

failed and were soon abandoned. A successful clearing-house for the exchange of kidneys for transplantation was implemented in New England at the suggestion of game theory. Algorithms like the deferred-acceptance algorithm have been designed for many-to-one choices, such as the assignment of students to schools, and have been successfully implemented, for example, in New York and Boston.

All in all, the ideas pioneered by Gale and Shapley have made an impact in the solution of matching problems in the actual world, and probably would make a greater impact if they were more widely known. This is an encouraging example of the potentiality of social mechanism design.

2. GRADING TEAM PROJECTS

Dr. Schönsinn[6] requires the students in his Game Theory class to do projects as teams, with three students in each team. As a game theorist, Dr. Schönsinn worries that his students may face an effort dilemma, especially if they all get the same grade for the project. Consider, for example, Team Technology, a group of three freshman engineering majors. These three students, Augusta, Bill, and Cecilia, each can choose to make a big effort (work) or a slight effort (shirk) on the team project. Their common grade will depend on the average effort. A student's payoff is the grade minus the subjective cost of effort if she or he chooses "work." The payoffs are as shown in Table 19.5. Grades are shown on a numerical scale of D = 0, C– = 1, C = 2, C+ = 3, B = 4, A = 5. (Dr. Shönsinn's university only gives plus or minus grades for C's). This game has a dominant strategy equilibrium in which everybody shirks and the group get a C–.

Dr. Schönsinn wants to set up grading rules for his class that will lead them to choose the cooperative solution at the upper left, where everyone works. There are two difficulties. First, effort is

---

[6]This example was suggested by Amoros, Pablo and Luis C. Corchon and Bernardo Moreno, The scholarship assignment problem, *Games and Economic Behavior*, **38**(1) (January, 2002), pp. 1–18.

Table 19.5. A Student Effort Dilemma.

		Cecilia			
		Work		Shirk	
		Bill		Bill	
		Work	Shirk	Work	Shirk
Augusta	Work	4,4,4	3,5,3	3,3,5	0,4,4
	Shirk	5,3,3	4,4,0	4,0,4	1,1,1

unobservable. All that anyone can observe is the results of the effort — how great a contribution each student makes to the project. Dr. Schönsinn assumes that a student who works will always make a bigger contribution than a student who shirks. Thus, if he can rank the students in the team by their contributions to the project, and give extra points to those who make the bigger contributions, that will create an incentive for the students to work rather than shirk. The second problem, though, is that Dr. Schönsinn doesn't know which students make the bigger contribution in any given team. Only the students in the team know that. But we will set the second problem aside, for now, and focus on the first. The students are ranked according to their contributions, and the student who is first in the ranking gets 3 points, while the student ranked second gets 2. The student ranked last gets nothing. Now suppose that all of them choose "work." Then each one has a $1/3$ chance at the 3-point bonus and a $1/3$ chance at a 2-point bonus, so the expected value payoff is $4 + (1/3)3 + (1/3)2 = 5\ 2/3$. The payoffs with the bonus system are shown in Table 19.6.

We see that in the revised game, "work" is a dominant strategy. The game is now cooperative dominant — just what we want for a mechanism to lead to efficient results!

But now we have to deal with the second problem: Dr. Schönsinn doesn't even know which student makes the bigger contribution to the project. He only gets their completed paper. The students

Table 19.6. The Student Effort Dilemma with Bonus Points.

		Cecilia			
		Work		Shirk	
		Bill		Bill	
		Work	Shirk	Work	Shirk
Augusta	Work	$5\frac{2}{3},5\frac{2}{3},5$	$6\frac{1}{3},5,6\frac{1}{3}$	$6\frac{1}{3},6\frac{1}{3},5$	$3,5,5$
	Shirk	$5,6\frac{1}{3},6\frac{1}{3}$	$5,5,3$	$5,3,5$	$2\frac{2}{3},2\frac{2}{3},2\frac{2}{3}$

themselves know who has made the bigger contributions, so that means Dr. Schönsinn has to get the information from them. The problem is that each one has an incentive to distort her report. By ranking themselves best, regardless of what they really contributed, each could increase their chances of getting the bonus points anyway. Dr. Schönsinn will have to design a mechanism that can solve that problem.

Here is his plan. Each student will rate the other two students, but not themself. Dr. Schönsinn believes that his students agree that the person who makes the biggest contributions should get the biggest rewards, if their own rewards are not affected.[7] So each student will honestly rate the other two students on their relative contribution: Greater or less. As a first approximation Dr. Schönsinn assigns each student the minimum of the rankings the other two students have given them. But there may be ties at this stage. If there is a tie between two of the students, then Dr. Schönsinn rates them as they are rated by the third student, not involved in the tie.

---

[7]This could be upset by collusion, if two students were to agree to get the third to do all the work, and then rank one another as first. This would be a cooperative arrangement between the two, and Dr. Schönsinn is following the standard practice in social mechanism design: Relying on non-cooperative equilibria to predict the actions of the people in the game. However, in this case and in some others, it might be helpful to allow also for collusion and irrationality.

In Team Technology, Augusta and Bill worked, but Cecilia shirked.[8] Augusta and Bill drew lots, and Augusta got to write the more important section of the paper, so hers was the biggest contribution, while Bill's was second and, because of her shirking, Cecilia's was least. Thus, their rankings are as shown in Table 19.7, where the students doing the ranking are listed vertically on the left and the students being ranked are listed horizontally across the top. Thus, for example, student B (Bill) ranks student A (Augusta) first and student C (Cecilia) second, not counting himself. The diagonal cells are blank because nobody is allowed to rank themselves. At the bottom row are the minimum rankings for each student, the column minimum, and as a first step, Dr. Schönsinn applies those rankings. That means Augusta will get the 3 point bonus — as she should. But Bill and Cecilia are tied. Dr. Schönsinn breaks the tie by looking at the rating by the third student, the one not in the tie, Augusta. Since Augusta ranked Bill first (relative to Cecilia) Dr. Schönsinn demotes Cecilia to third ranking, and Bill gets the 2 point bonus — as he should.

Thus, Dr. Schönsinn has succeeded in his objective. He has designed a grading system that (1) eliminates the effort dilemma and makes "work" a dominant strategy; (2) elicits

A Closer Look: Information

As a rule, any social mechanism or authority will need information, and the information may have to come from the players in the game. The players could have incentives to "game the system" by lying or withholding information. In the grading example, the students could benefit by ranking themselves first, if they were allowed to do so. To make the mechanism successful, the designer must take care that players have incentives truthfully to reveal the information they have. This is an important principle of social mechanism design.

---

[8]She must not have understood the unit on dominant strategies! Remember, this is just an illustration, for the sake of the example, to make the rules clear.

Table 19.7. Rankings of Students by Students on Team Technology.

		Ratings of student		
		A	B	C
Ratings by student	A		First	Second
	B	First		Second
	C	First	Second	
Minimum		First	Second	Second

from the students information that they have in order to do this; and (3) does not give students any opportunity to enhance their own grades by falsifying the information they give.

As this example illustrates, information is often a key to successful mechanism design. Agents have information that may be needed to direct the mechanism — but their decisions to release that information, and to release it truthfully, are strategic decisions! If lying or withholding information is a best response, then some people (if not all) will lie or withhold information. A successful mechanism needs to have safeguards against this tendency.

3. A GAME OF TYPES

Ann, Bob and Carol are to work together on a project that requires that three distinct jobs are to be done. Ann, Bob and Carol are different "types," both in that they have different aptitudes for the three jobs, which may complement one another or clash, depending on the jobs they are assigned to, and in that they have different preferences. Using their initials, we say that Ann is type A, Bob type B, and Carol type C. The payoffs for this three-person game are shown in Table 19.8.

As we see in the table, the cooperative solution (with transferable utility) is for the jobs to be taken in the order A:1, B:2, C:3. This is the middle cell in the right panel, top row. However, the agents

Table 19.8. A Game with Players of 3 Types.

		C								
		1			2			3		
		B			B			B		
		1	2	3	1	2	3	1	2	3
A	1	2, 2, $\underline{2}$	0, 0, $\underline{11}$	0, $\underline{3}$, $\underline{7}$	2, 2, 2	1, 2, 4	8, $\underline{4}$, 1	2, 2, 1	10, 10, 10	0, $\underline{11}$, 0
	2	$\underline{3}$, 0, $\underline{8}$	$\underline{4}$, 0, 4	$\underline{3}$, $\underline{3}$, 3	$\underline{4}$, 2, 2	$\underline{2}$, 2, 2	9, $\underline{3}$, 0	$\underline{4}$, 2, 7	$\underline{11}$, 0, 0	$\underline{2}$, $\underline{4}$, 2
	3	2, 2, $\underline{4}$	2, 7, $\underline{4}$	0, $\underline{8}$, $\underline{3}$	2, 2, 2	1, 2, 2	2, $\underline{4}$, 2	2, 4, 2	2, 2, 2	0, $\underline{5}$ 0

have inefficient preferences, and in fact each one has a dominant strategy that differs from the cooperative solution. For A the dominant strategy is 2; for B, 3; and for C, 1. The best-response strategies are underlined in the table, and we see that there is a unique Nash equilibrium in pure strategies, a dominant strategy equilibrium, at 2, 3, 1. The dominant strategy equilibrium is very inefficient, paying 3, 3, 3 while the cooperative outcome pays 10, 10, 10.

Accordingly, Ann, Bob, and Carol agree to have a neutral authority decide which jobs they are to do. The problem for the authority is to assign the three workers efficiently. If he knows that Ann is of type A, Bob of type B, and Carol of type C, then the problem is simple enough: Assign them to jobs 1, 2, 3. But it is not that simple, because *the authority does not know which worker is of which type.* Indeed, this is a very common problem in social mechanism design. In order to design a satisfactory mechanism, the authority needs to predict the way that the agents in the game will respond to the mechanism. Different types of agents will respond differently to the same mechanism, so *predicting how an agent will respond* is equivalent to *finding out the agent's type.*

One possibility is to ask them. But that creates a new game in which the agent's strategy is the type she or he claims to be. In this case, Ann is really of type A, but her dominant strategy will be to report that she is type B in the hope that she is assigned to her dominant strategy job, job 2. Similarly, Bob's dominant strategy is to report that he is of type 3, and Carole of type 1. If the jobs are assigned according to the types that the agents report, we are back at the dominant strategy equilibrium, with inefficient payoffs 3, 3, 3. In short, the project of social mechanism design will have failed.

It is a fundamental problem of social mechanism design to create a mechanism that will induce the agents to reveal their types accurately, in their own interest. If it does not do this, then the mechanism cannot be *incentive compatible*, that is, cannot succeed when people act non-cooperatively.

But perhaps the problem can be solved in this case. The authority knows that there is exactly one person of each type among the three, and knows the efficient assignment of types to jobs. Instead of simply asking, the authority establishes the following mechanism:

1. The three agents each request that they be designated as a particular type and assigned to the corresponding job according to *A*:1, *B*:2, *C*:3.
2. Only one person can be designated as a particular type, i.e., assigned to a particular job. Accordingly,

    a. If two claim to be of the same type, both are penalized one point and they must bid to be assigned the type designation and job that they have both requested.

        i.   The third player is designated and assigned as they requested.
        ii. The loser gets the remaining type designation and job.
        iii. If neither will bid for the job they both claimed, then they instead compete for the remaining type and job.
        iv. If still no one will bid, types and jobs not already assigned are distributed at random.
        v.   The auction is an ascending Dutch auction, in which the authority announces prices of 1,2, ... and continues until one bidder drops out.
        vi. Penalties and bid payments go to the uninvolved third party.

   b.   If all three claim the same type, there are two auction stages.

        i.   At the first they compete for the type they claim.
        ii. Payment is divided between the two losers.
        iii. At the second stage they compete for the lower-num-
             bered job, and corresponding type designation.
        iv. If no one bids, at any stage, the auctions are revised
             along the lines in a. iii. and iv. above.

   c.   If there is no conflict then the type designations are made as
        requested.
   d.   Once the type designations are settled by this bidding pro-
        cess, the jobs are assigned as A:1, B:2, C:3, using the type
        designations allocated by the mechanism.

The three workers will now be playing a much more complicated
game, with two or more stages. The first stage is the type designation
and job that they request, but the subsequent stage or stages will be
their bidding strategies. The bidding strategies will be subgames of
the larger game. Thus, we can use backward induction to reduce the
game to another game much like Table 19.8, with the type and job
requests as the three workers' strategies, but with the payoffs that
result *after* the mechanism has been applied. This is a very complex
game, since there are twenty-seven strategy combinations at the *first*
stage and some may have more than one auction stage. Thus, we will
not be able to show the backward solutions of all of them, and will give
only a few examples. The reduced game is shown in Table 19.10.

First, suppose the type designations requested by A, B, and C
respectively are 2, 2 and 3. That is, B and C choose the cooperative
(truthful) strategies but A deviates by choosing what, in the original
game, would be their dominant strategy. Then A and B must compete for the designation as type 2. If A wins, the order is 2, 1, 3, and
(from Table 19.8) the payoffs then are 2, 1, 7. If B wins, then the
order is 1, 2, 3, the cooperative outcome with payoffs 10, 10, 10.
Winning can only make A worse off, so she will not bid, while B
could bid as high as 8 and still be better off. B wins with a bid of 1,
leaving net payoffs of 9, 8, 13.

Suppose the requests are 2, 2, 2. (This is, of course, purely hypothetical.) Then the three must compete to be designated as type 2. (a) Suppose A wins the first round. Then B and C are expected to compete to be designated as type 1. If B wins this second round, then the payoffs (before bids are paid) are 4, 2, 7, while if C wins the second round, they are 3, 3, 3. Neither will bid, since each is worse off if she or he wins. Accordingly, they compete instead to be type 3. Now the payoffs are reversed, and B will bid one, but C will win with 2, so the payoffs (before bid payments) are 4, 2, 7. (b) Suppose B wins the first round. Then A and C compete to be designated as type 1. A win for A means payoffs of 10, 10, 10, while for C it means 2, 7, 4. Therefore, C will not bid, A wins with a bid of 1, and the payoffs before bid payments are 10,10,10. (c) Suppose C wins the first round. Then A and B compete and the payoffs for a victory by one or the other are 8,4,1 or 2,2,2 respectively. B will bid one, but A wins with 2, so the payoffs before bid payments are 8,4,1. Table 19.9

Table 19.9. Payoffs Depending on the First-Round Winner.

First round winner	Payoffs before payments	After second-round payments	After first-round payments
A	4,2,7	6,3,5	5,3.5,5.5
B	10,10,10	9,11,10	9.5,10,10.5
C	8,4,1	6,4,3	6.5,4.5,2

Table 19.10. Payoffs in the Reduced Game with the Mechanism.

		C=1, B=1	C=1, B=2	C=1, B=3	C=2, B=1	C=2, B=2	C=2, B=3	C=3, B=1	C=3, B=2	C=3, B=3
A	1	10, 9.5, 10.5	8, 13, 9	4, 9, 0	6, 3, 4	13, 8, 9	8, 4, 1	8, 9, 13	10, 10, 10	13, 9, 8
	2	8, 1, 4	1, 4, 8	3, 3, 3	2, 5, 6	9.5, 10, 10.5	5, 9, 0	4, 2, 7	9, 8, 13	8, 1, 4
	3	5, 6, 2	2, 7, 4	1, 4, 8	2, 2, 2	5, 6, 2	7, 2, 4	2, 5, 6	9, 13, 8	9.5, 10.5, 10

shows the payoffs and bid transfers in the case of a first-round victory by each of the three. Clearly, B's victory is dominant, so that A and C will not bid at the first stage and the payoffs are 9.5,10 and 10.5.

Suppose that the requests are 1, 3, 2. Then, there being no conflict, the jobs are assigned as requested and the payoffs are 8, 4, 1.

In Table 19.10, once again, the best responses are underlined. What we see is that each agent again has a dominant strategy, and that is to truthfully reveal their type, so that the dominant strategy equilibrium in this carefully designed game is the cooperative equilibrium of the original game. (In the reduced game shown in Table 19.10, the cooperative strategies are strictly dominant, and in the bidding subgames the equilibrium strategies are weakly dominant). If the players act with non-cooperative rationality, there will in fact be no bidding and shifting of assignments, jobs and payoffs — instead, the three workers will simply report truthfully, and be assigned accordingly. But the designed game mechanism, with its authority, bidding and shifting, creates the incentives to them to report truthfully and act cooperatively.

This example cannot be generalized in any simple way. The results depend very much on the game we began with in Table 19.8; had we begun with a different game the same mechanism might have produced very different and less successful results.

Nevertheless, this example illustrates some important points about social mechanism design.

- Since the cooperative solution of the original game is a dominant strategy equilibrium with the mechanism, we can say that the mechanism *implements the cooperative solution in dominant strategies*. This is the gold standard of mechanism design. The objective will always be to make the cooperative strategies dominant when the mechanism is applied, because, in that case, the players need not guess what strategies the other players will choose in order to determine their own best response. If the cooperative solution is a Nash equilibrium with the mechanism but not a dominant strategy equilibrium, then the agents may have to learn what the other agents will do. In that case, we would want a Nash equilibrium that is stable in the presence of learning.

- Bidding, penalties and similar arrangements are common features of mechanism design, since they can provide incentives to agents to truthfully reveal their "type." As we will see in the next section, this is central to a widely supported government policy in the USA, the "cap-and-trade" approach to limiting pollution.

- This mechanism may be disrupted by collusion. For example, suppose that A and C form a coalition and make their initial requests 1 and 2 respectively. Then B's best response is 2, and after the bids and payments are made, the payoffs are 13, 8, and 9; so that A and C together get 22 at the expense of B, whose payoff is reduced to 8. A will have to make a side payment to C to make it worth C's while — a side payment of 2 would leave the colluders sharing their ill-gotten gains equally with final payoffs of 11, 8, 11. Thus, C must trust A; there must be "honor among thieves" — but that is usually necessary for successful collusion, and sometimes collusion happens anyway.

- We have assumed that the mechanism is costless, despite the time and resources that really would be required to carry out a bidding process and administer and enforce the payments. That doesn't matter (much) if the workers in the example play non-cooperatively, since then they play 1, 2, 3 and there is no bidding or transferring to administer. But presumably the authority would have to command resources enough to make the mechanism credible, and if there is collusion as in the previous bullet point, then the whole mechanism must be put into action. Thus, it is not likely really to be free, and we must hope that it costs only a little by comparison with the benefits of increased efficiency that it can bring about.

4. CAP AND TRADE

Some of the most important potential applications of mechanism design are in the field of environmental economics, and we have some experience in this field. Inefficient environmental pollution and the depletion of environmental resources such as water and fish are recognized problems for public policy. These problems are easily understood in terms of non-cooperative game theory. The cooperative arrangement in these cases is efficient restraint of the pollution or the non-depletion of the resource. This does not mean that pollution or depletion are completely eliminated, as a rule — there is an efficient rate of pollution or depletion, but it is not a non-cooperative equilibrium in the "game" of choosing rates of pollution or depletion. As a rule, inefficiently excessive rates of pollution and depletion are the consequences of dominant strategies, so that the "game" of pollution or depletion is an n-person social dilemma, a "tragedy of the commons." It is now widely accepted — even among many free-market conservatives — that government action is appropriate to restrain the pollution or depletion. The difficulty is to design a public mechanism that will do this efficiently.

There are two problems of mechanism design here. One arises from the demonstrated fact that governments cannot always be relied on to promote efficiency, and may indeed be an important source of inefficiency. We might put it this way: A "cooperative" solution in the game of government is a government that acts in ways that advance the mutual interests of the citizens. How, then, can we design a political mechanism that assures us a government that will do its best to act as an efficient instrument for the mutual benefit of the citizens? We have discussed this in Chapter 18 and need say little more here except that the design of voting and political systems is a branch of social mechanism design.

Here we are more concerned with the second problem. Supposing that the government is indeed committed to bring about an efficient rate of pollution or depletion, it will need some information to do it. For example, some "types" of companies and people may be able to reduce their pollution and depletion very cheaply, while other "types" can do so only at very great cost. It is efficient if those who can cut back at little cost cut back a great deal, while those who can cut back only at great cost are required to cut back only a little. (This may not seem just, but for now, we are concerned with efficiency rather than justice. We may aim to attain both efficiency and justice, but first, let us see if we can attain efficiency at all.)

Suppose that the authority asks the companies and people to tell which of them are "low cost types" and which are "high cost types." Acting non-cooperatively, everyone will claim to be a "high cost type," in hope of not being asked to cut back very much.

The "cap-and-trade" scheme addresses this. In a "cap-and-trade" scheme, agents of a particular category, such as fishermen or companies that generate electric power, are all required as a group to reduce their pollution or depletion in some proportional way — for example, fishermen may be allowed to catch 75% as many fish as they caught on the average in the last 5 years, or factories may be required to cut back their polluting emissions by 50%. This establishes a "cap" on the polluting or depleting activity of each individual or organization in the category. Corresponding to a cap is an allowance: Each polluter or depleter is allowed to continue to emit pollutants or to extract resources up to the level of the cap. But if the individual is a "low cost type," he may choose not to pollute up to the level of the cap, or to extract all the resources that the cap would allow. Instead, he can sell all or part of his allowance to "high cost types" who may then add the bought allowance to their own and so legally emit pollutants or extract resources beyond what would otherwise be the cap. The low-cost types exchange with the high-cost types, each trading away the item that they can give up with least sacrifice: For the low-cost types, pollution or depletion allowances, and for the high-cost types, money. Thus, the imposition of caps is followed by trading: Cap-and-trade.

Looking at this from the point of view of economic theory, we may suppose that there will be a supply of pollution or depletion allowances from low-cost types and a demand for them from high-cost types. If the price is just high enough so that the quantity supplied is equal to the quantity demanded, then we have a market equilibrium in the market for allowances. Moreover, we have some reason to expect that this market equilibrium is efficient, with the reductions in emission or extraction being carried out by those who can do so at least cost.

Regulations based on the "cap-and-trade" approach have been applied since the 1980's, when the United States adopted this method of reducing sulphur dioxide emissions that were among the causes of "acid rain." For greenhouse gasses, "cap-and-trade" emissions limits were mandated by the Kyoto Protocol and as a consequence have been adopted in a number of countries. There have also been applications to the allocation of subsurface water and fishery resources, among others. Here are some lessons from this experience.

- The "learning period" may be rather long. For the earliest applications, at least, it seemed to take years before the decision-makers subject to the regulations began the trading that would be necessary for the program to be a success.
- Given time, cap-and-trade regulations can be effective in reducing inefficiencies, and the cost of reducing pollution and depletion by efficient methods can be considerably less than a one-size-fits-all approach that does not allow for differences among those being regulated.
- Usually, the costs of reducing the polluting or depleting activities have been less than expected. This would be no surprise, since simple non-cooperative behavior would lead low-cost types to represent themselves as high-cost. However, there may also have been some bias in even honest estimates — low-cost types may have mistakenly thought that they were of the high-cost type, until they actually tried to reduce their pollution or depletion.

The "cap-and-trade" approach is indeed an instance of mechanism design, but not of game theory. The reasoning, and especially the expected efficiency of the market equilibrium, are based on the theory of "perfectly competitive" markets and market failure. Market equilibrium is an instance of Nash equilibrium in some market games, but markets may be distorted by monopoly power. Indeed, there is reason to believe that the market for sulphur dioxide emissions was sometimes distorted by imperfect competition, i.e., monopoly or oligopoly pricing of the allowances.

5. SUMMARY

Games are defined by their rules, which in turn influence the non-cooperative equilibria of rational players of the game. In social mechanism design, we turn this around, first identifying the target non-cooperative equilibrium, and then adjusting the rules (so far as possible!) to achieve this equilibrium. For some applications, including auctions and matching individuals to opportunities or to one another, there have been some important practical successful applications. For some matching problems, a deferred acceptance procedure can lead to a match in the core of the game, because it allows individuals to make decisions without guessing whether they might do better by waiting. Very often the trick is to design a mechanism that will give individuals an incentive to reveal the information they have, which might be information about themselves — what "type" of agent they are — or about one another. The highest standard is a set of rules that make the cooperative and informative strategies dominant strategies in the game that has been designed. In that case, the mechanism becomes self-enforcing and penalties, bribes and other enforcement mechanisms need not actually be put into effect, so long as people act with self-interested rationality.

Q19. EXERCISES AND DISCUSSION QUESTIONS

Q19.1. Building a Levee

Landowners Victoria and Wanda own adjacent plots along Winding River, and the plots may sometimes be flooded. Each can choose to build or not to build a levee, and if they both build a levee (at a cost of 1) both of their plots will be protected from floods. If either of them fails to build, they will both suffer flood losses of 5. The payoffs are shown in Table 19.11.

Table 19.11. Floodplain Payoffs.

		Wanda	
		Build	Don't
Victoria	Build	−1, −1	−6, −5
	Don't	−5, −6	−5, −5

This game has two equilibria, and one of them is inefficient. We want a social mechanism that assures efficient provision against floods. Now suppose the public authority proposes a flood insurance program that will pay all flood losses. The charge for the insurance is 1 if the landowner builds, 2 for a landowner who does not build if the other landowner builds, and 3 each if neither landowner builds.

Draw the payoff table if both are insured. Draw a tree diagram for the imbedding game in which the two landowners have the option to insure or not. Determine a subgame perfect equilibrium of the imbedding game using both backward and forward induction. Will the equilibrium be efficient? Will the insurance fund cover its costs?

Q19.2. Laissez Faire

A market system can be thought of as a mechanism for the allocation of scarce resources to their most productive uses. According to economist Friedrich von Hayek, information about the availability of resources is widely distributed in the population, with no central group having any large proportion of that information. Hayek argues that market systems are superior to other systems for allocation of resources. Interpret Hayek's claim in terms of the ideas of this chapter.

CHAPTER 20

Games, Experiments, and Behavioral Game Theory

Game theory and the experimental method both begin from an assumption of uniformity. In the experimental method, the assumption is that nature is uniform, so that the regularities observed in

To best understand this chapter, you need to have studied and understood the material from Chapters 1–11.

the experimental laboratory will also be applicable in other times and places. Throughout the 20th century, experimental methods were increasingly applied to human behavior. When experimental methods are applied to human behavior, the assumption of uniformity becomes an assumption about human behavior: The assumption that what we observe of human behavior in the laboratory will also be observed outside the laboratory. This is not as compelling an idea as the idea that inanimate nature is uniform. Context does matter for human behavior, and the laboratory context may modify behavior in important ways. Nevertheless, the experimental method has been a powerful source of insights across the human behavioral sciences.

In game theory, the assumption is again about the uniformity of human behavior, so that what we learn from human behavior in games can be extended to other kinds of interactions. This is closely related to the assumption of uniformity in experimental work: Indeed, we could think of a game like Nim as a laboratory for a certain kind of human behavior. Here again, context counts for human behavior, and we need to be a bit cautious with this assumption of uniformity; but at the same time the parallel between games and experiments is important, and a foundation of game theory.

It should be no surprise, then, that there has been experimental work in game theory from very early days. Indeed, experimental game theory began in 1950.

1. A PRISONER'S DILEMMA EXPERIMENT

An experiment with a modified Prisoner's Dilemma was almost certainly the first experiment in game theory, and in the first few decades of the development of game theory, the Prisoner's Dilemma attracted much of the attention of the experimentalists. In that first experiment, in January of 1950, Merrill Flood and Melvin Dresher of the RAND Corporation were the experimenters. The Flood-Dresher experimental game is shown in Table 20.1. (The table used in the experiment was different, and was designed to be a little confusing, since part of the purpose of the experiment was to find out if the subjects could figure out what the equilibrium strategies would be.) The experimental subjects were Armen Alchian, an economics professor at UCLA, and John Williams, head of the mathematics department of the RAND Corporation. As Table 20.1 shows, the game is modified from the Prisoner's Dilemmas we have studied in earlier chapters in that it is unsymmetrical — Williams does better than Alchian in three of the four strategy combinations, including those in which both cooperate or both defect. This proved to be a complication in the experiment.

In the Flood-Dresher experiment, Alchian and Williams played the game 100 times in succession. A record was kept not only of their strategies but also of their comments — a "talking through" experimental protocol that has been widely used in experiments by cognitive scientists in more recent decades and which gives some insights on their thought processes.[1] It is clear that they started out

---

[1] These are reported in William Poundstone's 1992 book, *Prisoner's Dilemma* (New York: Doubleday, 1992), pp. 108–116.

with different expectations, and to some extent retained quite different expectations.

Alchian expected Williams to defect, while Williams tried to bring about a cooperative solution by starting cooperatively and playing a trigger strategy (that is, retaliating for Alchian's non-cooperative play by himself playing non-cooperatively for one or more rounds). Williams tried playing according to a Tit-for-Tat rule. Alchian initially didn't "get it" and assumed that Williams was playing a mixed strategy. (Williams commented that Alchian was a dope.) Eventually Alchian got the idea that Williams was signaling for cooperative play. As the victim of an unsymmetrical payoff table, however, Alchian thought Williams ought to equalize the payoffs somewhat by allowing him, Alchian, to defect now and then. At the end of 100 rounds of play, Alchian worried that there would be no cooperation on the last play (since there would be no next play to retaliate on) and considered defecting earlier in order

HEADS UP!

Here are some concepts we will develop as this chapter goes along:

Games as Experiments: We can often check ideas from game theory by having real people play the game, and observing the strategies they choose.

Bounded Rationality: If people do not always succeed in choosing the best response or in maximizing their payoffs, but do try to come close they are said to be "boundedly rational."

Reciprocity: When people deviate from self-interested rationality to return favors or wrongs, they act with reciprocity.

Ultimatum Game: A game in which one player offers to divide an amount in proportions the player chooses, and both get the payoffs only if the other player agrees. This game has been widely played in experiments.

Centipede Game: A game in which one player can claim the bigger share of the benefits, but get an even bigger share if neither player takes this opportunity. This game, too, has played an important role in experiments.

Table 20.1. A Modified Prisoner's Dilemma.

		John Williams	
		Cooperate	Defect
Armen Alchian	Cooperate	(1/2, 1)	(−1, 2)
	Defect	(1, −1)	(0, 1/2)

to gain an advantage. In fact, both cooperated on plays 83-98 and both defected on play 100.

William Poundstone writes[2] "Alchian talks of Williams' unwillingness to 'share.' It is unclear what he meant by that." Let's pause to try to understand what Alchian had in mind. He seems to have continued to think in terms of mixed strategies and to have had in mind something like the following cooperative mixed strategy: If there could be an enforceable agreement, the two might have agreed to a joint mixed strategy in which they jointly play a mixture of (C,C) and (D,C)[3] with probability p for (C,C) and $(1 - p)$ for (D,C). The plays of (D,C) would enable Alchian to get some of his own back, at some cost in terms of total payoff. Since the total payoff per play is $1.5p$, any value of p less than 1 is inefficient, but Alchian's payoff is $1 - 0.5p$, so a reduction in p helps him out. Williams' payoff is $2p - 1$, and the expected value payoffs are equal when $p = 4/5$.[4] Alchian expected Williams to purchase his (Alchian's) cooperation with some reduction of p toward $4/5$, although perhaps not all the

---

[2] *Prisoner's Dilemma*, p. 107.

[3] Alchian's strategy is listed first. C means cooperate and D means defect.

[4] This cooperative mixed strategy is not an equilibrium, since the probabilities are not best responses. The fact that it does not maximize the total payoff suggests that it is not efficient, and therefore not a cooperative solution either, but that is not quite right. Since there is no "transferable utility" — side payments cannot be made — the correlated mixed strategy is efficient in the sense that neither player can be made better off without making the other worse off, and this is the way an economist like Alchian would have understood efficiency. So, the joint mixed strategy could be a cooperative solution to this game.

way. The cooperative strategy should be relatively easy for Williams to play — he always plays *C*, after all — but that puts Alchian in control of *p*, and Williams would have to observe Alchian's play for several rounds to estimate the probability Alchian was playing, and only then retaliate. Alchian noted, however, that Williams always retaliated, and interpreted that as a selfish unwillingness to share the benefits of cooperation.

Despite all this confusion, the two players managed to cooperate on 60 of the 100 games. Mutual defection, the Nash equilibrium, occurred only 14 times. Flood and Dresher showed their result to John Nash, and Nash pointed out that it was not really a test of the one-off Prisoner's Dilemma, since the repeated play changes the analysis considerably. Nash was right, of course, but we have since learned that the "subgame perfect" Nash equilibrium in a game like this is to defect on every round. Neither player did that. Alchian started out expecting the game to go that way, but revised his expectation and learned a quite different — and equally non-equilibrium — approach. Williams knowingly played a non-equilibrium strategy in the hope of getting the benefits of mutual cooperation.

2. BEHAVIORAL GAME THEORY

The results of this original experiment exemplify the results of many experiments on the Prisoner's Dilemma. In general, experimental subjects do not always play the dominant strategy equilibrium, and often they do play the cooperative strategy pair. This lends itself to two interpretations, and there has been controversy over which is the better interpretation. The interpretations are

1. People are not really as rational as game theory assumes, and fail to play the dominant strategy equilibrium because they do not understand the game.
2. People are better at solving social dilemmas than game theorists suppose, perhaps because they do not always base their actions on self-interest.

The controversies over interpretation of the Prisoner's Dilemma often assume that only one of these interpretations is true. But, as the Flood-Dresher experiment already showed, there could be a measure of truth (and falsehood) in each of them. Over the years, the experimental study of decision-making in games and game-like situations has grown into a distinct subfield called behavioral game theory. Methods have been borrowed from psychology and experimental economics, and experimental protocols developed specifically for the situations of game theory, where the results depend on the interaction of the strategies chosen. Protocols to allow for risk aversion have also been developed. With this specialized tradition of study, some important results have been accumulated.

Results from many experiments can be summed up as follows:

- Real human rationality is "bounded rationality." People do not spontaneously choose the mathematically rational solutions to games, but think them through in complex and fallible ways. People tend to play according to heuristic rules, or "rules of thumb" such as the trigger strategy rule Tit-for-Tat, which work well in many cases but are fallible.
- People can often find their way to the solutions of simple games, especially with some opportunity to learn through trial and error. In many games, play approaches the equilibrium with experience and learning, but in the Prisoner's Dilemma, at least, learning can have the opposite effect, in that people learn to cooperate (as Alchian did in the Flood–Dresher experiment).
- Different people may approach the social dilemma with different motivations as well as different approaches to solving the problem. Several experiments indicate that there is more than one *type* of decision-maker in game experiments.
- Other experimental studies have suggested a result that Flood and Dresher could not have obtained because everyone involved in that experiment was of the same gender. It is often found that women play differently than men.[5]

---

[5]Somewhat cautiously, it seems likely that females are, on the whole, more likely to try for and arrive at the cooperative outcome than males are. Some recent

We first consider a representative experimental study that addresses many of the difficulties that arise in some of the early studies. To illustrate how non-self-serving motives may enter into game theory and how experiments can reveal this, we will consider experiments with the Ultimatum and Centipede games, and apply some of the ideas that arise from those studies to a business case. To illustrate the role that bounded rationality can play, we will then explore *Level-k theory*, which has motivated a number of important recent experimental studies. This is a very selective review of behavioral game theory, and there are very many other ideas and experiments to draw on; but it will have to be enough for this chapter.

3. A MIXED EXPERIMENT

To illustrate the development of experimental game theory over the half-century following the Flood–Dresher experiment, let us consider one fairly general example, an experimental study[6] done in the late 1980's at the University of Iowa. The experimental subjects were business students. They were randomly matched to play the games and played them anonymously, via a computer network, without being able to see one another. This design assured that they were playing the one-off games the experimenters designed, and not some more complex repeated game. The payoffs were adjusted to offset risk aversion and similar considerations so that the

---

experiments by neurologists, involving brain imaging, indicate that (in female experimental subjects) brain activity associated with cooperative play in Prisoners' Dilemma experiments is consistent with the hypothesis that cooperation is emotionally rewarding. This brain activity is observed when the other player is human, but not when the subject plays against a computer. Angier, Natalie, Why We're So Nice: We're Wired to Cooperate, *New York Times*, Section F (Tuesday July 23, 2002), pp. 1–8. On the other hand, one early study reported in Rapoport, Anatole and Albert M. Chammah, *Prisoner's Dilemma* (University of Michigan Press, 1965), found women less likely to cooperate.

[6]Cooper, Russell W., Douglas V. DeJong, Robert Forsythe, and Thomas W. Ross, Selection criteria in coordination games: Some experimental results, *American Economic Review*, **80**(1) (March, 1990), pp. 218–233.

numerical payoffs should correspond closely to subjective payoffs.[7] The authors write, " ... there is strong support for the dominant strategy prediction ... and the Nash Equilibrium prediction...."[8]

The games they played were, in many cases, designed to mix elements of a social dilemma with elements of a coordination game. An example is given as Table 20.2. The upper left four cells in this game define a coordination game somewhat like Heave-Ho, in that the two equilibria, (1, 1) and (2, 2) can be ranked in the same way by both players, (2, 2) better than (1, 1). In these circumstances, it has seemed that (2, 2) might be a Schelling focal point and therefore the equilibrium most likely to occur. However, (2, 2) is not the cooperative solution. Strategies (3, 3) make both players better off still. However, strategy 3 is dominated, as the cooperative strategies are in a social dilemma.

In these experiments, Nash equilibria were usually observed. A few players played the cooperative strategy, 3. But the most usual Nash equilibrium in the game in Table 20.2 was (1, 1), not the better equilibrium at (2, 2). The experimenters judged that (1, 1) was chosen because the subjects assigned a positive probability to the possibility that the other player would play the cooperative strategy 3. If *Q* plays the cooperative strategy, 3, and *P* plays 1, *P* makes a killing with 1,000 points. Or *P* might reason that *Q* will choose 1 because *Q* thinks *P* will choose 3! To test this hypothesis, the experimenters also tried a slightly different game shown in Table 20.3, in which (3, 3) is no longer the cooperative solution. In the Table 20.3 game, (2, 2) is the cooperative solution, so there is no conflict between cooperative and better-ranked Nash solutions. In this game, (2, 2) was almost always chosen.

The experimenters concluded that the cooperative solution can influence the choice of Nash equilibria, even though the cooperative strategies are dominated. This experiment is a good example of

---

[7]The actual payoffs were determined by a lottery with payoffs and probabilities based on game points, so that risk aversion should have no effect on the choice of strategies. The details are too advanced for this introductory text.

[8]*Op. cit.*, p. 223.

Table 20.2. An Experimental Three-Strategy Game.

		Q		
		1	2	3
P	1	350, 350	350, 250	1,000, 0
	2	250, 350	550, 550	0, 0
	3	0, 1,000	0, 0	600, 600

Table 20.3. Another Experimental Three-Strategy Game.

		Q		
		1	2	3
P	1	350, 350	350, 250	1,000, 0
	2	250, 350	550, 550	0, 0
	3	0, 1,000	0, 0	500, 500

much experimental work in game theory, showing that many of the difficulties with the Dresher–Flood experiment can be addressed. It is consistent with many results that show that Nash and dominant strategy equilibria can be realized in experiments, especially when care is taken that the game is what it is meant to be and when the game is fairly simple and there is no conflict between Nash and cooperative outcomes. With respect to the Shelling focal point theory, the glass is at least half full, since the players found a Schelling focal point when they were not distracted by conflict between the cooperative and equilibrium solutions, and there really is no Schelling focal point when such a conflict exists.

4. ULTIMATUM GAMES

The Ultimatum Game shows the contrast between equilibrium and experimental results in an even stronger form. A certain amount of money (let us say, $50) is to be divided between two players. One player, the Proposer, offers to split the amount with the other player, the Responder. The Proposer suggests an amount that the Responder is to get, and the Responder can only say yes or no. There is no negotiation and no repetition.[9] If the Proposer and the Responder agree on the proportions in which the amount is to be split, then they each get the amount they have agreed upon. If they do not agree — that is, if the Responder answers "no," — then neither player gets anything.

To keep things a little simpler, let's say that the Proposer can only offer whole-dollar amounts to the Responder. Then the Proposer has 51 strategies: 0, 1,... 50. Figure 20.1 shows just two of those strategies in a game in extensive form: "offer $1" and "offer r." The unknown amount r can stand in for any strategy other than "offer $1," so we can analyze the game with this diagram. We can see that, if the offer is $1, the Responder's best response is to accept, since $1 is better than nothing. First, suppose r is more than $1. If the Proposer offers r, the Responder's best response is again to accept. Suppose, on the other hand, that r is 0. With an offer of 0, the Responder has nothing to lose by rejecting. If the probability of a rejection in that case is more than 2%, the Proposer will maximize his payoff by offering just one dollar — the minimum positive offer — and that is the subgame perfect equilibrium of this game.

The experimental results are quite different, however. On the one hand, Responders will usually reject offers of less than 30%. This is a sacrifice of self-interest, because, as we have pointed out, $1 or $5 is better than nothing, and nothing is what the Responder gets if he rejects the offer. Conversely, Proposers usually offer more than the minimum share to the Responders. That could, in principle, be rational self-interest. Knowing that rejections of small offers are probable, the Proposers might choose offers that maximize the

---

[9]Some experimenters have studied repeated Ultimatum Games, but here we want to focus on the one-off game, which has a clear non-cooperative equilibrium.

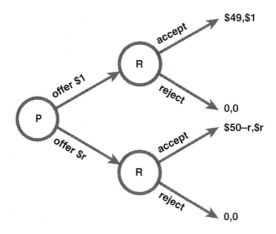

Figure 20.1.   An Ultimatum Game.

expected value of their payoffs — just enough to balance the gain from making a smaller offer against the increased risk of a rejection resulting from it. But there is evidence that Proposers often give the Responders a bigger share than the share that would maximize the expected value of payoffs, and 50–50 splits are very common.

The ultimatum game has been studied in some non-western cultures, and there is evidence of differences of detail across cultures as well as differences between genders within Western cultures. In qualitative terms, however, the results are similar and are as described in the previous paragraph.

How can we account for these results? Altruism, in its simplest form, will not do it. An altruist who tries to maximize total payoffs would never turn down any offer, even a zero offer. On the other hand, if the decision-makers have a preference for fairness, that could be consistent with the experimental results. There is evidence that perceived fairness influences the outcomes in many games. Nevertheless, we need to dig a little deeper.

Some recent studies are based on the hypothesis of *reciprocity*. The reciprocity hypothesis says that people will deviate from self-interest to return perceived favors or to retaliate for perceived slights. Thus, for example, a Responder in the Ultimatum Game who is offered $5 sees the offer as a wrong, since the Proposer keeps 90% of the payoff, and responds by sacrificing that $5 payoff to retaliate by leaving the Proposer with nothing. This is called "negative reciprocity." Sacrificing one's own payoffs to reward a good deed by the other is called "positive reciprocity." The reciprocity hypothesis says that real human behavior will often deviate from self-interest in the direction of reciprocity, both positive and negative. On the whole, the results for the Ultimatum Game seem consistent with reciprocity.

> **Definition:** *Reciprocity* — When a player gives up a greater payoff in order to reward a perceived self-sacrifice by the other player, or to retaliate for perceived aggressiveness by the other player, this is called **reciprocity**. In the first case, reward, it is called *positive reciprocity*. In the second case, retaliation, it is called *negative reciprocity*.

Another example of the reciprocity hypothesis is found in experimental studies of the Centipede Game.

5. CENTIPEDE GAMES AND RECIPROCITY

In Chapter 9, we used the Centipede Game as an example to illustrate subgame perfect equilibria. The simplest kind of Centipede game, a two-step game, is shown in Figure 20.2. Recall, the game begins with a payoff pot of 5 available. Player A can grab 4 of that, leaving 1 for player B, or pass the pot to B. With the pot in his possession, B can grab or pass. Whenever the pot is passed it grows larger, so that B can grab 6, leaving 2, for a total of 8, but if the pot is passed a second time the two players split a pot of 10 equally, for 5, 5.

We recall that the subgame perfect equilibrium is to grab at the first play, and this is true even if we extend the Centipede to 100 steps or more. In experimental studies, it is common for the pot to

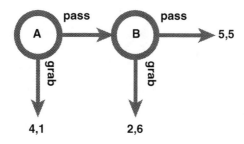

Figure 20.2.   A Simple Centipede Game.

be passed to the end point of equal division, and grabs at the first step are uncommon. In experiments with more than two steps, there are some cases in which the pot is grabbed in late stages of the game.

Some studies explain these results by the reciprocity hypothesis, relying mainly on positive reciprocity. Player B understands that player A can grab the larger part of the pot, and if Player A passes the pot, player B sees this as a favorable act, and is likely to reciprocate the positive act by passing as well, reducing his payoff from 6 to 5 but allowing Player A a payoff of 5 rather than the 4 player A has passed up. If Player A is motivated by reciprocity, he may anticipate player B's self-sacrifice, and reciprocate an expected good turn on the part of Player B. If both are motivated by reciprocity, we could see the pot passed to the end of the game. Even if Player A plays strictly according to self-interest, if Player A believes Player B is motivated by reciprocity, Player A would pass in the expectation that the pass would be reciprocated by B. (However, self-interested A might grab the pot at a later stage of a multi-step Centipede.)

EXERCISE: Suppose that A and B play a four-stage Centipede Game with plays by A, B, A, and finally B. Suppose A plays strictly according to self-interest and A believes B is motivated by reciprocity. What is A's best strategy?[*]

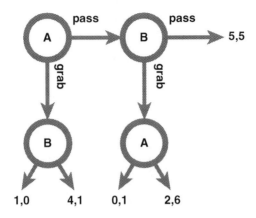

Figure 20.3.   A Centipede Game with Retaliation.

Thus, the Ultimatum Game gives a lead role to negative reciprocity, that is, costly retaliation, while the Centipede Game gives a
lead role to positive reciprocity. But negative and positive reciprocity
can reinforce one another. Consider the modified Centipede game
shown in extended form in Figure 20.3. After the "grab" stage, the
other player has an opportunity to retaliate, by choosing "left," or
not to retaliate, choosing "right." Retaliation makes both players
worse off — the retaliator gives up even his smaller proportion of
the pot, much as a Responder who turns down a positive offer gives
up his smaller proportion in the Ultimatum game. In both cases, in
effect, the retaliators give up some of their own (potential) payoffs
to punish the other player.

But does it make a difference in the Centipede game? Notice
that retaliation is never a best response in this game. Therefore, the
option to retaliate is irrelevant to the subgame perfect equilibrium.
At the first stage, "grab and don't retaliate" is the subgame perfect
equilibrium. Therefore, we would expect (based on the subgame
perfect equilibrium) that there would be no difference, in experiments, between the games in Figure 20.2 and those in Figure 20.3.
In actual experiments, however, "pass" strategies are more common

in games like Figure 20.3 than in games like Figure 20.2. Evidently, the threat of self-sacrificing negative reciprocity reinforces the promise of self-sacrificing positive reciprocity in inducing players to pass rather than grab.

6. BUSINESS APPLICATION: RECIPROCITY IN THE EMPLOYMENT RELATIONSHIP

Economists traditionally have taken the assumption of rational self-interest as given, and so there are few economic studies based on the reciprocity hypothesis. However, one study of the relationship between employers and employees by an economist[10] argued that reciprocal gifts are at the heart of successful employment relationships.

George Akerlof began from the observation that employers often pay more than the going wage in the labor market. This gives the employees something to lose if they should lose the job. After all, an employee who is paid just what he can get in the market has nothing to lose by being dismissed — he can get another job at the same wages tomorrow.[11] The idea is that this increases productivity: Employees who have something to lose will work harder, and productivity and profits will be improved as a result. When employers pay above-market wages and this maximizes profits by enhancing productivity, the higher wage is called an *efficiency wage*, rather than a market wage. But how does this increase in productivity happen? Akerlof argues that employees usually work harder than they really have to. Monitoring cannot be constant, and detailed observation in a case study indicated that employees worked more than the amount

---

[10] Akerlof, G. A., Labor contracts as a partial gift exchange, *Quarterly Journal of Economics*, **98**(4) (November, 1982), pp. 543–570.

[11] In more developed countries, there are a very few labor markets in which this is literally true. Examples might be day labor markets and hiring hall labor markets in which a person may work for a different employer every day. These are more common in less developed countries. It is not clear what the growth of temporary employment might mean for this argument.

of work needed in order to avoid being *caught* shirking. Akerlof's reasoning is that the employees are responding to their employer's perceived generosity with positive reciprocity. Akerlof drew from anthropological studies that had interpreted the observed behavior of people of various cultures as evidence of reciprocity. In many of these cases, reciprocity takes the form of mutual gift-giving. Akerlof saw a parallel to the anthropological studies in that the employees perceived their higher-than-market wages as a "gift" and responded with the "gift" of increased effort. But this is equally consistent with game theoretic experiments that point toward a tendency to sacrifice in order to return a perceived favor (or retaliate for a perceived slight).

What this suggests is not a new idea, but nevertheless one that is often overlooked. It is that the employment relationship is more profitable if it is not simply a case of conflicting interests. "Disgruntled" employees are less profitable employees, so smart employers find ways (including, but not only, money) to "gruntle" their employees.

7. LEVEL *K*

In most of game theory, we assume not only rationality but common knowledge of rationality. That means that there is no point in trying to outsmart anybody. If I know that the other players are rational *in just the same sense that I am*, then there is no point in my trying to outsmart anybody. The results of this analysis can be a good approximation to the play of people who have experience with the game and the responses of other players. In many cases, though, people have to choose their strategies without experience (if the game is played only once, or is being played for the first time) or other information or training. In cases of that kind, the common knowledge of rationality approach may be a poor approximation, for at least two reasons. First, the evidence shows that real human rationality is bounded. The solutions to games with common knowledge of rationality can be quite complex and beyond the capability of real people to solve, without experience or training. Second, we have evidence that suggests that people do often try to outsmart one another. It could make sense for me to outsmart my rival, if I know that his rationality is bounded and that I might be able to be a little "more rational" than he is. In short, it seems that people often act "as if" there were different types of players, some "more rational" than others.

This idea is central to the "Level k" approach to strategic thinking in one-off games. In this approach, decision-makers in games are of (at least) the following types:

Level 0: Players at Level 0 do not do any strategic thinking at all, but choose their strategies without much thinking, perhaps at random.
Level 1: Players at Level 1 choose the best response to the decisions of players at Level 0.
Level k: Whenever $k > 0$, a player at Level k chooses the best response to a player at Level $k - 1$. For example, a Level 2 player chooses the best response to a player at Level 1.

In addition, there may be two other player types: "equilibrium" players, who simply play a Nash equilibrium, and sophisticated players, who try to estimate the odds of being matched against a player of one of the other types and choose their response to get the best expected value payoff on the basis of the estimate.

As an example, let us consider the Location game from Chapter 4, Exercise 4.4. The payoffs for that game are shown for convenience in Table 20.4. We recall that two department stores, Mimbel's and Gacy's, have to choose locations for their flagship stores. The unique Nash equilibrium for this game is for both to play Center City.

We suppose that Level 0 players choose among the four locations at random, with a probability of ¼ for each of the four strategies. Now suppose Mimbel's believes that Gacy's is a Level 0 player. Then the expected values for Mimbels' strategies are as shown in Table 20.5. We see that, as a Level 1 player, Mimbels' best response is Center City, for an expected value payoff of 116.5.

Table 20.4. Payoffs in the Location Game (Repeats Table 4.14, Chapter 4).

		Gacy's			
		Uptown	Center City	East Side	West Side
Mimbel's	Uptown	30,40	50,95	55,95	55,120
	Center City	115,40	100,100	130,85	120,95
	East Side	125,45	95,65	60,40	115,120
	West Side	105,50	75,75	95,95	35,55

Table 20.5. Expected Value Payoffs for Mimbel's if Gacy's is Level 0.

			Total
Mimbel's	Uptown	30/4 + 50/4 + 55/4 + 55/4	47.5
	Center City	115/4 + 100/4 + 130/4 + 120/4	116.25
	East Side	125/4 + 95/4 + 60/4 + 115/4	98.75
	West Side	105/4 + 75/4 + 95/4 + 35/4	77.5

Table 20.6. Expected Value Payoffs for Gacy's if Mimbel's is Level 0.

			Total
Gacy's	Uptown	40/4 + 115/4 + 125/4 + 105/4	96.25
	Center City	95/4 + 100/4 + 65/4 + 75/4	83.75
	East Side	95/4 + 85/4 + 40/4 + 95/4	78.75
	West Side	120/4 + 95/4 + 120/4 + 55/4	97.5

Now suppose that Gacy's assumes that Mimbel's is a Level 0 player and so will choose among the strategies at random. Then Gacy's will expect payoffs as shown in Table 20.6. We see that Gacy's best response to Level 0 play is a West Side location for an expected value payoff of 97.5.

We have seen that if Mimbel's is a Level 1 player, then they will play Center City, while if Gacy's is a Level 1 player, they will play West Side. What about level 2? If Mimbel's is a Level 2 player, they will play their best response to Gacys' Level 1 strategy, and that is (again) Center City. If Gacy's is a Level 2 player, they will play their best response to Mimbels' level one strategy, which is Center City. At Level 3, again, each will play its best response to Center City — which is Center City; and so, the same is true at every higher level. Since the Nash equilibrium for this game is for both to play Center City, if both players play at least at Level 2, they will play the Nash Equilibrium. We cannot distinguish between an equilibrium player and a player at Level 2 or above. As for sophisticated players, we cannot tell without more information, since that will depend on the proportion of Levels 0, 1 and higher players.

The Level-k theory is a theory of play for one-off games, that is, games that are played only once or played for the first time. In experiments, typically, the experimental subjects will be matched to play the games with rotation so that each pair plays only once. This design assures that they are playing the one-off games the experimenters designed, and not some more complex repeated game. The strategies chosen on each play are recorded. If strategies corresponding to Levels 1 and 2 play, or other levels, are much more common than would occur at random, then the evidence tends to favor the Level k theory. On this basis, the theory has good support, and play at Levels 1 and 2 seem most common, while some equilibrium players and sometimes a few Level 3 and sophisticated players are observed. Very few if any Level 0 players are observed. In the Level k theory, Level 0 is not so much a theory of the play by any decision-maker as it is a hypothesis about the way some players model the play of their opponents or partners. The Location Game has not been studied by experimental methods, but if we were to do so, we would expect to see quite different play on the part of the subject in the role of Gacy's and Mimbel's. For Mimbels's, we would expect the equilibrium strategy Center City to be far the most common strategy choice. However, for Gacy's, we would expect to see the West Side and Center City locations both chosen more

Table 20.7. Hawk vs. Dove
(Repeats Table 5.8, Chapter 5).

		Bird B	
		Hawk	Dove
Bird A	Hawk	−25,−25	14,−9
	Dove	−9,14	5,5

commonly than the other two, with a relative frequency that would give us a clue as to the proportion of Level 1 players. If we were to observe that, it would tend to confirm the Level *k* theory; while quite different results, such as roughly equal proportions choosing other strategies or no difference between Gacy's and Mimbels' players, would lead us to doubt the Level *k* theory.

We see that in some games (such as the Location Game) the Nash equilibrium will be played if the players make decisions at a sufficiently high level. But that's not true in general, and to illustrate that, we can return to another familiar game, the Hawk vs. Dove game from Chapter 5, Section 8. For convenience, the payoff table for this game is repeated in Table 20.7.

As usual we assume that a Level 0 player chooses between the two strategies at random with equal probabilities. If Bird A is a Level 1 player, then, Bird A expects a payoff of $-\frac{11}{2}$ for the hawk strategy and $-\frac{4}{2}$ for the dove strategy. (Don't misunderstand the *negative* numbers: $-\frac{4}{2}$ is greater than $-\frac{11}{2}$). Therefore, a Level 1 player will play Dove. A Level 2 player will play the best

Definition: *Cognitive Salience* — In any problem of decision or inference, there may be one alternative that attracts attention because of some special characteristic or property. That alternative is said to be *cognitively salient*. The term comes from the Latin for a jump: the cognitively salient alternative is the one that jumps to mind.

response to Dove, which is Hawk; and a Level 3 player will play the best response to Hawk, which is Dove, and so on — all odd-level players will play Dove and all even-level players will play Hawk. We cannot say what an equilibrium player will play, since either strategy can correspond to an equilibrium. As for sophisticated players, we know that Dove is the best response if there are more than about 36% Hawks in the population, and Hawk otherwise. Accordingly, for this game a sophisticated player will play Hawk only if he thinks that even level players are less than 36% of the population, and otherwise will choose Dove.

A difficulty with the Level k approach is to determine what the Level 0 players will do — what it means to say that they do no strategic thinking at all. In some games, it does not seem that choosing the strategy at random really is what an unthinking player would do. Some strategies may have some special property that attracts attention to them, so that the unthinking decision-maker would tend to choose that strategy. The term for a property like this is *cognitive salience.* "Salience" is another word for prominence, for a feature that makes a particular strategy stand out, as a mountain or a peninsula does on a map, so "cognitive salience" is a feature that makes the strategy stand out at a simple stage of thought. The unresolved question is whether the unthinking player will choose at random or will choose a cognitively salient strategy (if there is one) with a high probability.

Here is a game to illustrate the problem. There is no story to go with it, but it will show how greed can make a difference, so we will call it the Greed Game. Players P and Q choose among four strategies 1, 2, 3, 4; and the payoff table is shown as Table 20.8. We see that this game has two Nash equilibria, corresponding to strategy pairs (2,2) and (3,3).

First, suppose that a Level 0 player chooses among the four strategies at random. Then a Level 1 player will estimate his expected value payoffs as shown in Table 20.9, and will accordingly play strategy 2. A Level 2 player will respond with strategy 2, and so will all higher level players. We expect to see all players above Level 0 playing the equilibrium at strategies 2,2.

Table 20.8. Payoffs in a Greed Game.

		Q			
		1	2	3	4
P	1	0,0	0,400	0,500	1,000,0
	2	400,0	300,300	0,0	400,0
	3	500,0	0,0	100,100	0,0
	4	0,1,000	0,400	0,0	0,0

Table 20.9. Expected Value Payoffs against a Random Player in the Greed Game.

			Total
P or Q	1	0/4 + 0/4 + 0/4 + 1,000/4	250
	2	400/4 + 300/4 + 0/4 + 400/4	275
	3	500/4 + 0/4 + 100/4 + 0/4	150
	4	0/4 + 0/4 + 0/4 + 0/4	0

However, the payoff of 1,000 at strategy pair 1, 4 or 4, 1 is twice as large as any other payoff in the game, something of a jackpot payoff. This might catch the attention of a Level 0 player. In this case, the large payoff might stand out because it appeals to the emotion of greed. Suppose, then, that instead of choosing at random, a level 0 player "goes for the gold" by choosing strategy 1. Then the best response, chosen by a Level 1 player, is strategy 3; and a Level 2 or higher player responds with strategy 3, playing the Nash equilibrium at strategy pair 3,3.

In this game, as we see, different models of Level 0 play lead to quite different predictions. In fact, each predicts the play of a Nash equilibrium, but they predict different equilibria — (2,2) if Level 0 chooses at random and (3,3) if Level 0 chooses according to cognitive salience. On the one hand, the Level *k* theory certainly would be more specific if we could always identify a single model of Level 0 play. On the other hand, we can let the evidence speak for itself. An experiment with the Greed Game — an experiment that has yet to be done — might show which model of Level 0 play is better: if (2,2) were much more common than (3,3) we would infer that the random interpretation of Level 0 play is the better one, for this particular game. Conversely, if strategies 1 and 4 were often chosen, this would be evidence against the Level k theory.

The Level k theory has the advantage of representing bounded rationality in the choice of strategies in a way that directly reflects the strategic interdependence of the players, and their models of other players; and that assumes that players may be of different types. Most of the experiments testing Level k theory have been more complex than these examples. In these studies, we find considerable evidence for the theory, and particularly for the existence of different types of play.

Alternative theories also have some support in other experimental studies. An important family of alternative models for one-off games assumes that decision-makers tend to choose best responses, but make errors with some probability. Usually, the more the decision-maker stands to lose by making the error, the less likely she or he is to make the error. At the same time, if players know that other players can make mistakes, that will influence their own choices of strategies. These assumptions lead to rather complex mathematical models of equilibrium that must be beyond the scope of this introductory text. Models of this kind have the advantage of allowing for learning, as a process of error-reduction.

However we model it, the evidence for bounded rationality is very strong. In behavioral game theory, we model boundedly rational decision-making in games and game-like situations and rely on experimental evidence to select the best models. This remains one of the most important frontiers of research and application in game theory.

8. WORK TO BE DONE (FRAMING)

Experimental game theory has adopted experimental approaches from other fields, particularly psychology and experimental economics, and in some cases has extended the results of those studies. One body of experimental studies (primarily in psychology) has focused on a phenomenon called framing. What these studies say is that human decisions may depend on the way the question is put. That is, the decision depends on the way the decision is "framed." For example: Would you rather buy a product that is 95% fat free or one that is 5% fat? Of course, the two phrases mean the same thing, but when did you see a product advertised as 5% fat?

Framing is illustrated by a classical experiment by psychologists Amos Tversky and Daniel Kahneman. It is not a game theory experiment, but an experiment about risk perception, a closely related field. The experiment is based on a hypothetical disease threat: an epidemic of an obscure tropical disease is expected to kill 600 Americans if no precautions are taken. There are two programs of precautions that might be taken, but they are inconsistent, so only one of the two can be taken. Experimental subjects were asked to choose between the two programs.

- One group of experimental subjects were told that
  - with Program A, 200 people will be saved,
  - while with Program B, there is a 1/3 probability that 600 will be saved and a 2/3 probability that none will be saved.
- The other group were told that
  - with Program A, 400 will die,
  - whereas with Program B, there is a 1/3 probability that nobody will die and a 2/3 probability that 600 will die.

Notice that these two descriptions actually describe the same events, but the first one calls attention to the fact that, with Program A, 200 people will live with certainty, whereas Program B calls attention to the fact that 400 will certainly die with the same program. The choice is the same either way: a certain 200/400 split of living and dead, on the one hand, and an all-or-nothing risk with slightly bad odds on the other.

Despite the fact that the two descriptions add up to the same, the two experimental groups responded very differently. The ones who were told that 200 would live opted for Program A by a 72-28 margin, while those who were told that 400 would die chose Program B by a 78-22 margin! Evidently the decisions depended very much on how the decision was framed — in a way positive or negative to program A.

Experimenters in game theory have only recently begun research on the role of framing in game theory. The reason for the delay probably is found in the roots of game theory in rational action theory. In traditional game theory models, we assume rationality and common knowledge of rationality — so the only frame the decision needs is the fact that the other players are rational. But when decisions are influenced by perceived reciprocity, framing may be important, since the perception of reciprocity may depend on the way the game is framed. Self-interest, altruism and bounded rationality, too, may give rise to decisions that depend on the way the decision is framed.

Hopefully, with further experimental work we will learn much more about the role of framing in games.

9. WHERE WE HAVE ARRIVED

All in all, these experimental studies raise some deep questions about some models of non-cooperative game theory *as a theory of human strategic behavior.* It seems clear that

- Real human rationality is bounded, not unbounded as the common knowledge of rationality assumption implies.
- Real human behavior in games sometimes approximates cooperative rather than non-cooperative outcomes.
- Real human behavior in games is influenced by non-self-interested motives, such as reciprocity.
- It is probable that real human behavior in games is influenced by framing.
- Real human behavior in games can be very complex and, for practical purposes, a considerable part of it is random.

However,

- There are a wide range of experimental and observational cases in which non-cooperative game models do describe observed behavior. Among these are many games in which people have plenty of opportunity to learn and refine their behavior.
- Non-cooperative and cooperative game solutions together can explain a great deal of real human behavior.
- Game theory is only partly a theory of human behavior. It is also partly a "ideal" theory, that is, a theory that explains how people would behave if they were (in one sense or another) rational.

To summarize, there are two ways to interpret the disagreement between theory and experiment in some important, exemplary games. One way is to interpret it as evidence against the rationality hypothesis and the general applicability of game theory. The other is to interpret it as a step toward extending the theory beyond the rationality assumption to a more general and realistic theory of strategic behavior. The second interpretation seems preferable (to the author of this textbook) in part because it allows us to retain the wide range of games and applications in which theory, experiment and observation agree well.

10. SUMMARY

Games and experiments are similarly based on assumptions about the uniformity of that which is observed, and games lend themselves well to experimental work. Thus, there has been extensive experimental work in game theory, beginning in the earliest days of game theory at the RAND Corporation. In some ways, the earliest experiments (which gave rise to the Prisoner's Dilemma) gave a foretaste of what was to come. First, there can be ambiguity about just what is tested. For example, experiments with repeated play cannot be used to directly test hypotheses about simple one-off games. But even when all precautions are taken to remove ambiguity, the results clearly go beyond the limits of non-cooperative equilibrium theory.

There are equilibrium examples that are confirmed by experiments, but there are others in which the experimental results raise questions about both unlimited rationality and self-interest. When there is a conflict between the non-cooperative and the cooperative solutions to an experimental game, the observed behavior is often pulled strongly toward the cooperative solution or otherwise influenced by it. It may be necessary to allow for a wide range of distinct non-self-interested and boundedly rational players, but it may be that a wide range of human strategic behavior can be understood as a boundedly rational expression of self-interest modified by reciprocity.

Q20. EXERCISES AND DISCUSSION QUESTIONS

Q20.1. Road Rage

Recall the "road rage" game from Chapter 9, Exercise 9.1. Al's strategies are "aggress, don't aggress" and Bob's strategies are (if Bob aggresses) "retaliate" and "don't." The payoffs are shown in Table 20.10.

In the subgame perfect equilibrium of this game, Bob aggresses and Al does not retaliate. However, drivers do not always aggress, and retaliation is sometimes observed when they do, even to the point (rarely!) of firing weapons at the aggressor. Explain these facts using concepts from this chapter. How would Bob and Al respectively play if they were motivated by reciprocity? The Washington State Police adopted a policy to discourage road rage by increasing the penalty for aggressive driving. Does this make sense in the light of the discussion in this chapter?

Table 20.10. The Road Rage Game.

		Bob	
		Aggress	Don't
Al	If Bob aggresses then retaliate; if not, do nothing	−50,−100	5,4
	If Bob aggresses then don't; if not, do nothing	4,5	5,4

Q20.2. Payback Game

Player 1 gets a sum of money. He can keep all of it or pass all or some of it on to Player 2. If some of the money is passed on to Player 1, the experimenter matches the amount, so that the amount passed to Player 2 is double the amount given up by Player 1. Player 2 can then pay back to Player 1 all or part of the money he has received. Any money passed back is again matched by the experimenter, so that Player 1 receives 1.5 times the amount passed back. Thus, if Player 1 starts with \$10 and passes it all, Player 2 gets \$20, and if he passes all of it back, Player 1 gets \$30.

What is the subgame perfect equilibrium of this game? Would it make a difference if the players are motivated by reciprocity? Drawing on the information in this chapter, how would you expect the experiment to come out?

Q20.3. An Environmental Game

Prof. Greengrass is interested in the motives for conservation of environmental resources such as forest land and underground water. These resources may be of benefit to future generations, but any generation also has the option to use them up, for example, by cutting the forests or polluting the underground water. In that case the resources are of less or no use to the future generations. For experimental purposes, Prof. Greengrass will set up a game along these lines: there are N participants, with N equal to two or more, and they will play in order. The first player is given a "resource" (a piece of paper) and may pass it to the second or return it to the experimenter. If it is passed, each player in turn has the same choices, while the game ends when the paper is returned to the experimenter. Each player who passes the paper to the next player gets one point, while the player who returns it to the experimenter gets two (unless he is the last in order, who can get only one point).

Draw the extended form diagram for this game, assuming there are 3 players in the game. Assuming that the players are rational and self-interested, what would you expect would be the result? What do you think would happen? Why? What difference would it make if the number of players were larger?

Q20.4. Effort Dilemma

An effort dilemma, recall, is a social dilemma in which the cooperative strategy is to work (with a strong effort) and the non-cooperative strategy is to shirk. The theory of indefinitely repeated games tells us that, if the probability of another play is large enough, cooperative play may occur. Design an experiment to test the influence of an increasing probability of repetition on the outcome of an effort dilemma.

Q20.5. Level *k*

Assuming that Level 0 play is random with equal probability for all strategies, determine the Levels 1, 2, and 3 play for the following exercises: Chapter 4, Exercises 4.4 and 4.5; Chapter 5, Exercises 5.2 and 5.3, Chapter 16, Exercises 16.2 and 16.3.

Endnote

[*]Answer to the exercise on page 487. Assume that, as in Figure 20.2, the cooperative payoff after four passes dominates the payoffs if A grabs at the first stage. If A believes that B is motivated by reciprocity, then A will pass and, if B passes, pass again. Anticipating that B will "return the favor" by passing, and so A will get the larger cooperative payoff. Thus, the knowledge that other players are not "rational and self-interested" leads the player who is rational and self-interested to act in ways that would not be rational otherwise.

CHAPTER 21

Evolution and Adaptive Learning

Think back to the Hawk versus dove example in Chapter 5. In this example, two birds get into conflict and each bird has to choose between an aggressive strategy and a strategy of retreat. Yet birds are

To best understand this chapter, you need to have studied and understood the material from Chapters 1–12, 14, and 20.

not the calculating rational creatures that we suppose human beings to be, in the context of game theory. In birds, much of the behavior we observe seems to be genetically determined. Hawks are aggressive, not because they choose to be, but because their genetic heritage makes them aggressive. Similarly, doves retreat because their genes have programmed them to retreat.

Nevertheless, game theory has had an influence on evolutionary biology. Biologists reason that evolution eliminates the strategies that do not work, so that game theory may be able to predict the result of this evolutionary process. The biologist John Maynard Smith, in particular, proposed the concept of evolutionarily stable strategies (ESS) as a game theory approach appropriate for evolutionary biology.

But even in the case of human beings, the assumption of perfect rationality made in classical game theory may go too far. Many observers of human behavior would say that human rationality is "bounded." Because human beings have limited cognitive capabilities, we may not really be able to come up with some of the more complicated computations for best response strategies described in game theory. One possibility is that "boundedly rational" human beings usually act according to habit, routine, and "rules of thumb." Even so, people learn. We are rational creatures in that we can learn and eliminate our mistakes. So, human learning may not be so very different from evolution, and the ideas the biologists have brought to game theory may be equally applicable to boundedly rational human beings.

This chapter explores the concept of ESS, along with its applications to both biology and boundedly rational learning.

1. HAWK VS. DOVE

The game of "Hawk vs. Dove" was given in Chapter 5 as an

HEADS UP!

Here are some concepts we will develop as this chapter goes along:

Population Games: If members of a population are matched at random to play a game, with different types playing different strategies, the whole sequence of play is a population game.

Replicator Dynamics: In a population game with replicator dynamics, each type increases or decreases relative to the total population in proportion as the payoff to the strategy played by that type exceeds or falls short of the average payoff.

Evolutionarily Stable Strategy: A Nash equilibrium that is stable under the replicator dynamics is an evolutionarily stable strategy equilibrium (ESS).

---

example of a two-by-two game with two equilibria. The payoff table is shown as Table 21.1. The birds' payoffs are changes in their inclusive fitness, that is, changes in their rates of reproduction relative to the average rates of reproduction in their populations. In population biology, the assumption is that the two birds are matched at random and face a conflict in which each bird can, in principle, choose two strategies: aggressive (hawk) or evasive (dove) Of course, the birds' strategies are determined by their genetic makeup, but if (for example) there are 25% hawks in the population and

Table 21.1. Hawk vs. Dove
(Repeats Table 5.8, Chapter 5).

		Bird B	
		Hawk	Dove
Bird A	Hawk	−25,−25	14,−9
	Dove	−9,14	5,5

75% doves, then the individual bird faces very much the same situation as if her opponent chose between the hawk and dove strategies with a mixed strategy, with 25% probability of choosing "hawk" and 75% probability of choosing "dove." In that case, a bird has three times the probability of being matched with a dove than with a hawk, so the payoffs to hawks are likely to be relatively high — with few fights and a lot of easy lunches.

Since the payoffs depend on probabilities, as with mixed strategies, we have to compute the expected values of payoffs. Suppose z is the proportion of hawks in the population (0.25 in the example in the previous paragraph). Therefore $(1 - z)$ is the proportion of doves (0.75 in the example). The expected value of the payoff for a hawk is

$$EV(\text{hawk}) = (-25z) + 14(1{-}z) = 14 - 39z$$

since the hawk's payoff is −25 when matched with another hawk and 14 when matched with a dove. When $z = 0.25$, as in the example, $EV(\text{hawk}) = 4.25$. The expected value payoff for doves is

$$EV(\text{dove}) = -9z + 5(1 - z) = 5 - 14z.$$

since the dove's payoff is −9 when matched with a hawk and 5 when matched with another dove. In the example, with $z = 0.25$, $EV(\text{dove}) = 1.5$.

Remember, these payoffs are proportional to rates of reproduction above replacement, so both populations will be growing. However, the population of hawks will be growing faster. The (weighted) average payoff is $1.5(1 - z) + 4.25z = 2.1875$. Thus, the population of hawks is growing $4.25/2.1875 = 1.9$ times as fast — nearly twice as fast — as the population of birds as a whole, while the population of doves is growing only about two thirds as fast.

These relative rates of reproduction determine what the population will be, in the "long run." Figure 21.1 shows the expected value payoffs to the two kinds of birds as the

A Closer Look: John Maynard Smith 1920–2004

John Maynard Smith was born in 1920 in London. Originally educated as an aeronautical engineer at Cambridge, England, he studied zoology at University College, London. He was on the faculty of University College, London 1951–1965 and then at the University of Sussex in England, where he retired in 1999. His contributions to game theory include applications to evolutionary theory, especially the concept of the evolutionarily stable strategy (ESS) as a factor in evolution.

proportion of hawks varies. The proportion of hawks, z, is on the horizontal axis. The broad gray line shows the expected value payoff for doves, as z varies from zero to one, and the dark line shows the expected value payoff for doves. At the intersection, the population is stable. The expected value payoffs are equal — as in a mixed strategy Nash equilibrium. With a little algebra we can see that this equilibrium is at $z = 9/25 = 0.36$.[1]

This is an Evolutionarily Stable Strategy, an ESS. According to Ferdinand Vega-Redondo, "a strategy ... is said to an ESS if, once

---

[1] $5 - 14z = 14 - 39z$
$(39 - 14)z = 14-5$
$25z = 9$
$z = 9/25 = 0.36$

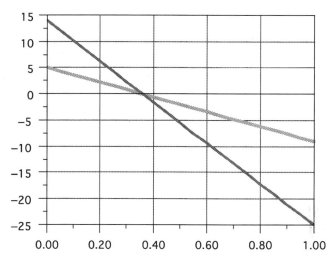

Figure 21.1. Relative Rates of Reproduction in a Hawk–Dove Game.

adopted by the whole population, no mutation ... adopted by an arbitrarily small fraction of individuals can 'invade,' (i.e., enter and survive) by getting at least a comparable payoff."[2] To apply this concept to the Hawk vs Dove game, we have to interpret the 9/25 equilibrium proportion as a mixed strategy — as if each individual bird

Definition: *Replicator Dynamics* — According to the *replicator dynamics*, the proportion of the population who play each strategy (in a population game) increases in proportion to the difference between the payoff to that strategy and the average payoff.

adopted a mixed strategy with 9/25 probability of playing Hawk. We then ask, if a small population were to adopt a different probability, would they get higher payoffs? The answer is no. The mixed strategy with $z = 9/25$ is a Nash Equilibrium, which means that each individual, and therefore every small population, is adopting its "best

---

[2]Fernando Vega-Redondo, *Evolution, Games and Economic Behavior* (England, Oxford: Oxford University Press, 1996), p. 14.

response" to the rest. Any "mutation" will not be a best response and will do worse still.

The dynamic idea underlying the ESS is the "replicator dynamics." Again quoting Vega-Redondo, in the replicator dynamics, "the share of the population which plays any given strategy changes in proportion to its *relative* payoff (i.e., in proportion to its deviation, positive or negative, from the average payoff)."[3] In the discussion of the Hawk-Dove example above, we can apply the replicator dynamics, noting that a) whenever $z < 9/25$, hawks have higher payoffs than the population average, and so their proportion, z, will increase, but b) whenever $z > 9/25$, hawks have lower payoffs than the population average, and so their proportion, z, will decline. Visually, therefore, whenever z is different from $9/25$, z will be moving toward $9/25$, and only $z = 9/25$ is stable under the replicator dynamics. Stable states under this dynamics are identical with ESS.

We recall that the Hawk vs. Dove Game also has two pure-strategy equilibria. In the population game, these Nash equilibria correspond to cases in which all members of the population are doves or are hawks. Are these equilibria evolutionarily stable? They are not. Think of an island with a population only of doves. If a few doves were to mutate and become hawks — or, more realistically, if a small flock of hawks were blown onto the island by a storm — then the population of hawks would grow toward the equilibrium proportion. Of the three Nash equilibria in the Hawk vs. Dove Game, only the mixed strategy equilibrium is an ESS.

We may draw two conclusions.

1. An ESS is a Nash equilibrium. To say that "no mutation ... adopted by an arbitrarily small fraction of individuals can 'invade,' (i.e., enter and survive) by getting at least a comparable payoff." is to say that every creature is playing its best response when the population is at an ESS.
2. Not all Nash equilibria are ESS, since not all Nash Equilibria are stable under the replicator dynamics.

---

[3] *Ibid.*, p. 45.

We will consider one further biological application of game theory.

2. A SEWAGE GAME

Escherichia coli, E. coli for short, is a common sewage bacterium. According to the New York Times, three strains of *E. coli* play something like a Rock, Paper and Scissors game.[4] In the Rock, Paper and Scissors game, two children choose among three strategies: announce "rock," "paper," or "scissors." Then paper "covers" (defeats) rock, rock smashes (defeats) scissors, and scissors cut (defeat) paper. Since any predictable strategy can be exploited, a mixed strategy with probabilities of all strategies at 1/3 is the only Nash equilibrium.

The three strains of *E. coli* bacteria are as follows: Strain 1 produces both a lethal poison and a protein antidote to protect itself from its poison. Strain 2 produces only the antidote. Strain 3 produces neither. However, the production of each chemical costs the bacterium something, and slows down the reproduction of the bacteria, so that Strain 1 reproduces less rapidly than Strain 2, and (in the absence of the poison) Strain 2 reproduces less rapidly than Strain 3. What could an ESS look like?

First, a population of 100% of one strain will not be an ESS. Suppose, for example, that the population is all Strain 1. Then a mutation creating a small population of Strain 2 could invade, and outgrow the Strain 1 population and take over. Suppose the population were all Strain 2. Then a mutant population of Strain 3 could invade and become predominant. Finally, if the population were all Strain 3, a mutant population of Strain 1 could invade, kill off Strain 3 with the poison, and become predominant.

When the experimenters grew the bacteria on a plate, the bacterial culture formed small clumps of each type, with competition between strains only at the boundaries. Where Strain 1 competed

---

[4]Henry Fountain, Bacteria's 3-Way Game, the *New York Times* (Tuesday, July 30, 2002), Section F, p. 3.

with Strain 2, Strain 2 outgrew and crowded Strain 1 so that Strain 1 pulled back from those borders, while on borders where Strain 1 competed with Strain 3, Strain 3 retreated from the poison produced by Strain 1, so Strain 1 advanced. Predominant numbers of either two strains similarly led to advances and declines of their two sorts of competitors. Thus, the boundaries were constantly shifting. However, the proportions of the three strains remained about the same, at 1/3 each. Here is why. Suppose that A is the proportion of Strain 1 in the population, B of Strain 2, and C of Strain 3. Then suppose A, B, and C are 40, 30, 30. Strain 3 will decline, since the probability that a Strain 3 bacterium will be matched with Strain 1 (where Strain 3 loses) is 1/3 larger than the probability that it will be matched with Strain 2 (where Strain 3 wins). But, for the same reason, Strain 2 will be increasing. From the point of view of Strain 1, both of these developments are bad news. Both the decline in Strain 3, which Strain 1 will beat, and the rise of Strain 2, which beats Strain 1, reduce the opportunities for Strain 1, so Strain 1 will begin to decline. The dynamics is complicated, but it leads to the conclusion that the population will be stable if, and only if, $A/B = B/C = C/A$. That means each strain is 1/3 of the population — just as in Rock, Paper and Scissors.

There are many other applications of game theory in evolutionary biology. It appears, then, that game theory can be usefully applied to non-humans. But the key concepts for the biological applications, ESS and the replicator dynamics, can also be applied to human beings and the evolution of the strategies they habitually choose.

3. BOUNDED RATIONALITY

In neoclassical economics and much of game theory, it is assumed that people *maximize*, or infallibly choose their *best response*. Others argue that real people cannot do that, but rather that real human rationality is *bounded*. There are several ways of expressing this. One of the earliest suggested that, if people have a satisfactory solution to a problem, they will look no further — rather than maximizing, they "satisfice." Another approach is expressed by the concept of *production systems* — that people act according to rules, though the rules may be very complex. This idea comes from studies in artificial intelligence,[5] but fits well with strategy rules like "Tit-for-Tat." We can put those ideas together, and say that people act according to rules that are not the ideal rules that would maximize their payoffs, but rather satisfactory rules. (Such rules are also known as heuristics.) As we saw in Chapter 20, there is a good deal of evidence to support this idea.

This is not to say that people don't learn. Learning is important in games, and game theory includes some models of perfectly rational learning. But a perfectly rational being would use all available information, would think in terms of probabilities, and would apply Bayes' rule to keep his estimated probabilities in line with the evidence. Here we are concerned with bounded rationality. Boundedly rational creatures make decisions on the basis of "rules of thumb," heuristic methods, and learn much less systematically. Nevertheless, boundedly rational learning means that boundedly rational human beings do change their strategies on the basis of their experience, even if they do not use all available evidence and apply Bayes' rule. They will eliminate heuristic rules and strategies that, in their experience, lead to lower payoffs. This is *adaptive learning*. They can also learn by imitating one another. If I see that my neighbor plays Tit-for-Tat in social dilemmas and my neighbor gets better payoffs than I do when I play always cooperate, I can switch and begin to play Tit-for-Tat. Taking both of these points into account, we might think of boundedly rational learning as an evolutionary process in which the relatively unfit rules are eliminated and the fittest rules survive. The fittest rules are, of course, the ones that

---

[5] Nobel Laureate Herbert Simon was an innovator of both of these interpretations, and thus is the most important figure in the origin of the idea of bounded rationality. See Simon, H. A., A behavioral model of rational choice, *Quarterly Journal of Economics*, **69**(1) (February, 1955), pp. 99–118 on satisficing, among many other writings, and Simon, Herbert A., The Information-processing Explanation of Gestalt Phenomena, *Computers in Human Behavior*, **2**, pp. 241–255 for an example of the latter.

yield the highest payoffs on the average. Accordingly, the replicator dynamics can be treated as a simple model of adaptive learning. We will explore this point with two examples.

4. EVOLUTION AND A REPEATED SOCIAL DILEMMA

Suppose a population of agents play the social dilemma shown in Table 21.2. Two agents will be randomly matched on each round of play. When they are matched to play the game, the agents will play the game repeatedly with no definite number of repetitions. The discount factor, allowing for both time discount and the probability that there will be no next round of play, is $5/8 = 0.625$.

The population who play this game is boundedly rational, in that they play according to one of three rules:

- always C
- always D
- Tit for Tat

The discounted present value payoffs for these three strategy rules are shown in Table 21.3.

At a given time, there is a small probability that an agent may switch strategies before the next match. Such an agent will shift from strategy R to strategy S with a probability

> **Definition:** *Heuristic Rules —* *Heuristic rules* for problem solving are rules that are fast and usually reliable, but informal or inconclusive because they can fail in unusual cases. Tit-for-Tat is an example.

that is proportionate to the payoff to strategy R relative to the payoff to strategy S. For example, if "always C" pays 2 on the average and "Tit-for-Tat" pays 2.5, the ratio of the payoffs is 2.5:2 = 5:4, so the probability of a shift from C to TfT is proportionately greater than the probability of a shift from TfT to C. Then more will shift to TfT than will shift away from it, and the proportion playing TfT will

Table 21.2. A Social Dilemma.

		\multicolumn{2}{c}{Q}	
		C	D
P	C	4,4	1,5
	D	5,1	2,2

Table 21.3. Strategy Rules and Payoffs in a Repeated Social Dilemma.

		Q		
		Always C	Always D	Tit-for-Tat
P	Always C	$10\frac{2}{3},10\frac{2}{3}$	$2\frac{2}{3},13\frac{1}{3}$	$10\frac{2}{3},10\frac{2}{3}$
	Always D	$13\frac{1}{3},2\frac{2}{3}$	$5\frac{1}{3},5\frac{1}{3}$	$8\frac{1}{3},4\frac{1}{3}$
	Tit-for-Tat	$10\frac{2}{3},10\frac{2}{3}$	$4\frac{1}{3},8\frac{1}{3}$	$10\frac{2}{3},10\frac{2}{3}$

increase. In general, the rules that give the higher payoffs will be played by an increasing proportion of the population. This is an instance of adaptive learning, and thus we can apply the replicator dynamics.

Since the agents are matched at random, the probability of being matched with a C player, a D player, or a Tit-for-Tat player depends on the proportions of C players, D players and Tit-for-Tatters in the population. This is complicated because we have two proportions to consider. However, "always C" is dominated, so we will keep it simple

Definition: *Adaptive Learning* — When agents make decisions according to heuristic rules, but try new rules from time to time and eliminate the rules that perform relatively poorly in their experience, the agents are said to learn *adaptively.*

by considering only a few cases with the proportion of C players at zero with various proportions of Tit-for-Tatters. The cases are shown in Table 21.4. Expected value payoffs for the three strategy rules are shown in the middle three columns. What we see is that the

Table 21.4. Some Tendencies in a Social Dilemma with Replicator Dynamics.

Proportion of Titfortatters	EV(always C)	EV(always D)	EV(TfT)	What happens
0.9	9.87	8.03	10.03	The population is increasingly dominated by TfTers and D players disappear.
0.7	8.27	7.43	8.77	Ditto
0.5	6.67	6.83	7.5	Ditto
0.3	5.07	6.23	6.23	Relative proportions of D and TfT players remain steady.
0.1	3.47	5.63	4.97	The population is increasingly dominated by D players and TfTers disappear.

equilibrium depends on the starting point — if there are very few Titfortatters at the beginning, then D will dominate and TfT disapprear, but with any proportion of Titfortatters above 0.3, Tit-for-Tat will win out and "always D" will disappear.

Thus, there are two stable equilibria with the replicator dynamics. One gives rise to cooperative outcomes and the other does not. The one that gives rise to cooperative behavior via a Tit-for-Tat rule is more probable than the other, in the sense that the population gets to it from more starting points. Thus, it seems that there is hope for boundedly rational learners to reach the cooperative equilibrium in some cases. Of course, the outcome of this example depends very much on the specific numbers and the probability of finding the cooperative equilibrium could be different if the numbers were different.

5. INFORMATIONALLY (ALMOST) EFFICIENT MARKETS

Many economists and financial theorists have argued that financial markets are informationally efficient. This means that the current price of a share in the XYZ corporation (for example) reflects all information about the profitability and risks of investment in XYZ that is available to the public. Because it reflects all available information, the stock price will change only in response to new information. That in turn means that the price of XYZ stock is not predictable — it will change in the future as a "random walk." This is the "efficient markets theory." Thus, according to this view, there is no point in studying individual stocks to learn all you can about them. You may as well buy stocks at random — or buy an "index" fund, which is a bundle of all the stocks that make up some popular stock price index. And, in fact, index funds have become quite popular with small-scale investors in recent decades.[6]

---

[6]This idea was set out for the popular reader by Burton Malkiel, *A Random Walk Down Wall Street*, 7th edition (New York: W. W. Norton and Co.), June 2000.

While the efficient markets theory may be approximately true, Grossman and Stiglitz argue[7] that it cannot be exactly true. For if it were true, no-one would bother to do any market research, and in fact the available information would not be reflected in the stock prices. For, after all, **available** information is not **free**. Effort is required to read corporate annual reports and financial news reports, and effort is a cost. There may be money costs as well, for the best information. Why would anyone bear all that cost if he could just as well buy at random or imitate the decisions of others for no cost?

The efficient markets theory is a logically correct statement about the situation of an individual investor in a world in which other investors are informed. However, if all investors act as if they were in a world of informationally efficient markets, then the markets cannot be informationally efficient. Not only

A Closer Look: Robert Axelrod and the Axelrod Tournament

In 1980, Robert Axelrod held a tournament of various heuristic rules for choosing strategies in the Prisoner's Dilemma. The tournament included simple rules like Tit-for-Tat and Always Defect, and also more complicated rules. He invited prominent game theorists to submit programs and computer-simulated the play of the strategy rules against one another. In a widely celebrated report, he found that Tit-for-Tat was best all around, although there is no perfect rule that will "beat" all the others.

Born in 1943 in Chicago, Axelrod attended the University of Chicago and Yale University, and is Walgreen Professor in the Department of Political Science and the Gerald R. Ford School of Public Policy at the University of Michigan.

---

[7]S. Grossman, and J. Stiglitz, On the impossibility of informationally efficient markets, *American Economic Review*, **70**, pp. 393–408.

Table 21.5. A Two-Person Investor Game.

		George	
		Informed	Uninformed
Warren	Informed	7,7	7,8
	Uninformed	8,7	4,4

that. We know from observation that, while some investors buy index funds and do not do difficult market research, other investors commit a great deal of time and money to market research.

We might interpret this in game theoretic terms. An individual investor has at least two strategies: he can be informed or be uninformed.[8] As a first step, suppose there are only two investors, so that we have a two-by-two game. If at least one investor informs himself, both investors can get a rate of return of 8% on their investments. However, the effort and cost of getting the information is equivalent to a reduction of the rate of return by 1%, so the net payoff to the investor who chooses to be informed is 7%. If neither becomes informed, however, they both get a rate of return of only 4%. This gives rise to the payoffs in Table 21.5.

This game has two Nash Equilibria in pure strategies — and each occurs where one investor is informed and the other is not. So, in this simplified game, we find an equilibrium at which just half the investors are informed, and those who are not informed come out with a higher payoff.

In the real world, there are many more than just two investors. If most of them are informed, then we would expect that the uninformed ones could do as well by investing at random or buying

---

[8]R. Cressman and J. F. Wen, whose ideas suggested this example, point out that the investor can also choose from a range of mixed strategies, but we will ignore these for simplicity. See R. Cressman and J. F. Wen (October 3, 2002), *Playing the Field: An Evolutionary Analysis of Investor Behavior with Mixed Strategies*, working paper, GREQAM, Centre de la Ville Charite, 2 rue de la Charite, 13002 Marseille, France.

index funds. In this sort of world, a person who buys an index fund is, in effect, imitating the more informed investors, and doing it very cheaply. If there are enough informed investors, the uninformed may do as well this way as they could if they made the effort to become informed, or even better. In fact, while we don't know for certain, it seems likely that even a minority of informed investors could push the prices of the stocks to their efficient levels. After all, the decisions of the informed are likely to have more impact than the actions of the uninformed — the decisions of the informed are targeted, while the decisions of the uninformed are not. We may also suppose that the investors learn adaptively to choose the best strategy, and accordingly we apply the replicator dynamics to this model.

In place of the two-by-two game, in Table 21.5, it would be better to treat this game as a proportional game, like the commuter game in Chapter 14, Section 3. Figure 21.2 shows payoffs for such a game, with the payoffs to uninformed investors shown by the solid curve. The payoffs in the figure are a little different from those in Table 21.5, in that the payoffs to an uninformed investor depend on how many investors are informed. An uninformed investor can get the maximum payoff of 8 only if he is the only uninformed investor in a population of informed investors.

At the other extreme, if all investors are uninformed, they make the 4% available to the truly random investor. However, as the proportion of informed investors increases, the quality of information in the market increases rapidly, so that the returns to uninformed investors who invest at random or buy index funds rises quite steeply and then levels off when $1/4$ or more are informed.[9] Informed investors can always get a return of 7%, net of the cost of information, as shown by the dashed horizontal line. Thus, it pays better to

---

[9]The leveling off is an instance of "diminishing returns" — in this case diminishing returns to investment research. It is computed as $4 + 4p^{0.25}$ where p is the proportion of investors who are informed and $p^{0.25}$ is the fourth root (square root of the square root) of p. This equation is an arbitrary one chosen for the sake of the example, but chosen to illustrate the diminishing returns to market research.

Payoffs to Informed and Uninformed Investors

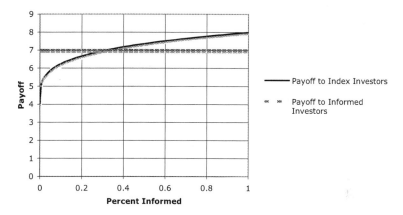

Figure 21.2.   Payoffs to Investors According to Proportion Informed.

be informed whenever the proportion of investors who are informed is less than 0.316, and it pays better to be uninformed when more than 31.6% of investors are informed.

Thinking of this as a proportional game, then, the equilibrium condition is that 31.6% of investors are informed. There are a large number of Nash equilibria — depending on which investors are informed and which are not — but every equilibrium has 31.6% informed and the rest uninformed.

Is this evolutionarily stable? It is. The replicator dynamic says that the proportion choosing to be informed will increase, or decrease, in proportion with the relative payoff. Figure 21.3 shows the payoff to being informed, as a proportion of the average payoff for the whole population. With the replicator dynamic, this quotient will determine the change in the proportion informed from one period to the next. Clearly, whenever the proportion is different from the equilibrium at 31.6%, it will be moving in the direction of 31.6%. Figure 21.4 shows two paths toward equilibrium, beginning from proportions of 0.7 (dashed) and 0.1 (dotted) over ten successive periods.

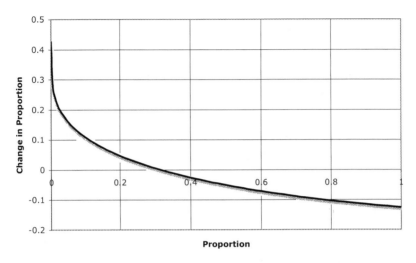

Figure 21.3.   The Relative Payoff of Being Informed.

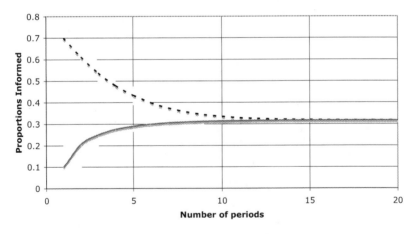

Figure 21.4.   Two Paths to Equilibrium.

In equilibrium, then, we have two "kinds" of investors. One "kind," the minority, does extensive market research and invests with care. They get a net payoff of 7%. The other kind buy index funds or invest at random. Because there are about 31.6% of investors who do research, the prices in the marketplace are fairly highly efficient and the inactive investors get a payoff of 7%. This sounds a bit like the real world we live in. Who is right? They both are, and in fact it doesn't matter who is inactive and who does research. There are no real differences between these "kinds" of investors except their strategies. All that is necessary for equilibrium is that 31.6% of them do research and are informed.

6. TIT-FOR-TAT, RECIPROCITY, AND THE EVOLUTION OF THE HUMAN SPECIES

The forgiving trigger strategy known as Tit-for-Tat is based on rational self-interest, but, as we have seen, it can bring about cooperative solutions in many cases when mutually unprofitable non-cooperative equilibria might occur otherwise. As we saw in Chapter 20, experimental evidence suggests that people do not always act in self-interested ways, and a number of studies are consistent with the idea that people often depart from self-interest in the direction of reciprocity. Some scholars believe that this tendency toward reciprocity may be part of the human genetic heritage — that our genes program us for some tendency to act with reciprocity, or to learn to act with reciprocity. In this context, we notice that although Tit-for-Tat is consistent with self-interest in many cases, it is consistent with reciprocity in even more.

Why should our genes predispose or "program" us for reciprocity? The argument of some scholars in "evolutionary psychology" is that our human and prehuman ancestors have lived with social dilemmas for a long time, and that preagricultural humans and prehumans who were predisposed to reciprocity would solve those social dilemmas more readily, and so would be more likely to survive and reproduce.[10]

---

[10]Those who believe that human beings did not evolve, but rather were created as they are by an intelligent God, should have less difficulty in explaining why our heredity predisposes us to reciprocity or to other non-self-interested motives. It is

One argument is that teamwork is required for many of the hunting and gathering activities by means of which preagricultural people and prehumans survived. For example, some preagricultural people in Africa hunted antelope by chasing an antelope until it dropped dead of exhaustion. No one human could do that — antelope have more stamina and can run longer than humans. Running is the antelope's field of specialization. So, the humans would hunt in teams, and a team of human hunters would post themselves at points equal distances apart around a large circle, with one hunter in reserve. Try to visualize them standing at their stations around the circle. We can call the points where the people stand station A, station B, and so on. It works like this: the person at station A chases the antelope toward station B, and then the person at station B takes over and chases the antelope to station C, and so on. The person who has chased the antelope from station A to station B now gets a long rest at station B, and the reserve person takes the place at station A. Once the antelope comes round to station A, the reserve person takes over and chases it to station B, and that first person takes another relay, following a rest. In that way, the humans could chase the antelope indefinitely, and certainly until it drops. This works best if everyone gives it an honest effort and nobody shirks — an age-old effort dilemma.

But even without teamwork, there are other reasons for hunting and gathering people to share food, as many of them routinely do. Hunting, after all, is risky, and even a good hunter may come home at the end of the day with no food for his family. In many preagricultural peoples, it is expected that the successful hunter shares his catch with those who are less successful.

Here is an example to show what that means. Suppose we have two preagricultural hunters, known as Horse and Bull. Each is

evolution that poses the harder question, since evolution is inherently a selfish process — as biologist Richard Dawkins suggested by the title of his book, *The Selfish Gene* (Oxford University Press, 1976). For a useful study of applications of Dawkins' ideas to social evolution, see Kate Distin, *The Selfish Meme* (Cambridge: Cambridge University Press, 2005).

skillful, but on any given day, either of them may come home empty-handed. To keep the example simple, let us say that either Horse or Bull will be successful on any given day, and the other will not, and the probability that it is Horse is 0.5, and therefore the probability that Bull is successful is the same, 0.5. Table 21.6 shows the payoffs when Horse has been successful in the hunt, Bull has not, and each person chooses between the strategies "share" and "don't share." Since Bull has nothing to share, his decision makes no difference. The −20 payoff to Bull (if Horse does not share) reflects the possibility that Bull will be so weakened he will not be a successful hunter tomorrow, either, and that could threaten his life. Table 21.7 shows the payoffs if Bull is successful and Horse is not.

These are not social dilemmas, because they are so unsymmetrical — the unsuccessful hunter, having nothing to share, has nothing to lose by sharing. But, on any given day, no-one knows which of these games they are playing. Accordingly, we calculate the expected value payoffs, and they are shown in Table 21.8.

Table 21.6. A Sharing Dilemma when Horse Has Been Successful.

		Horse	
		Share	Don't
Bull	Share	5,5	−20,10
	Don't	5,5	−20,10

Table 21.7. A Sharing Dilemma when Bull Has Been Successful.

		Horse	
		Share	Don't
Bull	Share	5,5	5,5
	Don't	10,−20	10,−20

Table 21.8. The Expected Value Payoffs of the Two Sharing Dilemmas.

		Horse	
		Share	Don't
Bull	Share	5,5	–7.5,7.5
	Don't	7.5,–7.5	–5,–5

Table 21.8 is a social dilemma — both hunters are better off if they both share, but refusing to share is a dominant strategy equilibrium. Of course, hunting people face this dilemma time and time again, with no definite limit. Accordingly, they might arrive at the cooperative solution by a self-interested Tit-for-Tat or similar trigger strategy. However, even with a trigger strategy, the cooperative outcome is only one of multiple stable outcomes; and when people make mistakes, or when there are more than two people, outcomes are even less certain. By contrast, human beings with some innate tendency to act with reciprocity would be all the more likely to share reciprocally, to return sharing when it is initiated, and to initiate sharing in the hope that it would be reciprocated — and thus all the more likely to survive in high-risk forms of hunting and gathering.

Thus, it seems that game theory and evolution have yet another connection: human beings may well have evolved to play certain kinds of games.

7. SUMMARY

Game theory can be applied to the evolution of non-human animals if we think of the games as random matches of individuals from a population and interpret the payoffs of games as differences in the reproductive success of the "players" of the games. These ideas lead to the concepts of ESS and the replicator dynamics. The replicator dynamics is a dynamics in which the proportion of the population playing a strategy increases in proportion to the relative payoff for that strategy in the game. A Nash equilibrium that is stable under this dynamics is an ESS. In this case, the "strategy" can be a "mixed strategy," that is, the proportion of the population playing each strategy. These "games" sometimes have the same equilibria as familiar cases from basic game theory.

Recognizing that boundedly rational human beings may choose their strategies according to relatively simple, fallible rules, we might model boundedly rational learning by applying the ESS to a population of rules. This can lead to Nash equilibria and can narrow the range of Nash equilibria, in some cases, to cooperative or second-best solutions, depending on the details of the game.

Q21. EXERCISES AND DISCUSSION QUESTIONS

Q21.1. Frog Mating Game

Recall the Frog Mating Game from Chapter 6, Exercise 6.3. The frogs are all males who can choose between two strategies for attracting females: the call strategy (call) or the satellite strategy (sit). Those who call take some risk of being eaten, while those who sit run less risk. On the other hand, the satellites who do not call may nevertheless encounter a female who has been attracted by another call. So, the payoff to the satellite strategy is better when a larger number of other male frogs are calling.

The payoff table for the three frogs is shown in Table 21.9, with the first payoff to Kermit, the second to Michigan J, and the last to

Table 21.9. Payoffs for Eager Frogs.

		Flip			
		Call		Sit	
		Michigan J.		Michigan J.	
		Call	Sit	Call	Sit
Kermit	Call	5,5,5	4,6,4	4,4,6	7,2,2
	Sit	6,4,4	2,2,7	2,7,2	1,1,1

Flip. Suppose 10,000 male frogs are randomly matched in groups of three to play this game. Determine whether the game has an ESS and, if so, what it is.

Q21.2. El Farol

El Farol is a bar in Santa Fe, New Mexico, where chaos researchers from the Santa Fe Institute often hang out. It is said that the El Farol is at its best when it is crowded, but not too crowded. We saw a three-person example a little like that in Chapter 6. Now suppose that there are 100 chaos researchers who choose between going to the El Farol and staying home (those are the two strategies). Say the number of people who go to El Farol on a particular night is N and suppose the payoffs to each person who goes is

$$P = N - 0.13N^2.$$

Does this game have an ESS? Is it efficient? Why or why not?

Q21.3. Maximizing profits

Economists often assume that businesses maximize profits, an operation that can be expressed in terms of calculus. Yet it is well known that very few businessmen use calculus in deciding how and when to change their prices and the outputs of their firms. Discuss this point in the light of concepts from this chapter. Can you think of some decisions that businessmen might learn to make very accurately in an ESS? Any that would be difficult even in an ESS?

Q21.4. Banking

Bankers can choose between two strategies: (a) lend only to borrowers with excellent credit; or (b) lend to all borrowers with fair credit or better. When most bankers choose strategy (a), scarcity of credit causes a lot of business failures so (a) is the more profitable strategy; but when most choose (b), easy credit makes for good business con-

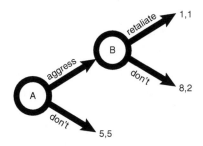

Figure 21.5. Game Y, a Retaliation Game.

ditions and (b) is the more profitable strategy. Specifically, if y is the proportion of banks choosing strategy (a), the payoff to (a) is $6 - 2y$, while the profit to (b) is $5 + y/2$. Analyze this example using the concept of an ESS.

Q21.5. Retaliation Game

Game Y, shown in extensive form in Figure 21.5, is a simple retaliation game.

Suppose a large number of agents are matched at random to play this game, sometimes as A and sometimes as B.

a. Express this as an *N*-person game. What is the instrument variable?
b. Could "retaliate" be an evolutionarily stable strategy?

Q21.6. A Fishy Dilemma

Fishermen in the Tethys Sea face an ironic social dilemma. The fish stocks in the Tethys Sea have already been depleted by overfishing. New technology enables the fishermen to catch an even greater proportion of the available fish in the Tethys Sea. On the one hand, those who do not adopt the new technology will catch fewer fish than those who do. On the other hand, the new technology will deplete the Tethys fishery even faster than the old technology. The more fishermen adopt the new technology, the fewer fish will be caught, regardless which technology one is using.

a. Express this as an N-person game.

    i.   What is the state variable?

    ii. Discuss or sketch how the payoffs to the two strategies will vary with changes in the state variable.

    iii. How would you find a Nash equilibrium?

b. Apply the Evolutionarily Stable Strategies model to this problem. Is it possible that continued use of the old technology could be stable?

    Explain your answers to all questions.

Index

Printed in the United States
by Baker & Taylor Publisher Services